RESEARCH AND DEVELOPMENT IN UNIVERSITY MATHEMATICS EDUCATION

In the last thirty years or so, the need to address the challenges of teaching and learning mathematics at university level has become increasingly appreciated by university mathematics teachers, and beyond, by educational institutions around the world. Indeed, mathematics is both a condition and an obstacle to success for students in many educational programmes vital to the 21st century knowledge society, for example in pure and applied mathematics, engineering, natural sciences, technology, economics, finance, management and so on. This breadth of impact of mathematics implies the urgency of developing research in university mathematics education and of sharing results of this research widely. This book provides a bespoke opportunity for an international audience of researchers in didactics of mathematics, mathematicians and any teacher or researcher with an interest in this area to be informed about state-of-the-art developments and to heed future research agendas.

This book emerged from the activities of the research project INDRUM (acronym for International Network for Didactic Research in University Mathematics), which aims to contribute to the development of research in didactics of mathematics at all levels of tertiary education, with a particular concern for the development of early-career researchers in the field and for dialogue with university mathematicians.

The aim of the book is to provide a deep synthesis of the research field as it appears through two INDRUM conferences organised in 2016 and 2018. It is an original contribution which highlights key research perspectives, addresses seminal theoretical and methodological issues and reports substantial results concerning the teaching and learning of mathematics at university level, including the teaching and learning of specific topics in advanced mathematics across a wide range of university programmes.

Viviane Durand-Guerrier is Professor Emeritus of Didactics of Mathematics at the University of Montpellier, France.

Reinhard Hochmuth is Professor of Didactics of Mathematics at the Leibniz University Hannover, Germany.

Elena Nardi is Professor of Mathematics Education at the University of East Anglia, United Kingdom.

Carl Winsløw is Professor of Didactics of Mathematics at the University of Copenhagen, Denmark.

EUROPEAN RESEARCH IN MATHEMATICS EDUCATION

ERME, the European Society for Research in Mathematics Education, is a growing society of about 900 researchers from all over Europe and beyond whose mathematics education research is shaped by a spirit of communication, cooperation, and collaboration.

The contributions in the ERME Series seek to understand and improve learning and teaching of mathematics at schools, colleges and universities, as well as in informal settings (e.g., related to street mathematics or to self-organized networks of teachers). The ERME Series puts an emphasis on reflecting the institutional, societal, and cultural contexts of learners, teachers and researchers and how this context shapes research and adopts a variety of perspectives on its research field.

The volumes are written by and for European researchers, but also by and for researchers from all over the world. An international advisory board guarantees that ERME stays globally connected. A rigorous and constructive review procedure guarantees a high quality of the series.

Volumes of the series:

(1) Developing Research in Mathematics Education
Twenty Years of Communication, Cooperation and Collaboration in Europe
Edited by Tommy Dreyfus, Michele Artigue, Despina Potari, Susanne Prediger and Kenneth Ruthven

(2) Working with the Anthropological Theory of the Didactic in Mathematics Education
A comprehensive casebook
Edited by Marianna Bosch, Yves Chevallard, F. Javier García and John Monaghan

(3) Professional Development and Knowledge of Mathematics Teachers
Edited by Stefan Zehetmeier, Despina Potari and Miguel Ribeiro

(4) Classroom Research on Mathematics and Language
Seeing Learners and Teachers Differently
Edited by Núria Planas, Candia Morgan and Marcus Schütte

(5) Research and Development in University Mathematics Education
Overview Produced by the International Network for Didactic Research in University Mathematics
Edited by Viviane Durand-Guerrier, Reinhard Hochmuth, Elena Nardi and Carl Winsløw

For more information about this series, please visit: https://www.routledge.com/European-Research-in-Mathematics-Education/book-series/ERME

RESEARCH AND DEVELOPMENT IN UNIVERSITY MATHEMATICS EDUCATION

Overview Produced by the International Network for Didactic Research in University Mathematics

Edited by
Viviane Durand-Guerrier, Reinhard Hochmuth, Elena Nardi and Carl Winsløw

Routledge
Taylor & Francis Group
LONDON AND NEW YORK

First published 2021
by Routledge
2 Park Square, Milton Park, Abingdon, Oxon OX14 4RN

and by Routledge
605 Third Avenue, New York, NY 10017

Routledge is an imprint of the Taylor & Francis Group, an informa business

British Library Cataloguing-in-Publication Data
A catalogue record for this book is available from the British Library

Library of Congress Cataloging-in-Publication Data
Names: Durand-Guerrier, Viviane, editor. | Hochmuth, Reinhard, editor. | Nardi, Elena, editor. | Winsløw, Carl, editor.
Title: Research and development in university mathematics education : overview produced by the international network for research on didactics of university mathematics / edited by Viviane Durand-Guerrier, Reinhard Hochmuth, Elena Nardi, Carl Winsløw.
Identifiers: LCCN 2020049745 (print) | LCCN 2020049746 (ebook) | ISBN 9780367365370 (hardback) | ISBN 9780367365387 (paperback) | ISBN 9780429346859 (ebook)
Subjects: LCSH: Mathematics--Study and teaching (Higher)
Classification: LCC QA11.2 .R443 2021 (print) | LCC QA11.2 (ebook) | DDC 510.71/1--dc23
LC record available at https://lccn.loc.gov/2020049745
LC ebook record available at https://lccn.loc.gov/2020049746

ISBN: 978-0-367-36537-0 (hbk)
ISBN: 978-0-367-36538-7 (pbk)
ISBN: 978-0-429-34685-9 (ebk)

Typeset in Bembo
by Taylor & Francis Books

CONTENTS

ILLUSTRATIONS

Figures

Tables

CONTRIBUTORS

Michèle Artigue
LDAR, Université de Paris
France

Caroline Bardini
Melbourne Graduate School of Education
Australia

Berta Barquero
University of Barcelona
Spain

Rolf Biehler
University of Paderborn
Germany

Marianna Bosch
Universitat Ramon Llull
Spain

Laura Branchetti
Università di Parma
Italy

Laura Broley
Concordia University
Canada

Stéphanie Bridoux
Université de Mons
Belgium

Faïza Chellougui
Bizerte Faculty of Sciences, University of Carthage,
Tunisia

Tony Croft
Loughborough University
United Kingdom

Viviane Durand-Guerrier
IMAG, CNRS, Univ. Montpellier
France

Yael Fleischmann
Norwegian University of Science and Technology
Norway

Helge Fredriksen
UiT - The Arctic University of Norway
Norway

Alejandro S. González-Martín
Département de Didactique, Université de Montréal
Canada

Nicolas Grenier-Boley
LDAR, Normandie University
France

Ghislaine Gueudet
CREAD, Université de Brest
France

Erik Hanke
University of Bremen
Germany

Thomas Hausberger
IMAG, CNRS, Univ. Montpellier
France

Reinhard Hochmuth
Leibniz University of Hannover
Germany

Emma Howard
University College Dublin
Ireland

Barbara Jaworski
Loughborough University
United Kingdom

Duncan Lawson
Coventry University
United Kingdom

Rafael Martínez-Planell
University of Puerto Rico at Mayagüez
Puerto Rico

Antoine Meyer
Université Gustave Eiffel
France

Elena Nardi
University of East Anglia
United Kingdom

Pedro Nicolás
Universidad de Murcia
Spain

Ann O'Shea
Maynooth University
Ireland

Birgit Pepin
Eindhoven University of Technology
The Netherlands

Johanna Rämö
University of Helsinki
Finland

Chris Rasmussen
San Diego State University
United States

Avenilde Romo-Vázquez
Instituto Politécnico Nacional
Mexico

Frode Rønning
Norwegian University of Science and Technology
Norway

Heidi Strømskag
Norwegian University of Science and Technology
Norway

Theresia Tabchi
Université de Reims Champagne-Ardenne
France

María Trigueros
Instituto Tecnológico Autónomo de México
México

Fabrice Vandebrouck
LDAR, Université de Paris
France

Megan Wawro
Virginia Tech
United States

Carl Winsløw
University of Copenhagen
Denmark

Michelle Zandieh
Arizona State University
United States

INTRODUCTION

Viviane Durand-Guerrier, Reinhard Hochmuth, Elena Nardi and Carl Winsløw

This book falls within the activities of INDRUM (acronym for International Network for Didactic Research in University Mathematics) that was created in 2015 by a group of scholars involved in this field.

In the last thirty years or so, the need to address the challenges of teaching and learning mathematics at university level has become increasingly appreciated by university mathematics teachers, and beyond by societies all around the world. Indeed, mathematics is both a condition and an obstacle to success for students in many educational programmes vital to the 21st century knowledge society, for example, in pure and applied mathematics, engineering, natural and social sciences, technology, economics, finance, management and so on. This breadth of impact of mathematics illustrates the necessity of developing research in university mathematics education and of sharing results of this research widely. A European and international community of research on university mathematics education has begun to emerge thanks to dedicated thematic working groups in international conferences, namely CERME (Congress of the European Society for Research in Mathematics Education), ICME (International Congress on Mathematical Education), EMF (Espace Mathématique Francophone) and RUME (Research in Undergraduate Mathematics Education). The idea of developing an International Network for Didactic Research in University Mathematics (INDRUM) was presented in Oberwolfach in December 2014, and the network was formally founded at CERME 9 in February 2015.

INDRUM aims to contribute to the development of research in didactics of mathematics at all levels of tertiary education, with a particular focus on building research capacity in the field and a concern for the development of early-career researchers in the field – and for dialogue with the mathematics community. The network has several objectives, including: elaboration of a scientific programme that will federate existing research in this field in Europe and beyond; development and emergence of international collaborations by applying to project calls at all levels (local, regional, European, international); reinforcement of the visibility of research studies to researchers in mathematics education and to the mathematics community; and, contribution to the development of doctoral programmes in this field. While the origin of the network is so far distinctly European, its scope and aims are clearly decidedly international. So far, two very successful conferences have been organized by the network: in 2016 (Montpellier, France) and 2018 (Kristiansand, Norway) with 100 and 120 participants respectively, from European and non-European countries. As this book was nearing completion in 2020, the third conference was fast approaching (Bizerte, Tunisia) and plans for the fourth and fifth were already under way. The conferences have attracted researchers with a wide range of experience on topics covering all levels of university education, including post-graduate studies. The INDRUM conferences offer room for deep discussion on selected topics and foster international collaboration.

INDRUM conferences are recognized as ERME Topic Conferences by the European Society for Research in Mathematics Education (ERME). Its organizers and scientific committee have been closely linked with the current and previous ERME thematic working groups on "University Mathematics Education", "Advanced Mathematical Thinking" and "Argumentation and Proof". The scientific committee of the INDRUM network is composed of thirty-two leading researchers in the field from fifteen countries.

The aim of this book is to provide a deep synthesis of the research field as it appears through the first two INDRUM conferences. It aspires to be an original contribution which highlights key research perspectives, addresses seminal theoretical and methodological issues and reports substantial results concerning the teaching and learning of mathematics at university level, including the teaching and learning of specific topics in advanced mathematics across a wide range of university programmes.

The book is organized in four sections. Part 1 consists of papers from the contributors to the two opening plenary lectures and the two plenary panels of INDRUM 2016 and 2018. The remaining two sections of the book provide a deep synthesis of research developments concerning the teaching and learning of specific mathematical topics (Part 2) and concerning teachers' and students' practices (Part 3). We now provide a short summary of the chapters in each of the three sections.

Part 1: Achievements and current challenges

Chapter 1: Mathematics education research at university level: achievements and challenges

Michèle Artigue gave the opening keynote lecture of INDRUM 2016 at the University of Montpellier in France. Chapter 1 reflects achievements and challenges of research in mathematics education at university level. In the first part of the chapter, Artigue gives an interesting insight into the historical development of the field based on her personal experience both as a researcher and as a university teacher. The narrative spans a bridge from interdisciplinary projects carried out by mathematicians and physicists in Paris in the early 1980s to a look at the socio-cultural turn through the lens of The Anthropological Theory of the Didactic. The second part outlines the current state of the field and present its strengths and promising developments as well as the challenges it faces.

Chapter 2: Lessons for mathematics higher education from 25 years of mathematics support

Chapter 2 by Duncan Lawson and Tony Croft is based on the keynote lecture by Lawson presented at INDRUM 2018 at the University of Agder in Norway. The article first clarifies what the authors understand by mathematics support, and describes the reasons for its development, which also led to the foundation of the Centre for Excellence in University-wide Mathematics and Statistics Support at Loughborough and Coventry Universities in 2005. After a sketch of reasons why mathematics support is currently more important than ever, the authors address the question of why mathematics support is so attractive to undergraduates in a range of disciplines. This reflection is based on data from a UK National Student Survey, focus groups and individual interviews. The conclusion puts mathematics support in an international context and addresses critical concerns regarding ways forward.

Chapter 3: Current interactions between mathematicians and researchers in university mathematics education

Chapter 3 by Caroline Bardini, Marianna Bosch, Chris Rasmussen and María Trigueros brings together contributions to the plenary panel at INDRUM 2016 which examined the interaction between researchers in mathematics and researchers in mathematics education (or didacticians). In three case studies from Australia, United States and Mexico, the nature and extent of different types of collaboration is discussed. The cases show the diversity of interactions, depending on not only the country and historical period, but also the educational level considered and the institutional context. In the conclusion, benefits and pitfalls of established relationships are highlighted. Last but not least, directions that seem important to strengthen are pointed out.

Chapter 4: Education and professional development of university mathematics teachers

In Chapter 4, Winsløw, Biehler, Jaworski, Rønning and Wawro set out from recognising that university mathematics teaching today must address the needs of highly varied student cohorts – in terms of background and specialism – and that it does so in a highly potent digital learning environment. Thinking more systematically about the professional preparation and development of university mathematics teachers includes presenting current practices (here, examples from the UK, USA, Germany and Norway are outlined) and proposing new ideas for discipline and context-specific teacher preparation and for identifying and rewarding quality teaching.

Part 2: Teaching and learning of specific topics in university mathematics

Chapter 5: Challenging issues in the teaching and learning of Calculus and Analysis

In Chapter 5, Trigueros, Bridoux, O'Shea and Branchetti address issues in Calculus and Analysis teaching such as sequences and one/two variable functions. The chapter highlights that, although student difficulties with one-variable functions have long been the object of research, the difficulties persist and further research is needed. Also, the case of multivariable functions has been under-researched, due to a common assumption that issues relating to their teaching and learning are a mere generalization of those concerning one-variable functions. Recent research presented at INDRUM 2016 and 2018 shows that it is not. The authors also emphasize the need to address further issues in more advanced topics in Analysis such as Topology and Complex Analysis.

Chapter 6: Task design in Calculus and Analysis

Calculus and Analysis task design remains an important issue for university mathematics education, both for students majoring in mathematics and those who are not, as well as for pre-service mathematics teachers. In Chapter 6, Vandebrouck, Hanke and Martínez-Planell present the various theoretical perspectives underpinning the different studies on task design presented at INDRUM 2016 and 2018. The authors emphasize that many of these studies consider the issue of the formalization of basic notions, such as real numbers, supremum, convergence of sequences, or continuity. The authors call for further exploration, documentation and discussion within the community on issues such as assessment and incorporation of technologies.

Chapter 7: Discrete mathematics, computer science, logic, proof, and their relationships

Discrete mathematics, roughly defined as concerning countable objects, recently gained importance both in research and education, mainly due to its foundational role in large parts of computer science. In Chapter 7, Chellougui, Durand-Guerrier and Meyer provide a synthesis of recent INDRUM research concerning the educational side of this "border area" between university mathematics and computer science, and especially the actual and potential significance of logic in both disciplines. The authors demonstrate the potential and interest of interdisciplinary didactical engineering in this area, and the need to pursue and deepen the epistemological analysis on which such research is founded.

Chapter 8: Abstract and Linear Algebra

Algebra at university level traditionally falls into two categories: Linear Algebra, taught to wide student cohorts, and Abstract Algebra, mainly taught to students of pure mathematics. In Chapter 8, Hausberger, Zandieh and Fleischman observe that research in these areas has remained somewhat fragmented and separated, with studies focusing on isolated notions or results, such as linear independence. But recent research presented at INDRUM has developed more overarching didactical perspectives on algebra, in terms of structural and discursive characteristics. This promises a more unified approach to future development of the didactics of algebra, which could be particularly important in enabling research-based innovation of university curricula beyond very local contexts such as particular course units.

Chapter 9: Mathematics and other disciplines, and the role of modelling

Most of the teaching of mathematics in higher education concerns students who are not specializing in mathematics. The challenges of defining, designing, motivating and assessing mathematics teaching and learning for such students is therefore an important field both at INDRUM and in similar venues. In Chapter 9, González-Martín, Gueudet, Barquero and Romo-Vázquez, setting out from recent research in this area presented at INDRUM, propose that interaction between different specialisms becomes increasingly important, both to make sense of actually observed teaching and learning, and to devise new and more relevant and efficient didactical approaches, for instance to help students engage in autonomous mathematical modelling activity.

Part 3: Teachers' and students' practices at university level

Chapter 10: Transitions to, across and beyond university

Transition to and across university mathematics has been attracting increasing research efforts. In Chapter 10, Hochmuth, Broley and Nardi offer a systematic review of the themes discussed in INDRUM 2016 and 2018 and of the major theoretical lenses used in this area. Three types of transition for university mathematics education are considered: from school to university; within and across university courses; and from university to the workplace. While the first type has been the object of research for a long time, research on the second and third is rather new in the field. There are therefore many unexplored areas that are worthy of consideration in future research.

Chapter 11: Students' in-class and out-of-class mathematical practices

In Chapter 11, Rasmussen, Fredriksen, Howard, Pepin and Rämö zoom in on students' in-class and out-of-class practices, use of resources out-of-class, role in assessment practices, responses to active learning initiatives, and in-class mathematical practices. They do so with a sharp theoretical focus on situating students' mathematical practices in relation to interactions with other students, the teacher, the mathematics, and resources. They conclude by considering how future research might engage in cross-pollination of the above themes and they highlight potential benefits from theoretical networking as well as from deploying theories that are appropriate for, but hitherto under-used in, examining students' in-class and out-of-class practices in complementary ways.

Chapter 12: Mathematics teaching practices at university level

In Chapter 12, Grenier-Boley, Nicolás, Strømskag and Tabchi characterise teaching practices with a sharp focus on teacher learning and teacher knowledge; frameworks of teacher knowledge; issues of communicating mathematics; and teacher-researchers' teaching practices. They consider particularly a quartet of approaches currently deployed for instructional/didactic design (Anthropological Theory of the Didactic, Theory of Didactical Situations, Realistic Mathematics Education and Lesson Study) and highlight developments with regard to instructional design for inquiry-based learning. They conclude by calling for stronger synergy between the research communities of mathematics and mathematics education in addressing the multifaceted challenges of teaching as well as for generating models that are either new or suitably transferable from research into mathematics teaching at other educational levels.

In this book, we offer an overview of the research presented and discussed in the two first INDRUM conferences, INDRUM 2016 and INDRUM 2018. Its chapters reveal the vitality and diversity of didactic research on university mathematics, highlight the need for further research in under-explored topics and propose paths for further research. The biennial INDRUM conferences will continue. INDRUM 2020 took place virtually from Bizerte (Tunisia) in September 2020, and INDRUM 2022 and 2024 are in preparation: they will both reveal new developments and kickstart new collaborations; they will strengthen and enlarge the INDRUM network, with the aim to contribute to the improvement of university mathematic education in its many and varied facets.

We hope the book will provide a useful opportunity for an international audience of researchers and university teachers with an interest in this area to be informed about state-of-the-art developments and heed future research agendas.

PART I

Achievements and current challenges

1

MATHEMATICS EDUCATION RESEARCH AT UNIVERSITY LEVEL

Achievements and challenges

Michèle Artigue

Introduction

This chapter has its origin in the introductory lecture I was invited to give at the first conference of the INDRUM network in Montpellier, in March 2016. As I pointed out at the beginning of the lecture, the demand made by the organizers of the conference was very challenging, for several reasons. The first reason is that this field, seen as an academic field of research, already has a rather long history which can be traced back to the seventies at least. To make sense of research achievements and of the limitation of these, one needs to be aware of some elements of this history. The second reason is that the field is a burgeoning field of research, addressing a huge diversity of questions and mobilizing an increasing diversity of theoretical approaches for their study (Winsløw & Rasmussen, 2019). Making sense of recent research achievements, through the multiplicity of existing discourses, and through the multiplicity of changing contexts where they are produced, is thus a very challenging task.

Reflecting on how I could take up the challenge and organize this lecture, I decided to structure it into two main sections. In the first, perhaps more directed towards researchers in the field not necessarily aware of the details of its historical development, I would try to provide an idea of the evolution of the field, through a reflective look at some important episodes of my personal trajectory. Then, in a second main section, I would present a more global vision of the current state of the field. I have decided to keep the same global organization for this chapter. However, I have updated the reflection developed in the second main section, taking into consideration the proceedings of the first INDRUM conference, the articles published in the recently created *International Journal of Research in Undergraduate Mathematics Education (IJRUME)*, and the synthetic presentations offered in (Biza, Giraldo, Hochmuth, Khakbaz & Rasmussen, 2017; Winsløw, Gueudet,

Hochmuth & Nardi, 2018) and the entries regarding University Mathematics Education (UME) research added in the second version of the *Encyclopedia of Mathematics Education.*

I should add that the vision presented is mine. This is a partial and subjective vision of the field, shaped by my personal trajectory and experience, and the reader is not obliged to agree with it. I hope, however, that it provides some useful elements to understand how this field of research has progressively matured and created its specific identity, what it has achieved, and to outline some perspectives for its future.

Some crucial episodes in a personal journey

Entering the field of university research: a mathematics-physics adventure

I entered research in university mathematics education at the turn of the eighties, when a group of mathematicians and physicists in my university, including researchers in mathematics and physics education, decided to create an experimental course for students entering the mathematics and science programme. Our main goal was to challenge the compartmentalization of the two disciplines. The design of this course was very innovative, with regular lectures on topics of common interest jointly prepared and given by one mathematician and one physicist, along with common tests and interdisciplinary projects (Artigue, 1981). The whole team met once a week and everyone attended the common lectures.

Everything worked fine with the exception of the planned common lecture on the notion of differential for which the mathematicians and physicists could not reach an agreement. This "differential clash" became a research issue that I first addressed jointly with the two didacticians of physics in the team, Laurence Viennot and Edith Saltiel, then with Marc Legrand and his colleagues in Grenoble, who were working on the teaching of the Riemann integral.

If I consider retrospectively the research we developed (Artigue, Menigaux & Viennot, 1990; Alibert et al., 1988), it is rather representative of some predominant characteristics of university research in mathematics and physics education at that time. The research has a clear cognitive orientation. It aims at understanding the conceptions of the differential developed by our students, the difficulties they meet with this notion and the associated processes in mathematics and in physics, and from a methodological perspective, it mainly uses a series of questionnaires and interviews.

This research also demonstrates a strong epistemological sensitivity, perhaps more characteristic of the French didactic community (Artigue et al., 2019). With the help of historians, indeed, we made specific efforts to understand the source of the "differential clash" observed, why mathematicians seemed so rigidly attached to their vision of the differential as a function, in fact a differential form, and why physicists seemed so rigidly attached to their pragmatic vision of differentials as very small, if not infinitesimal, increments. For that purpose, we worked on primary and secondary historical sources, and also systematically studied the traces of the differential/derivative educational debate.

This research also shows the strong desire of researchers to translate their research findings into educational action. The results of both the cognitive and epistemological analyses were used to make visible the negative effects of certain usual teaching practices to mathematicians and physicists. Students, for instance, declared in the interviews that it was better for them not to try to understand what a differential was, and to work mechanically, both in mathematics and in physics; they were not able to distinguish between situations requiring or not the use of differential processes, and they only succeeded in solving the tasks proposed to them because they had learned to detect the linguistic hints calling for the use of such processes in the texts, and to mobilize the respective rituals of each discipline. These results were also used to develop a compromise acceptable by the two teams. Moreover, math-physics tasks in line with this compromise were designed and implemented in specific workshops in the following years (Artigue, Menigaux & Viennot, 1988).

However, this research, and more globally the work carried out in this experimental course, shows that issues of ecology (Artaud, 1997) and sustainability were not really part of our agenda. In fact, when some years later the team moved to other projects and teaching activities, this experimental course disappeared.

Experiencing the "schizophrenia" of undergraduate university teachers: the teaching of differential equations

The second research I will evoke is the research I developed regarding the teaching of differential equations some years later. At that time, I was working on issues related to dynamical systems with some colleagues, with the support of a specialist of this domain, the mathematician Adrien Douady. I experienced the type of "schizophrenia" which is rather common to those who teach undergraduate courses: the complete disconnection between what is their lived experience as mathematicians and their lived experience as university teachers. The programme of the course on ordinary differential equations for second year students I had to teach was focused on solving algebraically by integration in finite terms some specific forms of equations, which led students to think that the goal of research in this domain was to progressively complete the book of recipes they were introduced to. Some exact solutions in terms of power series were also looked for but only advanced students could have access to other views.

With some mathematician colleagues, I began to reflect on the possibility of developing elementary ODE courses more in line with the epistemology of this domain, thanks to the use of computers (Artigue, Gautheron & Sentenac, 1988). Once again, the epistemological work was an essential dimension of the research, leading to the identification of three main historical strands, each one of them having its own *problématique* and development: the algebraic, the numerical and the geometrical-topological strand initiated much later than the two first ones by Henri

Poincaré at the end of the 19th century. Only the first strand, in its most elementary forms, was part of traditional courses (the theory of exact solving as initiated by Condorcet and Liouville was not considered).

In the mid-eighties, the creation of an experimental section for first year students at the University of Lille 1 offered new research possibilities. To address the issue at stake (the viability of a first ODE course reflecting the current epistemology of the field), we used a methodology of didactic engineering (Artigue, 2014), and relied on didactic constructs familiar to French didacticians – such as the notions of *setting* and *tool-object dialectics*[1] due to Regine Douady (1986) – together with fundamental constructs of the Theory of Didactical Situations (TDS) (Brousseau, 1997). This methodology was of course adapted to the specific context of university education. The didactic engineering was collaboratively designed with Marc Rogalski and his colleagues from the University of Lille 1, revised after its first experimentation, and then successfully implemented there for nearly a decade (Artigue & Rogalski, 1990). In fact, the first experimentation showed that the viability of such a course required a different institutional status for graphical representations than the limited heuristic status given to them in university courses; graphical representations had to be credited as a legitimate tool for reasoning and proof, of course in appropriate forms such as those developed in the didactic engineering (Artigue, 1992). Moreover, we discovered that such a change could not be limited to the teaching of differential equations; for evident reasons of coherence, this change impacted the didactic contract (Brousseau, 1997) of the whole Analysis course the teaching of ODE belonged to. This certainly helps to understand why, despite its repeated successful implementation, only the first situations of this design were more widely used. Today, conceptual tools such as the hierarchy of levels of didactic co-determinacy proposed by the Anthropological Theory of the Didactic (ATD) (Chevallard, 2002; 2019) help us to systematically consider the different conditions and constraints governing the possible ecology of didactic designs, beyond those situated at the level of the mathematical *theme* or *sector* directly addressed[2] – and to better anticipate their possible effects. We were less equipped to address these ecological issues thirty years ago. It must also be mentioned that, since then, research on the teaching and learning of differential equations has developed internationally, studying more systematically the accessibility and learning potential of teaching designs combining algebraic, numeric and geometrical-topological approaches as shown in the synthesis (Kwon, 2020).

These are just two examples among many others. They reflect the cognitive and epistemological focus of the research carried out at that time, and also the form that this focus was likely to take in the French didactic culture where TDS, with its underlying systemic perspective was the predominant theoretical approach. They also reflect the engagement of researchers in action, but without the conceptual tools that would have allowed them to seriously address dissemination and sustainability issues.

The international scene: the AMT Working Group of PME and the ICMI Study on the teaching and learning of mathematics at university level

On the international scene, the state of research at that time is well reflected by the work of the *Advanced Mathematic Thinking* (AMT) working group of the International Group for the Psychology of Mathematics Education (PME) which I entered in the late eighties, and the *AMT* book coordinated by David Tall (1991) resulting from this work. This book confirms the cognitive orientation, already mentioned, of research towards the study of students' learning processes, thinking modes, conceptions and difficulties, and also the strong influence of constructivist perspectives. One of the main aims of the working group was to elucidate the specific nature, if any, of what its participants called "advanced mathematical thinking", and thanks to this elucidation, to better understand what differentiated learning processes at university from those experienced before by students. Even if a definitive answer is not provided in the book, some criteria are proposed in terms of relationship to abstraction, symbolism and generalization, role of definitions, formal reasoning and proof. One can also observe the important role played in research by constructions that take the form of distinctions such as that between *concept definition* and *concept image*[3] due to David Tall and Shlomo Vinner (1981), or emerging theories such as the construction developed by Dubinsky based on Piaget's idea of reflective abstraction that would become APOS theory (Arnon et al., 2014) or that developed by Tall that led to his theory of cognitive development along three different worlds: the *conceptual-embodied world* based on perception, action and thought experiment, the *proceptual-symbolic world* of calculation and algebraic manipulation and the *axiomatic-formal world* of set-theoretic concept definitions and mathematical proof (Tall, 2013). One can also note the predominance of Calculus / elementary Analysis as a mathematical domain. At the time, many research projects, indeed, were motivated by the high level of failure in the Calculus undergraduate courses, a gateway to scientific studies in most universities. The concept of limit, considered as its foundational core concept, is especially addressed.

Another characteristic, well representative of educational research at that time, is the emphasis put on cognitive discontinuities and their epistemological sources in learning processes, and on the persistent difficulties generated by these discontinuities. Different theoretical constructs contribute to the conceptualization of these discontinuities – I just mention three of these: the notion of *epistemological obstacle* borrowed from Bachelard's epistemology (Bachelard, 1938; Brousseau, 1983) that shows the role played in students' resilient difficulties by forms of knowledge which have proved to be effective in other contexts (as shown by researchers such as Bernard Cornu (1991) and Anna Sierpinska (1985) for the concept of limit); the discontinuity between *proceptual* (Gray & Tall, 1994) and *formal thinking*; the discontinuity between concepts that emerge as necessary ingredients for the solution of specific problems such as the concepts of derivative and integral, and concepts that respond to unifying and formalization needs such as the concept of abstract vector space, named FUG (formalizing, unifying, generalizing) concepts by Aline Robert (1998), Jean-Luc Dorier (2000) and their colleagues.

Discontinuities are also identified between domains, for instance between Algebra and Analysis. I just mention below some of those which have proved especially challenging for students (Artigue, 1997):

- the change needed in the perception of equality as, in order to understand the mechanism of proofs in Analysis, it must be perceived as a sign expressing arbitrary level of closeness;
- the predominant role taken by inequalities over equalities, and more than that, the transition from global perspectives regarding the solving of inequalities to a subtle combination of local and global perspectives;
- the change induced by the transition from reasoning modes based on equivalence to reasoning modes based on the use of sufficient conditions, whose effectiveness requires the ability to lose information in a controlled way, taking into account both the respective orders of magnitude of terms in symbolic expressions and the local character of reasoning in Analysis.

Let me stress that what is more globally addressed here is the change and reconstructions needed in mathematical practices when moving from one domain to another – what today, referring to ATD, I would call a change in *mathematical praxeologies*[4].

As stressed earlier, research first focused on discontinuities, but progressively it became more sensitive to the essential role played by connections and flexibility in teaching and learning processes. Such an evolution has been supported by the increased attention paid to the semiotic dimension of mathematical activity in educational research, and also by the evolution of technology and its specific semiotic affordances. In UME research, this evolution is visible for instance in the presentation of the state-of-the-art of research about the teaching and learning of linear algebra co-authored by Jean-Luc Dorier and Anna Sierpinska in the ICMI Study 11 devoted to the teaching and learning of mathematics at university level (Dorier & Sierpinska, 2001). Their chapter underlines the complexity of connections at stake in linear algebra: connections between different languages (geometric, algebraic, abstract), between different registers of representations (graphical, algebraic, symbolic representations, tables), between Cartesian and parametric points of view, and between synthetic-geometric, analytic-arithmetic and arithmetic-structural modes of reasoning. Moreover, regarding teaching practices, they underline that university teachers, most often, jump without any precaution between these different representations and languages, underestimating the difficulties that these jumps provoke for their students. Of course, connections and flexibility are not a specificity of university mathematics, but what changes is their intensity, and the autonomy given to the students regarding their management.

I was involved in the ICMI Study 11 just mentioned, co-ordinated by Derek Holton, as a member of its International Programme Committee and this was a very interesting experience. As is the case for any ICMI Study, our collective work intertwined general reflection on the themes identified in the discussion document, syntheses of research advances, and the presentation and discussion of many innovative realizations carried out in different contexts. Published ten years after the AMT book, the ICMI Study 11 volume (Holton, 2001) shows the diversity of issues addressed by those interested in teaching and learning at university level, not just researchers in the field: curricular and assessment issues, teaching practices, relationships between mathematics and other disciplines, affordances of technology, and teacher education including that of university teachers still in an emerging state at that time. Issues related to the secondary/university transition are addressed in several chapters, and they are more widely approached than in the AMT book, considering the diversity of social and psychological moves that this transition entails for students. The research section, however, tends to show that the "social turn" in mathematics education research, as denoted by Steve Lerman (2000), has not yet substantially impacted research at that level.

The socio-cultural turn through the lens of ATD

For me, in fact this socio-cultural turn was tightly linked to the incorporation of ATD in my research perspectives. It first occurred with the supervision of Brigitte Grugeon's doctoral thesis on the transition between vocational high school and technological high school in France (Grugeon, 1995). This incorporation proved to be so productive that I engaged Frederick Praslon, another doctoral student of mine, to adopt ATD as his macro-theoretical framework for the study he was undertaking on the secondary/university transition regarding the concept of derivative and its mathematical environment (Praslon, 2000).

As expressed very well by Marianna Bosch and her colleagues (Bosch, Fonseca & Gascón, 2004), adopting such a perspective on the secondary/university transition represents a radical move. The lens is no longer directed towards the students and their cognitive functioning or development, but towards the institutional practices that condition and constrain, both explicitly and implicitly, what they have the possibility to learn or not, and to the associated systems of norms and values that remain partly tacit. In his pioneering work, Frederick Praslon in fact used ATD to question the common vision of this transition at the time: a transition from the proceptual to the formal mathematical world, from intuitive and pragmatic reasoning modes to rigorous mathematical ones. Carefully analyzing mathematical praxeologies in scientific high school and first university year through a diversity of institutional sources, he showed that contrary to what was often claimed by university teachers, a substantial universe around the notion of derivative was already built at the end of high school in France, at least for students in the scientific stream, but that a dramatic extension of the landscape was taking place in the first six months at university, which he visualized using concept maps. He also showed that the transition from secondary to mathematics-sciences programmes at university was not a radical move from the proceptual to the formal mathematical world, from an intuitive and algorithmic Calculus to the approximation world of Analysis; it was more like an accumulation of micro-breaches, less visible and thus not appropriately addressed by the institution. The main breaches he identified are the following:

- an increasing speed in the introduction of new objects;
- a greater diversity of tasks making routinization much more difficult;
- much more autonomy given in the solving process for similar tasks;
- a new balance between the particular and the general, the *tool* and *object* dimensions of mathematical concepts;
- objects more controlled by definitions, results more systematically proved, and proofs which are no longer "the cherry on the cake" but take the status of mathematical methods.

As he evidenced, the conjunction of these breaches created a substantial gap but university teachers were not aware of it in their great majority and tended thus to under-estimate the induced cognitive cost for their students. To make university teachers and students sensitive to these changes, Praslon designed a set of tasks situated in the gap between the two cultures: a priori compatible with high school knowledge but fully exotic in high schools, and at the same time not genuine university tasks.

I will not enter into more details in this section. Since that time, the use of ATD for the study of institutional transitions has substantially developed, with the incorporation of conceptual tools such as the hierarchy of levels of didactic co-determinacy not available at the time of Praslon's doctoral thesis, or the development of specific notions such as the notion of *completeness of praxeologies* (Bosch, Fonseca & Gascón, 2004). These developments have made possible the identification of new characteristics of the secondary/university transition: incompleteness and isolation of high school praxeologies, changes in the respective importance attached to the praxis and theoretical blocks of praxeologies, and in the topogenetic distribution of roles between teacher and students. Beyond that, the development of the design dimension of ATD within the paradigm of "Questioning the world" and the concept of Study and Research Path (SRP) (Chevallard, 2015; 2019) and the refinements of the theory induced by its increasing use with students in mathematics service courses, for instance engineering students, have opened new research perspectives (see for instance the pioneering theses by Berta Barquero (2009) and Avenilde Romo-Vázquez (2009) or more recently those of Rita Vázquez Padilla (2017) or Ignasi Florensa Ferrando (2018), and also Castela & Romo Vázquez, 2011)

Mentioning these more recent developments ensures a transition with the next section of this chapter in which I consider more broadly the state of the field, highlighting what I perceive as its main strengths and promising evolutions as well as the challenges it faces.

Strengths, promising evolutions and challenges

Strengths and promising evolutions

The field of UME research presents evident strengths. It has developed over more than four decades and is clearly a burgeoning field of research today. This is attested for instance by the increasing number of participants in the associated Topic

Working Group at CERME Conferences, the success of the first INDRUM Conferences held in 2016 and 2018, the creation of the INDRUM network itself, and also the creation of an international research journal especially devoted to UME in 2015: *The International Journal of Research in Undergraduate Mathematics Education* (*IJRUME*).

This is also a field where regular efforts of coordination and synthesis are made with effective results. I have already evoked two of these and the AMT and ICMI Study 11 books they led to. In the recent years, new syntheses have been produced, such as the chapter I co-authored with Carmen Batanero and Philip Kent for the second NCTM Handbook (Artigue, Batanero & Kent, 2007), the book *Amongst Mathematicians* (Nardi, 2008), the survey led by Mike Thomas for ICME-12 (Thomas et al., 2014), the chapter synthesizing the work carried out at the 10 CERME conferences (Winsløw, Gueudet, Hochmuth & Nardi, 2018), the special issue (Nardi, Biza, González-Martín, Gueudet & Winsløw, 2014) focusing on theoretical paradigms underlying UME research, the synthesis produced by the University Topic Study Group at ICME-13 (Biza, Giraldo, Hochmuth, Khakbaz & Rasmussen, 2017), and last but not least, the fact that 10 entries devoted to UME research have been added to the second version of the *Encyclopedia of Mathematics Education*[5]. All these syntheses show that a substantial amount of knowledge has been accumulated along the years, and also that structured, connected and coherent accounts of this knowledge are accessible to those interested in the field, which is not always the case in the field of mathematics education. The fact that the community working in UME research is still of reasonable size certainly favours coordination.

I have stressed the importance of epistemological reflection in the emergence of the field, which I also consider to be a strength, and I have evoked some forms this reflection has taken. This epistemological work is going on, efficiently accompanying the development of UME research. There is no doubt that the fact that researchers specializing in UME have generally a strong mathematics background and are used to collaborate with research mathematicians, at least for their teaching activity, supports this epistemological work. These conditions make it easier for them to take into account in their epistemological analyses the evolution of mathematical sciences and the diversity of praxeologies these nurture in an increasing number of disciplines and contexts today. Epistemological work also benefits from the advances of research in the history, epistemology and philosophy of sciences, and from the evolution of research perspectives in these areas. My laboratory being attached since 2000 to a doctoral school structured around philosophy, history, epistemology and didactics of sciences, I have a personal experience of such interactions. I observe for instance an increasing shared interest in the diversity of life forms that mathematics always had and still has, according to the institutions and cultures where it is produced and practised, which currently supports the development of more productive interactions between research communities.

The emergence of the field was also characterized by the domination of cognitive constructivist perspectives. I consider as an achievement of UME research the fact that it has succeeded in emancipating from these perspectives whose limitations are evident. Moreover, an interesting point is that this emancipation most often goes along with a process of reformulation and reconstruction of earlier research outcomes making possible their exploitation in the new research perspectives. I have personally experienced such reconstructions in the diverse doctoral theses I have supervised on the secondary-tertiary transition (Praslon, 2000; Bergé, 2004; Najar, 2010). These doctorate students, in fact, were able to incorporate constructs and results regarding the learning of Calculus and Analysis such as those identified in the AMT book for instance into research *problématiques* based on institutional and anthropological theoretical frameworks. I see also a sign of it in the current enterprise of networking between APOS and the ATD, two theoretical constructions that I would position at the extreme opposites of our field and its use in UME research (Bosch, Gascón & Trigueros, 2017; Vásquez, 2017).

Another strength of the field is certainly its move from investigations focusing on the student to a more balanced interest in both the student and the teacher. This move is not proper to UME research as is well known, but it seems to have been more difficult to achieve at this level of schooling. Today, however, this obstacle seems eventually overcome, and university teacher practices are becoming an object of study in their own (Nardi & Rasmussen, 2019). Research both investigates lecture-oriented practices still dominant in UME, their potential and limitation, and complements or alternatives to these. Researchers also invest in the design, implementation and analysis of projects aiming at the didactic acculturation of university teachers, extending the pioneering work of Barbara Jaworski and Elena Nardi at Oxford University years ago (Nardi, Jaworski & Hegedus, 2005), and benefiting from the potential of new theoretical perspectives, such as those offered by the theories of community of practice (Biza, Jaworski & Hemmi, 2014) and community of inquiry (Jaworski, 2008) (see (Jaworski, 2019) for a synthetic presentation).

More generally, theoretical evolution in the field is both promising and challenging. I have already evoked the increasing use in research of ATD, the potential of which for university research has been especially analysed by Carl Winsløw in his regular lecture at ICME-12 (Winsløw, 2014). Furthermore the *Research in Mathematics Education* Special Issue (Nardi, Biza, González-Martín, Gueudet & Winsløw, 2014) is especially insightful from this perspective, considering the affordances of a range of socio-cultural, institutional and discursive theories: ATD, TDS and the theories of Communities of Practice and Communities of Inquiry already mentioned, and also instrumental and documentational approaches (Gueudet, Pepin & Trouche, 2012) and the theory of Commognition (Sfard, 2008). There is no doubt for me that these theories offer evident potential for research at university level and at the transition between secondary and university education. Without entering into more details about this potential, I would say that the most challenging perspective for me is that offered by commognition. I still wonder about the consequences of adopting such a radical discursive approach, but the way it engages us to analyse communicative acts involving students and teachers (Nardi, Ryve, Stadler & Viirman, 2014), to think about the teacher role and the resilience of university practices such as lecturing (Sfard, 2014) is, without any doubt, really insightful. More globally, despite the efforts of clarification and networking undertaken, there is no doubt that the increasing diversity of the theoretical landscape in the field is by itself challenging.

This being said, there is no doubt that the field also presents some weaknesses and faces many other challenges, and in the next sub-section, I will discuss some of them.

Challenges

In the ICMI Study 11 book, it was pointed out that research concentrated too much of its efforts on the classical training of future mathematicians despite the fact that these represent only a very small percentage of university students being taught some mathematics. This was the reason why, when I was asked to lead the authorship of the chapter on learning mathematics at post-secondary level of the second NCTM Handbook (Artigue, Batanero & Kent, 2007), I proposed as co-authors Carmen Batanero and Philip Kent, who could help me to offer a more balanced perspective, paying due attention to stochastic and engineering education. Still today, there is no doubt that research remains biased both in terms of domains and in terms of population. Fields of increasing importance in mathematics – such as the stochastic field including probability and statistics, and more generally applied and computational mathematics – seem still under-investigated. For instance, in the 69 original articles published in *IJRUME* in the years 2015–2018, the great majority (45) either directly focus on the teaching and/or learning of Calculus/Analysis and Abstract/Linear Algebra, or use these domains to address more general issues, especially the teaching and learning of proof; only six focus on the learning of other mathematical domains (combinatorics, logic, probability) and only three address the specificity of service courses for non-mathematics disciplines. Not independently from that, still the practice of the mathematician researcher, and even the pure mathematician researcher, is the explicit or implicit reference in most research studies. The diversity of forms of professional relationship with mathematics for which university courses may prepare graduates is still not sufficiently investigated and taken into account, despite the fact that we are much better equipped for tackling such issues.

However, some positive evolution can be observed. As pointed out in the entry of the Encyclopedia of Mathematics Education on Service-courses in university mathematics (Hochmuth, 2020), "research on mathematical service-courses plays an increasing role at international conferences on university mathematics education like CERME (Winsløw et al., 2018), ICME (Biza, Giraldo, Hochmuth, Khakbaz & Rasmussen, 2017), INDRUM (Durand-Guerrier, Hochmuth, Goodchild & Hogstad, 2018) and RUME (Weinberg, Rasmussen, Rabin, Wawro & Brown, 2017) (in the designated working-groups, respectively)". And, among the 30 references associated with this entry, 21 correspond to publications after 2010.

UME research in the field of stochastics teaching and learning is also increasing, but perhaps in a way less visible from the mathematics education community because it has its specific channels of publication like the *Journal of Statistics Education* (JSE) or the *Statistical Education Research Journal* (SERJ). It is interesting to notice that, as pointed out in (Eichler & Zapata-Cardona, 2016), this research pays particular attention to the affective dimension of learning processes, an area of research well developed in mathematics education (Hannula, 2014) and also increasingly present in UME research. Moreover, research is also extending beyond the level of undergraduate courses, as pointed out in (Winsløw, Gueudet, Hochmuth & Nardi, 2018). The research presented in (Hausberger, 2018) on the development of structuralist thinking and the meta-concept of structure is a good example of such extension. However, achieving a better coverage by research into the diversity of tertiary mathematics is still a challenge for the field.

Another challenge in my opinion is the excessive predominance of small-scale qualitative studies, involving a very limited number of students or teachers. Moreover, reviewing submissions or reading papers, I am also often disappointed when, after reading in the methodological section that the authors have collected a huge amount of data, I observe then in their analyses that the evidence provided for supporting their claims reduces to the micro-analysis of some limited episodes. The a priori possible triangulation between different levels and dimensions of analysis that would make the results more convincing is at most evoked. The increasing attention paid to the semiotic and discursive dimensions of mathematical activity with the associated needs of micro-analysis seems to reinforce this trend, and there is no doubt that the strict limitation of space imposed to conference proceedings does not help. From this perspective, the contrast between conference papers, for instance those of CERME and INDRUM proceedings, and articles published in *IJRUME* not submitted to such limitations is interesting. However, we certainly need a better balance between small-scale and medium-scale studies at least in the field, between the quantitative and the qualitative in research methodologies, and also more triangulation between different levels and dimensions of analysis.

Another essential challenge is the still very limited influence of research outcomes on university teaching practices today. Of course, such difficulties are not specific to the field of UME research, but one could expect that, being themselves researchers, university teachers would be more open to considering research advances and what these can offer them to better understand their students, and to improve their teaching practices. Unfortunately, this does not seem to be the case in general, for many reasons which range from the low institutional value attached to teaching activities in comparison to research activities at university to the image of the didactic field itself in the mathematics community, in many countries. There is no doubt that making sense of research results in mathematics education and converting them into something useful in practice, is not an easy task. The activities of networking between theoretical frameworks I have been involved in through different projects in the last decade (Bikner-Ahsbahs & Prediger, 2014; Lagrange & Kynigos, 2014) have evidenced that, even for a researcher in mathematics education, to make sense of other research approaches and results, just by reading the associated literature and by attending seminars or conferences, is difficult. In these projects, appropriation resulted in fact from the collaborative building of networking praxeologies on top of our own research praxeologies (Artigue & Bosch, 2014), and the process was very progressive. The communication between mathematicians and didacticians does not face exactly the same problems, as extensively discussed in Fried & Dreyfus (2013). However, this experience has reinforced my conviction that to overcome the current limitations, we must not think in terms of dissemination of research results, but in terms of collaborative projects, building and negotiating, jointly with mathematicians and other university teachers, *problématiques* that make sense for all those involved, and meet their respective interests and needs. And then we should combine our respective knowledge and expertise in these projects through appropriate praxeologies. Of course, collaborative projects have existed for decades. The two experiences I mentioned at the beginning of this text were clearly collaborative projects; the ICMI Study 11 book (Holton, 2001) presents a range of examples. However, each of them more or less appears as a particular and isolated case. We can go further today; we have more powerful conceptual tools in order to approach ecological and institutional issues, to approach collaborative work and relationships between communities, to build, analyse and compare projects, and, last but not least, to consider the long-term dynamics necessarily at stake. Once again, some promising evolutions are observed as shown in Winsløw et al. (2018), but the issue remains a major challenge in many parts of the world.

To this can be added the challenge of connection between the living field of mathematics, so dynamic and diverse, and undergraduate mathematics education, both in terms of content and practice. There is no doubt that, contrary to graduate mathematics education, undergraduate mathematics education is poorly connected to the mathematics of today in most universities. It is generally argued that the limited mathematical background of undergraduate students makes the connection with the sophisticated world of current mathematics impossible. This may be the case if we consider that this connection must necessarily be expressed in terms of operational knowledge. However, if we consider mathematics as a part of human culture, we must admit that, as any cultural form, our mathematical culture is not reducible to its operational part. The distinction between different forms of relationship with mathematical objects and practices opens the landscape towards alternative didactic strategies and practices. Those developed and used by the very active community engaged in the popularization of mathematics could be a source of inspiration. However, as was evidenced by ICMI Study 16 which addressed this topic (Barbeau & Taylor, 2009), up to recently at least, didactic research has paid limited attention to popularization practices, and more generally informal mathematics education. Moreover, as far as I know, the didactic community still has limited contact with the community of research in science communication with which the collaboration could certainly be helpful. There are thus resources that could be more systematically explored for addressing this challenge.

When considering the connection with the living field of mathematics, one cannot forget that technological evolution has substantially impacted professional mathematical practices, in particular by providing much more powerful tools for supporting an experimental dimension of mathematical practices that has always existed, and by making this experimental work more visible and sharable (cf. for instance the journal *Experimental Mathematics*). However, in many places, undergraduate mathematics education seems still blind to this evolution, even when those in charge make extensive use of technology in their professional activity. Making visible the experimental dimension of mathematics tends to be perceived as an obstacle to the entrance into the deductive game of mathematics aimed at, leading in some first-year programmes to the banishment of any technological tool from algebra or analysis courses and/or from assessments. Overcoming such a limited epistemological view and its negative consequences is a challenge that researchers in mathematics education have faced for decades, but there is no doubt that the situation remains critical today at university level in many places, contributing to the rupture with mathematical practices in secondary education. Some perspectives are however promising, such as those associated with the experimentation of inquiry-based practices in university courses (see (Rasmussen & Wawro, 2017) for instance) but the situation remains very challenging in most contexts.

In fact, the changes induced by the evolution of technology are not limited to those just evoked, raising new challenges. The digital era which we have entered induces dramatic changes in the way we access information and build knowledge, in the way we communicate, interact and work. New pedagogical strategies develop, such as reversed pedagogy, MOOCs and diverse forms of hybrid pedagogy, which need to be studied. The number of online resources increases exponentially as well as the diversity of learning sources, and modalities of use. Once again, didactic research offers promising tools to take up this challenge, the documentational approach initiated by Ghislaine Gueudet and Luc Trouche being one of the most recent (Gueudet, Buteau, Mesa & Misfeldt, 2014; Gueudet, Pepin & Trouche, 2012).

Conclusion

Opening the first INDRUM Conference, I had tried to share with the participants the experience of a researcher who has been active in the field of university mathematics education for more than three decades. I organized my reflection around the historical evolution of this field of research, convinced that this could help understand its current state and better appreciate its achievements, identify its strengths and weaknesses, as well as challenges for future research. I have tried to make clear that substantial advances have been made, both from a theoretical and empirical point of view, that knowledge has progressively accumulated, and that, even if weaknesses still exist, we are today reasonably equipped to take up the many challenges that we have to face.

Three years after the first INDRUM conference, I observe promising evolutions that make me confident in the collective capacity of the growing community of

UME researchers to take up the many challenges it faces, many of these being shared with other communities in mathematics education research but also some being more specific to the UME context. As pointed out in the introduction, my vision of the field is partial and subjective, shaped by my own experience and by the research and educational cultures in which it has mainly developed. I hope, however, that it is not so singular and will be of interest for all those committed to the field, that it will refresh the memory of those who have been involved in the field for a long time while contributing to the acculturation of the newcomers who do not share this background.

Notes

1 A setting denotes the set of objects of a mathematical field, the relationships between them, their diverse formulations and the associated mental images. A mathematical concept is seen as a *tool* when one focuses on its use to solve problems, as an *object* when considered as a cultural object, element of scholarly knowledge. The *tool/object* dialectic is based on the fact that, in the historical development of concepts, the tool dimension emerges before the object one, and plays with their dialectic interactions.

2 The hierarchy of levels of didactic co-determinacy is structured around ten levels in interactions: Topic - Theme - Sector - Domain - Discipline - Pedagogy - School - Society - Civilization - Humanity. The theme "Qualitative approach of ODE" was part of the sector "ODE", itself part of the domain "Analysis" in the course created at the University of Lille 1.

3 These notions express a distinction between mathematical concepts as formally defined (concept definition) and the cognitive processes by which they are conceived, with the concept image denoting "the total cognitive structure associated with the concept, which includes all the mental pictures and associated properties and processes" (Tall & Vinner, 1981).

4 The concept of praxeology is a core concept in ATD, used to model all kinds of human practices. A praxeology is defined as a four-component structure (T, τ, θ, Θ), T being a *type of tasks,* τ a *technique*, i.e. a way of performing tasks of type T, θ a *technology*, i.e. a way of explaining and justifying τ, and Θ a theory providing background to θ.

5 In the first version, there was only one such entry entitled "Calculus Teaching and Learning" covering both research carried out at secondary and undergraduate levels. The new entries are: University Mathematics Education, Linear Algebra Teaching and Learning, Abstract Algebra Teaching and Learning, Analysis Teaching and Learning, Differential Equations Teaching and Learning, Logic in University Mathematics Education, Service-Courses in University Mathematics Education, Secondary-tertiary transition in mathematics education, Preparation and Professional Development of University Mathematics Teachers, Teaching Practices at University Level.

References

Alibert, D., Artigue, M., Courdille, J.M., Grenier, D., Hallez, M., Legrand, M., Menigaux, J., Richard, F. & Viennot, L. (1988). Le thème "différentielles" - un exemple de coopération maths-physique dans la recherche. In G. Vergnaud, G. Brousseau, & M. Hulin (Eds.), *Didactique des Acquisitions des Connaissances Scientifiques. Actes du Colloque de Sèvres* (pp. 7–45). Grenoble: La Pensée Sauvage.

Arnon, I., Cotrill, J., Dubinsky, E., Oktaç, A., Fuentes, S., Trigueros, M., & Weller, K. (2014). *APOS Theory: A Framework for Research and Curriculum Development in Mathematics Education.* New York: Springer.

Artaud M. (1997). Introduction à l'approche écologique du didactique. L'écologie des organisations mathématiques et didactiques. In M. Bailleul *et al.* (Eds.), *Actes de la IXème Ecole d'Eté de Didactique des Mathématiques* (pp. 101–139). ARDM

Artigue M. (1981). Une expérience de coordination des enseignements de mathématiques et de physique en DEUG SSM. *Cahier de Didactique et Pédagogie*, 22, IMAG. Grenoble.

Artigue, M., Gautheron, V., & Sentenac, P. (1988). Qualitative study of differential equations. In T.F. Banchoff, I. Capuzzo Dolcetta, M. Dechamps, M. Emmer, H. Koçak, & D.L. Salinger (Eds.), *ECM/87, Educational Computing in Mathematics* (pp. 135–144). Amsterdam: North Holland.

Artigue M., Menigaux J. & Viennot L. (1988). *Questionnaires de Travail sur les Différentielles. Brochure n°76.* IREM Paris 7.

Artigue M., Menigaux J. & Viennot L. (1990). Some aspects of students' conceptions and difficulties about differentials. *European Journal of Physics*, 11, 262–267.

Artigue, M, & Rogalski, M. (1990). Enseigner autrement les équations différentielles en DEUG. In Commission inter IREM Université (Ed.), *Enseigner Autrement les Mathématiques en DEUG A Première Année* (pp. 113–128). Lyon: LIRDIS.

Artigue, M. (1992). Functions from an algebraic and graphic point of view: cognitive difficulties and teaching practices. In E. Dubinski & G. Harel (Eds.), *The Concept of Function: Aspects of Epistemology and Pedagogy* (pp. 109–132). MAA Notes no. 25. Mathematical Association of America.

Artigue, M. (1997). Teaching and learning elementary analysis. In C. Alsina *et al.* (Eds.), *International Congress on Mathematics Education, Selected Lectures* (pp.15–30). Sevilla: SAEM Thales.

Artigue, M., Batanero, C., & Kent, P. (2007). Learning mathematics at post-secondary level. In F. Lester (Ed.), *Second Handbook of Research on Mathematics Teaching and Learning* (pp. 1011–1049). Greenwich, CT: Information Age Publishing.

Artigue, M. (2014). Didactical engineering in mathematics education. In S. Lerman (Ed.), *Encyclopedia of Mathematics Education* (pp. 159–162). New York: Springer.

Artigue, M. & Bosch, M. (2014). Reflection on networking through the praxeological lens. In A. Bikner-Ahsbahs, & S. Prediger (Eds.), *Networking of Theories as a Research Practice in Mathematics Education* (pp. 249–266). New York: Springer.

Artigue, M. *et al.* (2019). The French didactic tradition. In B. Werner, M. Artigue, M.A. Mariotti, R. Sträßer, & M. Van den Heuvel-Panhuizen (Eds.), *European Traditions in Didactics of Mathematics* (pp. 11–56). New York: Springer.

Bachelard, G. (1938). *La Formation de l'Esprit Scientifique.* Paris: Librairie Jean Vrin.

Barbeau, E. & Taylor, J.P. (2009). *Challenging Mathematics In and Beyond the Classroom. The 16th ICMI Study.* New York: Springer.

Barquero, B. (2009). *Ecología de la Modelización Matemática en la Enseñanza Universitaria de las Matemáticas.* Doctoral thesis. Universidad Autónoma de Barcelona.

Bergé, A. (2004). *Un Estudio de la Evolución del Pensamiento Matemático: El Ejemplo de la Conceptualización del Conjunto de los Números Reales y de la Noción de Completitud en la Enseñanza Universitaria.* Doctoral thesis. Universidad de Buenos Aires.

Bikner-Ahsbahs, A., & Prediger, S. (Eds.) (2014). *Networking of Theories as a Research Practice.* New York: Springer.

Biza, I., Giraldo, V., Hochmuth, R., Khakbaz, A., & Rasmussen, C. (2017). *Research on Teaching and Learning Mathematics at the Tertiary Level: State-of-the-art and Looking Ahead.* ICME-13 Topical Surveys Book Series. New York: Springer.

Biza, I., Jaworski, B., & Hemmi, K. (2014). Communities in university mathematics, *Research in Mathematics Education*, 16(2), 161–176.

Bosch, M., Fonseca, C., & Gascón, J. (2004). Incompletitud de las organizaciones matemáticas locales en las instituciones escolares, *Recherches en Didactique des Mathématiques*, 24(2–3), 205–250.

Bosch, M., Gascón, J., & Trigueros, M. (2017). Dialogue between theories interpreted as research praxeologies: the case of APOS and the ATD, *Educational Studies in Mathematics*, 95(1), 39–52.

Brousseau, G. (1983). Les obstacles épistémologiques et les problèmes en mathématique. *Recherches en Didactique des Mathématiques*, 4(2), 165–198.

Brousseau, G. (1997). *Theory of Didactical Situations in Mathematics*. Dordrecht: Kluwer.

Castela, C., & Romo Vázquez, A. (2011). Des mathématiques à l'automatique: étude des effets de transposition sur la transformée de Laplace dans la formation des ingénieurs, *Recherches en Didactique des Mathématiques*, 31(1), 79–130.

Chevallard, Y. (2002) *Organiser l'étude 3. Écologie & régulation*. In J.-L. Dorier, M. Artaud, M. Artigue, R. Berthelot, R. Floris (Eds.) *Actes de la 11e Ecole d'Eté de Didactique des Mathématiques* (pp.41–56). Grenoble: La Pensée Sauvage.

Chevallard, Y. (2007). La dialectique des medias et des milieux. In G. Gueudet & Y. Matheron (Eds.), *Actes du Séminaire National de Didactique des Mathématiques, Année 2007* (pp. 344–366). Paris: IREM Paris 7.

Chevallard, Y. (2015). Teaching mathematics in tomorrow's society: A case for an oncoming counter-paradigm. In Sung Je Cho (Ed.), *Proceedings of the 12th International Congress on Mathematical Education* (pp. 173–188). New York: Springer.

Chevallard, Y. (2019). Introducing the anthropological theory of the didactic: An attempt at a principled approach. *Hiroshima Journal of Mathematics Education*, 12, 1–44.

Cornu, B. (1991). Limits. In D. Tall (Ed.), *Advanced Mathematical Thinking* (pp. 153–166). Dordrecht: Kluwer.

Dorier, J.-L. (Ed.) (2000). *On the Teaching of Linear Algebra*. Dordrecht: Kluwer.

Dorier, J.-L., & Sierpinska, A. (2001). Research into the Teaching and Learning of Linear Algebra. In D. Holton (Ed.) (2001). *The Teaching and Learning of Mathematics at University Level. An ICMI Study* (pp. 255–274). Dordrecht: Kluwer.

Douady, R. (1986). Jeux de cadres et dialectique outil-objet [Games between setting and tool-object dialectics]. *Recherches en Didactique des Mathématiques*, 7(2), 5–32.

Durand-Guerrier, V., Hochmuth, R., Goodchild, S., Hogstad, N.M (2018). *Proceedings of the Second Conference of the International Network for Didactic Research in University Mathematics*. Kristiansand: University of Agder & INDRUM.

Duval, R. (1995). *Sémiosis et Pensée Humaine [Semiosis and human thinking]*. Bern: Peter Lang.

Eichler, A., & Zapata-Cardona, L. (2016). *Empirical Research in Statistics Education*. New York: Springer.

Florensa Ferrando, I. (2018). *Contributions of the Epistemological and Didactic Analysis: Question-Answer Maps in Engineering and in Teacher Education*. Doctoral thesis. Universitat Ramon Llull.

Fried, M.N., & Dreyfus, T. (Eds.) (2013). *Mathematics and Mathematics Education: Searching for Common Ground*. New York: Springer.

Gray, E., & Tall, D. (1994). Duality, ambiguity and flexibility: a proceptual view of simple arithmetic. *Journal for Research in Mathematics Education*, 26(2), 115–141.

Grugeon, B. (1995). *Etude des Rapports Institutionnels et des Rapports Personnels des Elèves à l'Algèbre Elémentaire Dans la Transition Entre Deux Cycles d'Enseignement: BEP et Première G*. Doctoral thesis. Université Paris Diderot – Paris 7.

Gueudet, G., Pepin, B., & Trouche, L. (2012). *From Text to "Lived resources": Mathematics Curriculum Material and Teacher Development*. New York: Springer.

Gueudet, G., Buteau, C., Mesa, V., & Misfeldt, M. (2014). Instrumental and documentational approaches: from technology use to documentation systems in university mathematics education, *Research in Mathematics Education*, 16(2), 139–155.

Hannula, M.S. (2014). Affect in Mathematics Education. In S. Lerman (Ed.) *Encyclopedia of Mathematics Education*. New York: Springer.

Hausberger, T. (2018). Structuralist praxeologies as a research program in the didactics of Abstract Algebra. *International Journal of Research in Undergraduate Mathematics Education*, 4(1), 74–93.

Hochmuth, R. (2020). Service-courses in university mathematics education. In S. Lerman (Ed.), *Encyclopedia of Mathematics Education* (second edition). New York: Springer.

Holton, D. (Ed.) (2001). *The Teaching and Learning of Mathematics at University Level. An ICMI Study*. Dordrecht: Kluwer.

Jaworski, B. (2008). Building and sustaining inquiry communities in mathematics teaching development: Teachers and didacticians in collaboration. In K. Krainer, & T. Wood (Eds.) *International Handbook of Mathematics Teacher Education: The Mathematics Teacher Educator as a Developing Professional* (vol. 3, pp. 335–361). Rotterdam: Sense.

Kwon, O.N. (2020). Differential equations teaching and learning. In S. Lerman (Ed.), *Encyclopedia of Mathematics Education* (second edition). New York: Springer.

Lagrange, J. B., & Kynigos, C. (Eds.) (2014). Special Issue: Representing mathematics with digital media: Working across theoretical and contextual boundaries. *Educational studies in Mathematics*, 85(3).

Lerman, S. (2000). The social turn in mathematics education research. In J. Boaler (Ed.), *Multiple Perspectives on Mathematics Teaching and Learning* (pp. 19–44). Westport, CT: Ablex Publishing.

Najar, R. (2010). *Effets des Choix Institutionnels Sur les Possibilités d'Apprentissage des Etudiants. Cas des Notions Ensemblistes Fonctionnelles Dans la Transition Secondaire/Supérieur.* Doctoral thesis. Université Paris Diderot – Paris 7 & Université virtuelle de Tunis. https://tel.archives-ouvertes.fr/tel-00564191.

Nardi, E., Jaworski, B., & Hegedus, S. (2005). A spectrum of pedagogical awareness for undergraduate mathematics: From "tricks" to "techniques". *Journal for Research in Mathematics Education*, 36(4), 284–316.

Nardi, E. (2008). *Amongst Mathematicians. Teaching and Learning Mathematics at the University Level.* New York: Springer.

Nardi, E., Biza, I., González-Martín, A.S., Gueudet, G., & Winsløw, C. (Eds.) (2014). Institutional, sociocultural and discursive approaches to research in university mathematics education. *Research in Mathematics Education*, 16(2).

Nardi, E., & Rasmussen, C. (2019). Teaching practices at university level. In S. Lerman (Ed.), *Encyclopedia of Mathematics Education* (second edition). New York: Springer.

Nardi, E., Ryve, A., Stadler, E., & Viirman, O. (2014). Commognitive analyses of the learning and teaching of mathematics at university level: the case of discursive shifts in the study of Calculus. *Research in Mathematics Education*, 16 (2), 182–198.

Oliveira, C. (2015). *Una Posible "Razón de Ser" del Cálculo Diferencial Elemental en el Ámbito de la Modelización Functional.* Doctoral thesis. Universidad de Vigo.

Praslon, F. (2000). *Continuités et Ruptures Dans la Transition Terminale S/DEUG Sciences en Analyse. Le Cas de la Notion de Dérivée et son Environnement.* Doctoral thesis. Université Paris Diderot – Paris 7.

Rasmussen C., & Wawro, M. (2017). Post-calculus research in undergraduate mathematics education. In J. Cai (Ed.), *Compendium for Research in Mathematics Education*, (pp. 551–581). Reston, VA: National Council of Teachers of Mathematics.

Robert, A. (1998). Outils d'analyse des contenus mathématiques à enseigner au lycée et à l'université, *Recherches en Didactique des Mathématiques*, 18(2), 139–190.

Romo-Vázquez, A. (2009). *La formation mathématique des futurs ingénieurs*. Doctoral thesis. Université Paris-Diderot. https://tel.archives-ouvertes.fr/tel-00470285/document.

Sfard, A. (2008). *Thinking as Communicating*. Cambridge: Cambridge University Press.

Sfard, A. (2014). University mathematics as a discourse – why, how, and what for? *Research in Mathematics Education*, 16(2), 199–203.

Sierpinska, A. (1985). Obstacles épistémologiques relatifs à la notion de limite. *Recherches en Didactique des Mathématiques*, 6(1), 5–67.

Tall, D., & Vinner, S. (1981). Concept Image and Concept Definition in Mathematics with particular reference to Limits and Continuity, *Educational Studies in Mathematics*, 12, 151–169.

Tall, D. (Ed.) (1991). *Advanced Mathematical Thinking*. Dordrecht: Kluwer.

Tall, D. (2013). *How Humans Learn to Think Mathematically: Exploring the Three Worlds of Mathematics (Learning in Doing: Social, Cognitive and Computational Perspectives)*. Cambridge: Cambridge University Press.

Thomas, M., de Freitas Druck, I., Huillet, D., Ju, M.K., Nardi, E., Rasmussen, C., & Xie, J. (2014). Key mathematical concepts in the transition from secondary school to university. In Sung Je Cho (Ed.) *Proceedings of the 12th International Congress on Mathematical Education* (pp. 265–284). New York: Springer.

Vázquez Padilla, R. (2017). *Diseño de actividades didácticas basadas en modelización para la formación matemática de futuros ingenieros*. Doctoral thesis. CICATA-IPN, México.

Weinberg, A., Rasmussen, C., Rabin, J., Wawro, M. & Brown, S. (Eds.) (2017), *Proceedings of the 20th Annual Conference on Research in Undergraduate Mathematics Education*. San Diego, CA.

Winsløw, C. (2014). Mathematics at university: the anthropological approach. In Sung Je Cho (Ed.) *Proceedings of the 12th International Congress on Mathematical Education* (pp. 859–875). New York: Springer.

Winsløw, C., Gueudet, G., Hochmuth, R., & Nardi, E. (2018). Research on university mathematics education. In T. Dreyfus, M. Artigue, D. Potari, S. Prediger, & K. Ruthven (Eds.), *Developing Research in Mathematics Education: Twenty Years of Communication, Cooperation and Collaboration in Europe* (pp. 60–74). London: Routledge.

Winsløw, C., & Rasmussen, C. (2019). University mathematics education. In S. Lerman (Ed.), *Encyclopedia of Mathematics Education* (second edition). New York: Springer.

2

LESSONS FOR MATHEMATICS HIGHER EDUCATION FROM 25 YEARS OF MATHEMATICS SUPPORT

Duncan Lawson and Tony Croft

Introduction

In 1995, the London Mathematical Society, the Institute of Mathematics and its Applications and the Royal Statistical Society (the three leading professional and learned bodies in the mathematical sciences in the UK) issued the report *Tackling the Mathematics Problem* (LMS, 1995). This report introduced the phrase "the Mathematics Problem" to describe the situation of an increasing number of new undergraduates, typically students of engineering or the physical sciences, arriving at university inadequately prepared for the mathematical demands of their course. The specific areas that the report identified as problematic were: lack of fluency and accuracy in numerical and algebraic calculations; decline in ability to solve multi-step problems; and a changed perception of the nature of proof.

This report was followed by other reports such as *The Changing Mathematical Background of Undergraduate Engineers* (Sutherland & Pozzi, 1995) and *Measuring the Mathematics Problem* (Hawkes & Savage, 2000), both published by the Engineering Council. Whilst much of the content of these reports was dependent on the judgement of academic staff, *Measuring the Mathematics Problem* did, as its name implies, seek to quantify the problem through the provision and analysis of relevant data. A vital contribution to this was the work of Lawson (2000). This work was based on analysis of the results of incoming undergraduates on a diagnostic test at Coventry University. The diagnostic test had been unchanged over the period under consideration. It covered a range of basic topics including arithmetic, basic algebra, lines and graphs, triangles, trigonometry, further algebra and basic calculus. These topics were all on the school mathematics syllabus and form a vital foundation for students entering engineering courses at university.

The results of students entering university with specific grades in A-level mathematics[1] were compared over the period from 1991 to 1999. These results showed that over this period of time the performance of students entering university with a specified grade in A-level mathematics declined. Furthermore, the analysis of the diagnostic test results revealed that students entering university in 1999 with a grade C (i.e. a middle-ranking pass grade) performed at a similar level to those students who had entered university in 1991 with a grade N (a narrow fail grade). Lawson (2003) subsequently extended this work and showed that the average performance of the 2001 A-level grade B cohort was almost indistinguishable from that of the 1991 A-level grade *N* cohort.

This evidence and further substantial lobbying by higher education institutions (HEIs) and professional bodies over subsequent years led the Government to commission an inquiry into post-14 mathematics education. The report of this commission *Making Mathematics Count* (Smith, 2004) identified a range of issues in mathematics education which could not be solved in the short-term and made a number of recommendations for resolving these issues, recognising that these would take several years to have an effect on universities. The report concluded that "In the short-term, the Inquiry believes that Higher Education has little option but to accommodate to the students emerging from the current GCE[2] process" (Smith, 2004, p. 95).

HEIs had reached this conclusion for themselves many years previously and started to make the required "accommodation". A range of measures were introduced including bridging courses to be taken before starting on the university course proper and the modification of the university mathematics curriculum to include revision of key aspects of the school mathematics curriculum which experience showed were not well understood by many incoming students. Several institutions chose to tackle the Mathematics Problem through the introduction of mathematics support provision.

One commonly used definition of mathematics support, given by Lawson, Croft and Halpin (2003, p. 9) is

> A facility offered to students (not necessarily of mathematics) which is in addition to their regular programmes of teaching through lectures, tutorials, seminars, problems classes, personal tutorials, etc.

In practice, the most common form of mathematics support was the drop-in centre. Various forms of drop-in centre were implemented in different institutions. Several variants of the drop-in centre model of mathematics support are presented in Marr and Grove (2010) and Lawson (2012).

The basic idea behind a drop-in centre is the provision of a location where students can "drop in" (that is, come at a time of their own choosing) to receive assistance with mathematical difficulties. Such drop-in centres set out to have a welcoming, non-judgemental atmosphere, providing a "safe" location separate from the normal teaching and assessment process where students can ask even very basic questions without fear of this counting against them in some way at a later stage in their education. Such centres typically provide an array of resources such as handouts and textbooks; in addition there will often be computers for students to access video materials and take self-assessment diagnostic tests. However, the resource that is most valued by students is the provision of one-to-one interaction between the student dropping-in and a mathematics tutor (Lawson, Halpin & Croft, 2002).

The Mathematics Problem may have begun as a UK issue, however it has not remained so. Similar issues have been reported in many developed countries, for example Australia (Matthews et al., 2012), Ireland (Gill et al., 2010) and across Europe (Alpers, 2008). In response, mathematics support centres have been introduced at the majority of institutions in Australia (MacGillivray, 2009b) and Ireland (Cronin et al., 2016). In continental Europe, the introduction of mathematics support centres has been a little slower. A few institutions in Norway, Slovakia and Switzerland, supported by established practitioners from the UK and Ireland, have developed mathematics support centres. In Germany, a broader range of mathematics support initiatives have been introduced, including a small number of universities which have set up mathematics support centres (as reported in Liebendörfer et al., 2017).

Development of mathematics support

In its early days, there were some who regarded mathematics support as rather peripheral and not a significant phenomenon in the learning and teaching of mathematics. Kyle (2010, p. 103) reflects thus on his views about mathematics support:

> Although I might not have put it in these terms at the time, I probably regarded mathematics support as a form of cottage industry practised by a few well meaning, possibly eccentric, individuals, who may themselves have been hard pushed to offer a credible rationale for this work.

However, such ambivalence towards mathematics support did not stop its development throughout the first decade of the 21st century.

In 2004, the Higher Education Funding Council for England (HEFCE) announced its largest ever teaching and learning development programme: Centres for Excellence in Teaching and Learning (CETLs). HEIs individually or collaboratively could bid to have elements of their practice recognised as a Centre for Excellence and receive substantial funding (amounts up to £4.5 million over 5 years were available for individual centres). At this time, two of the largest mathematics support provisions in England were the Mathematics Learning Support Centre at Loughborough University and the Mathematics Support Centre at Coventry University. Together these two centres submitted a collaborative bid to be recognised as a CETL. This bid was successful and the resulting Centre for Excellence was called **sigma**: Centre for Excellence in University-wide Mathematics and Statistics Support, or just **sigma** for short.

Individual CETLs had set out their plans of work in their bids and there was huge variation in the nature and focus of the CETLs. From the beginning, **sigma** set out to have an impact across the sector and did not focus its attentions primarily on activities within the two host HEIs, as was the case with many other CETLs. The original CETL bid had indicated that **sigma** would fully fund Leeds University to develop its own mathematics and statistics support (MSS) provision. The process of establishing this provision at Leeds University would be closely monitored so that a blueprint for setting up MSS at other HEIs could be produced. Further funding (allocated on a matched-funding basis) would be made available through a competitive bidding process to enable other HEIs to establish their own MSS provision.

Over the 5 years of the CETL programme, **sigma** spent substantial sums of money on facilitating the development of MSS in other HEIs. This was not only through the direct provision of funds to other HEIs to establish MSS, as described in the previous paragraph. In addition, a range of other measures were funded including the development of thousands of learning resources for students that were made freely available through the website www.mathcentre.ac.uk, and good practice guides for staff including: *How to set up a mathematics and statistics support provision, Gathering feedback on mathematics and statistics support provision* and *Tutoring in a mathematics support centre*[3]. Furthermore, **sigma** promoted professional development for those involved in the delivery of MSS through the provision of:

- training courses for those involved in tutoring in MSS centres;
- workshops focused on operational issues related to running MSS centres (such as evaluating the provision, publicising the provision and understanding the increasingly diversified pre-higher education mathematics curriculum);
- opportunities for secondments to **sigma** to work with Loughborough or Coventry colleagues on MSS projects;
- an annual conference, the CETL-MSOR[4] conference to enable the sharing of good practice and the development of scholarship in MSS.

By the end of CETL funding in 2010, **sigma** had been able to establish the **sigma** network, a free association of individuals in HEIs across England and Wales involved in the provision of MSS. In addition, **sigma** had worked with MSS providers in Scotland to help establish the Scottish Mathematics Support Network and with colleagues from Ireland to assist the on-going development of the Irish Mathematics Learning Support Network.

The success of **sigma** had been observed by the National HE STEM[5] Programme[6], a major HEFCE funding initiative aimed at increasing the number of graduates in STEM disciplines. This Programme commissioned **sigma** to continue its work of enabling the roll-out of MSS in HEIs across England and Wales. Following the scheduled ending of the National HE STEM Programme in 2012, HEFCE directly funded **sigma** to maintain the development of MSS nationally and to enable the continuation of a sustainable network. Alongside continuing the initiatives described above, **sigma** focused on activities that would strengthen the network of MSS practitioners through the creation of a community of practice. **sigma** provided funding to revive the journal *MSOR Connections* (which had ceased due to the demise of the Higher Education Academy Subject Centres), in which many practice-focused articles relating to MSS had previously been published. In addition, modifications were made to the by-now well-established competitive bidding process for match-funding. This was extended to not just receiving bids from HEIs without any MSS provision but also to enabling those with a small provision to bid for funding to enhance and extend their provision. All successful bids were allocated a mentor (an experienced MSS professional from another institution who was active within the **sigma** network). The role of such mentors was to provide assistance to the grant-holder in the delivery of their objectives. Furthermore, mentors encouraged colleagues delivering the new or enhanced provision to themselves become active participants in the **sigma** network. This latter aim was further strengthened by the expectation that grant-holders would either present at the CETL-MSOR conference or write an article for *MSOR Connections*.

By the end of the final tranche of HEFCE funding in 2016, **sigma** had directly helped facilitate the establishment of 36 new MSS provisions at HEIs across England and Wales. In addition, the **sigma** network was firmly established as a valuable community which could be sustained through contributions of time and expertise from those involved in MSS at HEIs across the sector. MSS had become a firmly established part of the HE infrastructure throughout the UK and Ireland. As Youdan (quoted in Fletcher, 2013, p. 49) put it:

> Mathematics and statistics support has now attained a critical mass and overcome the significant hurdle where universities worry whether offering such support is an indication of modest aspirations. The accepted position is now that it is a student's right to receive support with the mathematical content of their degree.

Surveys of the extent of provision of mathematics support across universities in the UK (Lawson, Croft & Halpin, 2003; Perkin & Croft 2004; Perkin, Croft & Lawson, 2013; Ahmed et al., 2018, Grove, Croft & Lawson, 2019) have shown that the number of institutions providing mathematics support has increased significantly during the 21st century.

MSS had also gained academic credibility. The previously held view of mathematics support as a "cottage industry" had changed. Kyle's (2010, p. 104) previously quoted appraisal of MSS continues:

> Now only a few years on, we see that the concept of mathematics support has not only become firmly embedded in UK Higher Education, but colleagues have moved on to gather data on the way students use such resources and look for optimal strategies for the delivery of this support, and this is perhaps the most convincing evidence of acceptance. Mathematics support came of age in the first decade of the 21st century. What might once have been described as a cottage industry now plays a respected and widely adopted role in Higher Education.

The growing maturity of the MSS community has led, amongst other things, to an increase in the volume of research and scholarship relating to MSS. A recently published literature review (Lawson, Grove & Croft, 2019) surveying publications since 2000 contains 117 references.

Increased demand for mathematics and statistics support

Whilst **sigma** played a significant role in establishing MSS as "respected and widely adopted", there were also external factors which contributed to this. The most important of these was the increasing quantification of many disciplines brought about by technological advances in data collection. As a consequence of this, the Mathematics Problem was no longer the preserve of engineering and the physical sciences. Professional bodies and learned societies from an ever-expanding range of disciplines highlighted both how their disciplines were becoming increasingly quantitative and also how students studying their disciplines were inadequately prepared for such quantitative material since the majority of them had not studied any mathematics since the age of sixteen.

The Biotechnology and Biological Sciences Research Council strategic plan for 2010–2015 (BBSRC, 2010, p. 15) identified that:

> As bioscience becomes increasingly quantitative there is also an urgent need **to raise the mathematical and computational skills of biologists** [original emphasis] at all levels.

The British Academy, a national body for social sciences and the humanities, issued a position statement *Society Counts* (British Academy, 2012) which highlighted weak quantitative skills in their disciplines, identifying a shortage of academic staff able to teach such skills. The Royal Society of Arts, in a report *Solving the Maths Problem* (Norris, 2012), whose title invoked previous reports from engineering and mathematical sciences bodies, concluded that (ibid, p. 11):

> English universities are side-lining quantitative and mathematical content because students and staff lack the requisite confidence and ability.

The nature of higher secondary education in the UK is such that mathematics is not compulsory after the age of 16. This makes the UK an outlier in terms of educational provision in developed countries. Hodgen et al. (2010), in a study of 24 educational jurisdictions, found only six where the study of mathematics was not compulsory after age 16. The six were Australia (NSW), England, Ireland, Northern Ireland, Scotland and Wales. In all six of these educational jurisdictions, mathematics support provision in higher education is widespread. A report by the UK's Advisory Committee for Mathematics Education (ACME, 2011) has sought to quantify the problem caused by the optionality of mathematics post-16 in the UK. Their report estimates that in any given year around 330,000 students enter higher education for the first time taking courses where there would be benefit from having studied mathematics or statistics at a level beyond GCSE[7]. However, only around 125,000 students per year undertake such study. The 205,000 students who comprise the gap between these two figures are likely to be in need of MSS once at university.

The introduction of the Teaching Excellence Framework (TEF), compulsorily in England and voluntarily in other countries in the UK, has caused HEIs to pay increasing attention to student retention rates since this is one of the key metrics that is assessed in determining the outcome of a TEF assessment (DfE, 2017). A National Audit Office report into student retention in higher education (NAO, 2007, p. 32) identified the fact that "Many students require some additional academic support, especially in mathematical skills" and throughout the report inadequate preparation in mathematics is identified as a major contributor to students not completing their courses. Because of the capacity of MSS to make a difference to retention and drop-out rates, increasingly MSS provision is being viewed by institutional senior management as a strategic issue for the institution and not a local issue for the mathematics department (Tolley & Mackenzie, 2015).

Although evaluation of the impact of MSS is not straightforward, there is substantial and growing evidence of its effectiveness. A literature survey of evaluation of mathematics support centres was published by Matthews et al. (2013); subsequent work by Dzator and Dzator (2018), whilst not a literature survey, cites several studies that have taken place since 2013, as does Lawson, Grove and Croft (2019). One of the most important studies in this area is O'Sullivan et al. (2014). They report on a multi-institutional large-scale student evaluation of mathematics support across the island of Ireland, presenting both quantitative and qualitative data. The qualitative data includes feedback from students reporting the difference that engaging with MSS has made to them in the following terms (O'Sullivan et al., 2014, pp. 38, 42 and 75):

"Made me see that it is not impossible to grasp a particular mathematical task, but that it takes practice and time."
"Maths isn't scary any more."
"I am even starting to enjoy maths now."

Students who had engaged with mathematics support were asked if they had considered dropping out because of difficulties with the mathematical elements of their course. Those students who answered this question affirmatively were asked a follow-up question of whether the availability of MSS had influenced their decision not to drop out. 22% of the students surveyed reported that they had considered dropping out and of these 64% indicated that the availability of MSS had been a factor in their decision to remain on course.

At its conception, MSS was designed to provide support to "at risk" students, i.e. those in danger of failing their course of study because of difficulties with the quantitative components of the course. The original aims could have been classified as remedial. However, MSS has not been restricted to "at risk" students and the voluntary nature of engagement with this support (for example, students choose to attend a drop-in centre, they are not compelled to do so) has made the provision available to all. MSS has not been promoted as a remedial service. Instead, MSS has sought to engage with all students who wish to improve their learning of mathematics and statistics. As a consequence, many of the students who avail themselves of MSS are not "at risk" students but able students who are seeking to improve their performance in mathematics and statistics from good to excellent (Pell & Croft, 2008).

Several quantitative studies, such as those of Gallimore and Stewart (2014), Dzator and Dzator (2018) and Jacob and Ni Fhloinn (2018), have consistently shown that students who engage frequently with MSS achieve better outcomes (in terms of pass rates and average marks on mathematics and statistics modules) than students with similar characteristics (typically entry qualifications) who do not use MSS. This raises the question about why some students do not engage with MSS, particularly those who are in danger of failing their mathematics or statistics modules. When such non-users of MSS are asked for their reasons for not engaging, the most commonly given reasons are such things as "Not being aware that MSS was available" or "Not knowing where the MSS support provision was located" (Symonds, Lawson & Robinson, 2008). Other studies of non-engagement with MSS, such as Mac an Bhaird et al. (2013), O'Sullivan et al. (2014) and Grehan et al. (2016) report similar findings.

Whilst most MSS providers devote considerable energy and resources to publicising their provision (using a range of "marketing" methods such as information in course packs, presentations during induction week, posters, branded merchandise, lecture shout-outs, prominent links on the institutional virtual learning environment, social media, etc.) it is of course possible that some students are unaware that they can access MSS. However, reasons such as "not knowing where the MSS provision is located" perhaps indicate that additional factors are influencing their non-engagement. If such students were motivated to engage with MSS, it would not be difficult for them to find its location (in many institutions it has a prominent position in the library or some other central student facility). A small number of students indicate that their non-engagement is due to deeper reasons such as feeling that they had too many problems (i.e. were beyond help), feeling demoralised or fearing embarrassment. It seems likely that such reasons may underlie the more superficial reason of "not knowing where the MSS provision is located". Securing engagement of such students will always be difficult where the model of MSS provision is one which requires students to "opt in" (such as a drop-in centre).

Mathematics and statistics support and specialist mathematical sciences students – part 1

Although the definition of MSS (Lawson, Croft & Halpin, 2003), quoted on page 2 of this chapter, states that mathematics support is for "students (not necessarily of mathematics)" it might have been a more accurate representation of the original intention had it said that mathematics support was for "students (not primarily of mathematics)". Drop-in centres originally targeted engineering and physical sciences students and, for the reasons described in the preceding section, the expectation is now that students from a much wider range of "user-disciplines" will avail themselves of MSS. Particularly given the original focus of MSS on students at risk of failure because of weaknesses in quantitative skills, it was not anticipated that many students studying for degrees in the mathematical sciences would engage with MSS. After all, such students have chosen to study mathematics or statistics and will have had to demonstrate a comparatively high level of competence in these disciplines to have been admitted to the course. However, experience has shown that students studying the mathematical sciences are often those most likely to engage with MSS. This phenomenon has been reported at several universities, for example by MacGillivray (2009a), Solomon, Croft and Lawson (2010) and Grove, Guiry and Croft (2019). Indeed both MacGillivray (2009a) and Solomon et al. (2010) report that mathematics students "colonised" the physical space of the drop-in centre in the universities in question.

This raises the question as to why specialist mathematical sciences students should make such extensive use of MSS provision, and in particular, the drop-in centre. One answer to this question may be found in the results of the National Student Survey.

The national student survey and mathematical sciences students

The National Student Survey is an annual survey of all final-year undergraduates at all publicly funded HEIs. The survey presents students with a sequence of statements relating to a range of different aspects of their experience at their university. Students are asked to indicate their level of agreement with each statement using a five-point Likert Scale: Strongly agree, Agree, Neither agree nor disagree, Disagree, Strongly disagree. The statements cover aspects such as the teaching that students have received, assessment and feedback, student support, learning resources and course organisation.

The results from the NSS feed into grading and ranking systems for universities including official measures such as the Teaching Excellence Framework (DfE, 2017) and unofficial (but nonetheless influential) ones such as league tables produced by The Guardian and The Times / Sunday Times newspapers. In such gradings, NSS data is typically analysed by measuring the percentage of students showing some level of agreement with a particular NSS statement (ie. respondents selecting the first two items on the Likert scale – Strongly agree and Agree – are combined into a single category and, by default, those selecting the other items are also combined into a single category). The complete set of NSS data is made publicly available by the Office for Students[8].

The results of the NSS give a way to compare students experience in studying different subjects in higher education in the UK. The Joint Academic Coding System (JACS)[9] is a way of classifying subjects. In this classification there are 21 high-level subjects such as Medicine & Dentistry, History & Philosophical Studies, Engineering & Technology and Mathematical Sciences. The analysis that follows takes the NSS results from 2016 and breaks them down by the 21 subjects at the highest level of JACS.

In the 2016 NSS there were 22 statements with which students were asked to indicate their level of agreement (this has since changed and the current version of the NSS contains 28 statements). With four of these statements, students from the mathematical sciences showed a higher level of agreement than students from any other subject. These statements are set out in Table 2.1.

The fourth statement in this list is not related to students' experience in learning and teaching but is more to do with the social side of university. It has been suggested that mathematical sciences students have the highest level of agreement with this question because they have lower expectations in this sphere than students of many other subjects.

The other three questions relate to what might be classified as process-focused statements. The second and third statements reflect the administrative efficiency of the course. These statements are not directly related to learning and teaching, but do relate to the environment in which such learning and teaching takes place. Since assessment is a major driver of student learning, the first statement is directly connected to learning. In part, mathematical sciences' high score here may be due to the nature of the discipline. Many assessment tasks taken by mathematical sciences students have a correct answer and students may therefore be more likely to perceive the marking of such tasks to be fair than students whose assessment tasks are more essay-based where students often feel that there can be elements of subjectivity in the marking and that the final mark awarded can be affected by who actually marks the work.

A set of statements on the NSS explores students' views of their experiences with teaching at their institution. With these statements, the level of agreement of mathematical sciences students is lower than for students of most other subjects. Table 2.2 shows the position of mathematical sciences in an ordered list of the 21 subjects for the three teaching-related statements.

Table 2.3 shows the last five subjects, with their level of agreement, in the ordered list for the statement "Staff have made the subject interesting".

It can be seen from Table 2.3 that the gap between the Mathematical Sciences in 18th position and the subjects below it in the list in positions 19–21

TABLE 2.1 Statements where mathematical sciences students have a higher level of agreement than students from any other high level JACS subject

Statement	*Level of agreement*
Assessment arrangements and marking have been fair	87%
Any changes in the course or teaching have been communicated effectively	88%
The course is well organised and running smoothly	89%
I am satisfied with the Students' Union at my institution	71%

TABLE 2.2 Position of mathematical sciences in a list of 21 subject areas ordered by level of agreement with the statement

Statement	*Position of Mathematical Sciences*
Staff are good at explaining things	15
Staff have made the subject interesting	18
Staff are enthusiastic about what they are teaching	14

TABLE 2.3 Subjects with the lowest levels of agreement with the statement "Staff have made the subject interesting"

Position	*Subject*	*Level of Agreement*
17	Mass Communication and Documentation	81%
18	Mathematical Sciences	78%
19	Business and Administrative Studies	77%
20	Computer Science	77%
21	Engineering and Technology	76%

(two percentages points cover all four subjects) is smaller than the gap upwards (three percentage points) to the subject in 17th position. This suggests that there is very little difference between these last four subjects in the list and there is a clear gap then to the next subject up (Mass Communication and Documentation). At the other end of this ordered list, in equal 1st position, 91% of students from Veterinary Sciences and 91% of students from Historical and Philosophical Studies indicated agreement with this statement. Academic staff who teach mathematical sciences would say that mathematical sciences is an extremely interesting subject but they are not managing to communicate this interest to their students as effectively as staff in most other subjects.

There are two statements in which the level of agreement of mathematical sciences students was lower than in any other subject. These are shown in Table 2.4.

Across all subject areas, the average level of agreement with the two statements shown in Table 2.4 were 81% and 84% respectively. The highest level of agreement with the first statement was 89% amongst students of both Medicine & Dentistry and Teacher Education, and for the second statement was 95% amongst students of

TABLE 2.4 Statements where mathematical sciences students have a lower level of agreement than students from any other high level JACS subject

Statement	*Level of Agreement*
The course has helped me present myself with confidence	69%
My communication skills have improved	68%

Medicine & Dentistry. These statements are related to personal development and cover areas that are important in terms of employability following university.

In summary, when comparing the experience of mathematical sciences students with those of students from other subjects, using data from the 2016 NSS (which is entirely student self-reporting), we find that their experience in process-related areas is excellent, in teaching is unsatisfactory and in personal development is very poor. These results echo the findings of Hewson (2011) who analysed the 2010 NSS results. His analysis showed that students of mathematical sciences had the highest levels of agreement of all subjects with the statements relating to the fairness of marking and course organisation, was below all subjects other than computer science for the statement relating to making the subject interesting and below all subjects for the statement about presenting oneself with confidence. In other words, over this six-year period, mathematical sciences has maintained its strengths but not improved its weaknesses.

Mathematics and statistics support and specialist mathematical sciences students – part 2

The results of the NSS outlined above, particularly those relating to the teaching of the mathematical sciences, perhaps throw some light on to why students from the mathematical sciences have a tendency to "colonise" mathematics support drop-in centres. The work of Solomon et al. (2010) and Solomon, Lawson and Croft (2011) reveals some of the reasons behind the comparatively low opinion that mathematical sciences students have of their teaching at university.

Relationships with lecturers and tutors are very important to many students. Their experience in the traditional lecture and tutorial settings is often not satisfactory for them. They report being too embarrassed or intimidated to ask questions in a lecture setting, which could often be in front of hundreds of other students. Even in smaller group tutorials, they have a fear of being patronised by being told that the answer to their question is "obvious" or that something they are struggling with is "simple". They have reservations about visiting staff in their offices even during times which have been publicised as "office hours"; such consultations take place in "their space" and the visit is either interrupting them from doing something else (if no other students are present) or is constrained by there being a queue of students who need to be seen during the available hour.

Research evidence shows that mathematics support centres completely alter the power dynamic (Solomon et al., 2010). Support centres are viewed as neutral ground that does not belong to either the tutor or the students. A tutor on duty in the support centre is away from his/her desk and clearly is there in order to be asked questions and so the sense of interrupting them from doing something else is not present. Perhaps because of the way MSS tutors are trained, certainly because of the intended ethos of MSS where no question is regarded as too basic, students should not experience discouraging remarks such as "this is easy" when seeking help in a drop-in centre.

As the NSS results show, mathematical sciences students have the lowest level of agreement with the statement "The course has helped me to present myself with confidence". Although this question is probably more focused on students' ability to make oral presentations (a graduate employability skill which should be developed in all undergraduate courses), it also raises the issue of confidence. Whilst mathematical confidence is different from more general personal confidence, a learning experience which undermines the mathematical confidence of students who previously identified themselves as good at mathematics is likely to impact on their more general confidence. Mathematical confidence is a major issue for some undergraduates in the mathematical sciences who find unsettling the transition from school mathematics, where the focus is on process and algorithmic exercises, to university mathematics with its emphasis on proof and rigour. In the study of Solomon et al. (2010), 26% of the mathematics undergraduates who participated in their survey reported that since coming to university "I realised that I am not very good at maths" compared to 80% who reported "I was better at mathematics than most in my class at school".

For students whose mathematical confidence is low, the MSS drop-in centre can provide a safety that they do not experience in lectures or tutorials. The one-to-one interaction with a tutor removes the sense of public exposure that is present in a group learning situation such as a lecture or tutorial, where students with low confidence would not "risk" asking a question. There is safety too in that the MSS tutor is (usually) not going to be involved in the summative assessment process and so students are more likely to reveal areas of lack of understanding knowing that this is not going to "be held against them" at some stage in the future.

A commonly-held caricature of a mathematics student is a male loner who is striving to be better than his contemporaries. This caricature does not resonate with many mathematics students (not only female ones) whose natural approach to learning is far more collaborative than competitive. In very simple ways, a mathematics support centre facilitates students in adopting a collaborative approach to learning. It is a physical space that is set aside for study rather than social activity, but it is not one where silence or even quiet is expected. The furniture is often round or hexagonal tables which make it easy for a small number of students to work together. These practical factors can lead to the formation of undergraduate student communities of practice, as demonstrated by the following comment by student "Roz" quoted in Solomon et al. (2010, p. 428):

> I used it [the mathematics support centre] a lot because a group of us who tend to get fairly good marks used it a lot. Other people sort of came in to work with us and got the help and so on and so … and it developed a real upspin, it was really kind of in a sense the place to be, and there was a lot of people, there was a lot of use.

It is not necessary to have an MSS drop-in centre to provide many of the attributes that students of the mathematical sciences indicate are valuable to them. Waldock et al.

(2017) describe the development of a purposefully designed informal learning space for mathematical sciences students which has many of these characteristics. However, the availability of such spaces is rare in HEIs and drop-in centres are much more prevalent. It is therefore not surprising that many mathematical sciences students tend to gravitate towards them. As Solomon et al. (2011, p. 580) conclude

> Support centres appear to have a significant impact on discourses of ability and learning; they lead in particular to an appreciation of, and emphasis on, collaborative work and, in consequence, to a shift in attitudes towards university mathematics as a community of enquiry as opposed to an individual performance-oriented pursuit.

Most MSS providers seek to support all students from across their institution - the original, full name of **sigma** was "Centre for Excellence in University-wide Mathematics and Statistics Support". Therefore supporting mathematical sciences students lies within the remit of MSS. However, where disproportionate use is made of MSS by such students steps may need to be taken to address their needs in other ways. Students of other disciplines, particularly those where the level of mathematical content is relatively low, can be intimidated from engaging with MSS if it seems to always be dominated by specialist students engaging in discussion about advanced topics. The two institutions studied in MacGillivray (2009a) and Solomon et al. (2010), where it was reported that mathematical sciences students had "colonised" the MSS provision, both introduced separate rooms solely for the use of such students. Such a measure is a short-term solution. There is a need for those teaching the mathematical sciences to reflect on and enhance their own practices, in the light of the NSS results and high levels of demand for MSS.

Conclusions

MSS has become a well-embedded feature of the HE infrastructure in many countries, notably the UK, Ireland and Australia. In other countries such as Germany, Norway, Slovakia and Switzerland, MSS is in its infancy but support centres have been established with assistance from colleagues in the UK and Ireland. In most institutions where MSS is established there is no shortage of students wanting to engage with the provision. Indeed, one of the most common responses received on student evaluation questionnaires, when asked if they have any suggestions for how the MSS provision can be improved, is that it should be available for more hours than it currently is.

As well as the growth in potential users due to the increasing quantification of many subjects, described in Section 3, in recent times MSS has seen increase in demand from another source. The origins of MSS were in facilitating the transition into university; the latest new demand comes from enabling the transition out of university. Many employers now use numerical reasoning tests as part of the selection process for graduate level jobs. For students who have not formally studied any mathematics since the age of 16, such tests can be extremely daunting and they look for MSS to help to prepare them.

Although MSS as it is currently delivered in HEIs around the world has many positive features, it should not be regarded either as "the finished article" or as a panacea – the final solution to the Mathematics Problem. The MSS community is aware that delivery of MSS will always be "a work in progress" and means and methods of delivery will need to adapt to both address existing shortcomings (such as securing engagement from those who seem not to be motivated to engage) and to respond to new demands (such as providing support for graduate selection tests).

With regard to solving the Mathematics Problem, it should also be remembered that MSS was part of universities' response to the conclusion of the Post-14 Mathematics Education Inquiry that "In the short-term, the Inquiry believes that Higher Education has little option but to accommodate to the students emerging from the current GCE process" (Smith, 2004, p. 95). An implication of this statement was that in the medium-term external actions, such as those recommended in the Inquiry's report, would alleviate the mathematics problem. MSS has clearly become something more than short-term, but this does not remove the need for others to play their parts too. Many incoming undergraduates would benefit from being better mathematically prepared on entry to university, particularly by choosing to study some mathematics post-16. The mathematical demands of courses could be made more explicit to prospective students, although for universities this is double-edged. On the one hand, it might influence students in their choice of subjects at school so that more of them do study mathematics post-16; but on the other hand, it might deter potential applicants and make it harder for universities to recruit to certain subjects. Improved communication about the Mathematics Problem to teachers and parents, as well as applicants, could tend to push students to the former rather than the latter of these courses of action.

The authors of this chapter do not suggest that MSS should be introduced in every university. We believe that the need (or not) for MSS is dictated by local circumstances and that there should be very clear reasons for introducing MSS and such provision should have specific aims. As set out above, at universities throughout the UK, there is a very clear need brought about by the widespread under-preparedness of many undergraduate and postgraduate students for the mathematical demands of their courses.

There are those who would criticise the provision of MSS. Some suggest that "We don't have the kind of student who needs support at our institution". If this is the case, then we would agree that establishing MSS is not necessary. However, Mackenzie et al. (2016), who interviewed senior managers from 23 institutions from across all sectors of UK higher education, found that in all institutions there was an awareness that there were many students who were struggling with the mathematical demands of their studies.

Others opine that "If students do not know **that** [a piece of elementary mathematics], then they should not be at university". Whilst this may be their opinion, the reality is that there are many students who are at university and who do not know "***that***". MSS is about dealing with this reality.

Still others suggest that the provision of MSS is a collusion with the inappropriate school curriculum and those who teach badly in higher education; they suggest that if there was no MSS then something would have to be done to remedy these failings. The authors' reply to this is that changing the school curriculum and improving poor teaching in higher education are laudable aims, but ones that will not be achieved quickly. As was quoted previously, the Smith Inquiry concluded "In the short-term, the Inquiry believes that Higher Education has little option but to accommodate to the students emerging from the current GCSE process" (Smith, 2004, p. 95). In the meantime, there are students who are struggling and it is our belief that they should not be regarded as collateral damage in the process of bringing about larger-scale system change. Today's students need help today, not an improved system in five or ten years' time.

Finally, we would repeat the findings of Solomon et al. (2011, p. 580) quoted earlier that

> they [support centres] lead … to a shift in attitudes towards university mathematics as a community of enquiry as opposed to an individual performance-oriented pursuit.

This is surely something to be welcomed. This does not mean that MSS is a panacea, but the outcomes of MSS have benefitted hundreds of thousands of students worldwide over the last 25+ years and have led to MSS becoming firmly established in many HEIs around the world. All the indications are that it will remain so for many years to come.

Notes

1 A-levels are national qualifications in the UK typically taken at age 18, ie. at the end of secondary education. In the 1990s, these qualifications were graded A-E (pass grades), N (narrow fail), U (unclassified).
2 At this time, most students took GCE (General Certificate of Education) qualifications (most frequently A-levels) as a means of qualifying for entry into university.
3 These guides, and several others, are available from the **sigma** network website www.sigma-network.ac.uk.
4 CETL-MSOR = Continuing Excellence in Teaching and Learning in Mathematics, Statistics and Operational Research.
5 STEM = Science, Technology, Engineering and Mathematics.
6 See https://www.birmingham.ac.uk/university/colleges/eps/STEM/National-HE-STEM-Programme/national-he-stem-programme.aspx (Accessed 17 March 2019) for further details.
7 GCSE level is the qualification taken at age 16, ie. at the end of compulsory education in mathematics.
8 https://www.officeforstudents.org.uk/advice-and-guidance/student-information-and-data/national-student-survey-nss/get-the-nss-data/ (Accessed 17 March 2019).
9 https://www.hesa.ac.uk/support/documentation/jacs (Accessed 17 March 2019).

References

ACME (2011) *Mathematical Needs: Mathematics in the Workplace and in Higher Education*. The Royal Society, London, UK.

Ahmed, S., Davidson, P., Durkacz, K., Macdonald, C., Richard, M. & Walker, A. (2018) The provision of mathematics and statistics support in Scottish Higher Education Institutions (2017): a comparative study by the Scottish Mathematics Support Network. *MSOR Connections*, 16 (3), 5–19.

Alpers, B. (2008) *Short Report on the Discussion Groups on "What are the Major Problems Facing Engineering Maths Education in Europe?"* Online report. European Society for Engineering Education (SEFI). Available at http://sefi.htw-aalen.de/Seminars/Short%20Report%20on%20the%20Discussion%20Groups%20on%20Major%20Problems%20dl.pdf (Accessed 17 March 2019).

BBSRC (2010) *The Age of Bioscience: Strategic Plan 2010–2015*, Biotechnology and Biological Sciences Research Council, Swindon, UK.

British Academy (2012) *Society Counts: Quantitative Skills in the Social Sciences and Humanities*, British Academy, London, UK.

Cronin, A.G., Cole, J., Clancy, M., Breen, C. & Ó Sé, D. (2016) *An Audit of Mathematics Support Provision on the Island of Ireland in 2015*, National Forum for the Enhancement of Teaching and Learning in Higher Education, Dublin, Ireland.

DfE (2017) *The Teaching Excellence and Student Outcomes Framework Specification, October 2017*, Department for Education, London, UK.

Dzator, M. & Dzator, J. (2018) *The Impact of Mathematics and Statistics Support at the Academic Learning Centre*, Central Queensland University, Teaching Mathematics and Its Applications, Advance online publication https://doi.org/10.1093/teamat/hry016.

Fletcher, L. (2013) *The Mathematics Support Community of Practice: A Report of the Achievements of sigma Within the National HE STEM Programme*, **sigma**, Loughborough, UK. Available at http://www.sigma-network.ac.uk/wp-content/uploads/2012/11/sigma-final-report.pdf (Accessed 17 March 2019).

Gallimore, M. & Stewart, J. (2014) Increasing the impact of mathematics support on aiding student transition in higher education, *Teaching Mathematics and its Applications*, 33 (2), 98–109.

Gill, O., O'Donoghue, J., Faulkener, F. & Hannigan, A. (2010) Trends in performance of science and technology students (1997–2008) in Ireland, *International Journal of Mathematical Education in Science and Technology*, 41 (3), 323–339.

Grehan, M., Mac an Bhaird, C. & O'Shea, A. (2016) Investigating students' levels of engagement with mathematics: Critical events, motivations and influences on behaviour, *International Journal of Mathematical Education in Science and Technology*, 47 (1), 1–28.

Grove, M., Croft, T. & Lawson, D. (2019) The extent and uptake of mathematics support in high education: results from the 2018 survey, *Teaching Mathematics and Its Applications*, Advance online publication https://doi.org/10.1093/teamat/hrz009.

Grove, M.J., Guiry, S. & Croft, A.C. (2019) Specialist and more-able mathematics students: understanding their engagement with mathematics support, *International Journal of Mathematical Education in Science and Technology*, Advance online publication https://doi.org/10.1080/0020739X.2019.1603407.

Hawkes, T. & Savage, M. (2000) *Measuring the Mathematics Problem*, The Engineering Council, London, UK.

Hewson, P. (2011) Preliminary analysis of the National Student Survey, *MSOR Connections*, 11 (1), 25–28.

Hodgen, J., Pepper, D., Sturman, L. & Ruddock, G. (2010) *Is the UK an Outlier? An International Comparison of Upper Secondary Mathematics Education*, Nuffield Foundation, London, UK.

Jacob, M. & Ní Fhloinn, E. (2018) A quantitative, longitudinal analysis of the impact of mathematics support in an Irish university, *Teaching Mathematics and Its Applications*, Advance online publication https://doi.org/10.1093/teamat/hry012.

Kyle, J. (2010) Affordability, adaptability, approachability and sustainability, in *Responding to the Mathematics Problem*, Marr, C.M. & Grove, M.J. (Eds.), pp. 103–104. Maths, Stats & OR Network, Birmingham, UK.

Lawson, D. (2000) Changes in the competency of A level students, in *Measuring the Mathematics Problem*, Hawkes, T. and Savage, M. (Eds.), pp. 6–9. The Engineering Council, London, UK.

Lawson, D. (2003) Changes in student entry competences 1991–2001, *Teaching Mathematics and Its Applications*, 22 (4), 171–175.

Lawson, D. (2012) *Setting up a Maths Support Centre*, National HE STEM Programme, University of Birmingham, UK.

Lawson, D., Croft, T. & Halpin, M. (2003) *Good Practice in the Provision of Mathematics Support Centres*, 2nd edition, LTSN Maths, Stats & OR Network, Birmingham, UK.

Lawson, D., Grove, M. & Croft, T. (2019) The evolution of mathematics support: a literature review, *International Journal of Mathematical Education in Science and Technology*, Advance online publication https://doi.org/10.1080/0020739X.2019.1662120.

Lawson, D, Halpin, M. & Croft, T. (2002) After the diagnostic test – what next? Part 2, *MSOR Connections*, 2 (1), 23–26.

Liebendörfer, M., Hochmuth, R., Biehler, R., Schaper, N., Kuklinski, C., Khellaf, S., Colberg, C., Schurmann, M. & Rothe, L. (2017) *A framework for goal dimensions of mathematics learning support in universities*. CERME10 Feb 2017, Dublin, Ireland.

LMS (1995) *Tackling the Mathematics Problem*, A joint report of the London Mathematical Society, the Institute of Mathematics and Its Applications and the Royal Statistical Society, London, UK.

Mac an Bhaird, C., Fitzmaurice, O., Ni Fhloinn, E. & O'Sullivan, C. (2013) Student non-engagement with mathematics learning supports. *Teaching Mathematics and Its Applications*, 32 (4), 191–205.

MacGillivray, H.L. (2009a). Learning support and students studying mathematics and statistics. *International Journal of Mathematical Education in Science and Technology*, 40 (4), 455–472.

MacGillivray, H.L. (2009b) *Learning Support in Mathematics and Statistics in Australian Universities - A Guide for the University Sector*, The Australian Learning and Teaching Council, New South Wales, Australia. Available at http://mathscentre.ac.uk/resources/uploaded/guide–altc-learning-support-in-maths-and-stats.pdf (Accessed 17 March 2019).

Mackenzie, H., Tolley, H., Croft, A.C., Grove, M.J. & Lawson, D.A. (2016) Senior management perspectives of mathematics and statistics support in higher education: moving to an 'ecological' approach, *Journal of Higher Education Policy and Management*, 38 (5), 550–561.

Marr, C.M. & Grove, M.J. (2010) *Responding to the Mathematics Problem: The Implementation of Institutional Support Mechanisms*, Maths, Stats & OR Network, Birmingham, UK.

Matthews, K.E., Belward, S., Coady, C., Rylands, L. & Simbag, V. (2012) *The state of quantitative skills in undergraduate science education – findings from an Australian study*, Australian Council of Deans of Science, Victoria, Australia.

Matthews, J., Croft, A.C., Lawson, D.A. & Waller, D. (2013). Evaluation of mathematics support centres: a literature review. *Teaching Mathematics and Its Applications*, 32 (4), 173–190.

National Audit Office (2007) *Staying the Course: The Retention of Students in Higher Education*, The Stationery Office, London, UK.

Norris, E. (2012). *Solving the Maths Problem: International Perspectives on Mathematics Education*, Royal Society of Arts, London, UK.

O'Sullivan, C., Mac an Bhaird, C., Fitzmaurice, O. & Ní Fhloinn, E. (2014) *An Irish Mathematics Support Network (IMLSN) report on student evaluation of mathematics support: insights from a large scale multi-institutional survey*, National Centre for Excellence in Mathematics and Science Teaching and Learning, Limerick, Ireland. Available at http://mural.maynoothuniversity.ie/6890/1/CMAB_IMLSNFinalReport.pdf (Accessed 17 March 2019).

Pell, G. & Croft, A.C. (2008) Mathematics support - support for all? *Teaching Mathematics and Its Applications*, 27 (4), 67–173.

Perkin, G. & Croft, A.C. (2004) Mathematics support centres – the extent of current provision. *MSOR Connections*, 4 (2), 14–18.

Perkin, G. Croft, T. & Lawson, D. (2013) The extent of mathematics support in UK higher education—the 2012 survey. *Teaching Mathematics and its Applications*, 32 (4), 165–172.

Smith, A. (2004) *Making Mathematics Count*, The Stationery Office, London, UK.

Solomon, Y., Croft, A.C. & Lawson, D.A. (2010) Safety in numbers: mathematics support centres and their derivatives as social learning spaces, *Studies in Higher Education*, 35 (4), 421–431.

Solomon, Y., Lawson, D.A. & Croft, A.C. (2011) Dealing with "fragile identities": resistance and refiguring in women mathematics students, *Gender and Education*, 23 (5), 565–583.

Sutherland, R. & Pozzi, S. (1995) *The Changing Mathematical Background of Undergraduate Engineers*, The Engineering Council, London, UK.

Symonds, R., Lawson, D.A. & Robinson, C. (2008) Promoting student engagement with mathematics support, *Teaching Mathematics and Its Applications*, 27 (3), 140–149.

Tolley, H. & Mackenzie, H. (2015) *Senior Management Perspectives on Mathematics and Statistics Support in Higher Education*, **sigma**, Loughborough, UK. Available at http://www.sigma-network.ac.uk/wp-content/uploads/2015/05/sector-needs-analysis-report.pdf (Accessed 17 March 2019).

Waldock, J., Rowlett, P., Cornock, C., Robinson, M. & Bartholomew, H. (2017) The role of informal learning spaces in enhancing student engagement with mathematical sciences, *International Journal of Mathematics Education in Science and Technology*, 48 (4), 587–602.

3

CURRENT INTERACTIONS BETWEEN MATHEMATICIANS AND RESEARCHERS IN UNIVERSITY MATHEMATICS EDUCATION

Caroline Bardini, Marianna Bosch, Chris Rasmussen and María Trigueros

A diversity of roles and interactions

One of INDRUM's main objectives is to contribute to the development of research in didactics of mathematics at all levels of tertiary education, with a particular focus on strengthening different forms of collaboration between all mathematicians involved in multifaceted university practices such as research in mathematics, research in mathematics education, university teaching, curriculum design, etc. These forms of collaboration might differ depending on the didacticians' institutional settings – education vs. mathematics or science departments. They can also be affected by the fact that, besides assuming various levels of interlocution with didacticians, mathematicians are part of the object of study of didactics research, which impedes a certain symmetry in the relationship. This complexity in the interactions between mathematicians and didacticians raise many questions, depending on the roles and institutional positions assumed by the former.

When mathematicians act as teachers, the largest question concerns the ways of collaboration between educational researchers and "practitioners", from the design, experimentation and data gathering of teaching practices, to the dissemination and impact of research results. This situation is quite similar to that of secondary and primary school teachers. Researchers in UME need to be aware of the difficulties university teachers encounter, and reformulate these teaching questions to integrate them into UME lines of research. At the same time, they need to consider the possible impact of research in the evolution of the university system, analysing the conditions needed for the reception and implementation of research results.

However, many factors make the situation very different from that of other educational levels. One is related to the lack of visibility of research in UME due to its relative youth as a research field and also to its weak presence in university teacher educational proposals. Another difference is the fact that university teachers usually participate in the design and evaluation of university programmes, being at the same time responsible for the creation, implementation and sometimes even evaluation of curriculum proposals. Finally, environmental factors related to the different positions of mathematicians and didacticians in the departments of mathematics or education, as well as the interactions between their respective research societies, seem to be crucial in shaping the relationships between both communities. All in all, it is clear that the status of the two research communities is far from being comparable, which leads to a non-balanced communication between mathematicians and didacticians.

This chapter presents three experiences from different countries (Australia, USA and Mexico) that illustrate different ways of collaboration between mathematicians and didacticians. The three cases were selected based on the respective authors' experiences and reflections on the current state of affairs in their respective countries and often insights into the challenges as well as prospects and possibilities for collaboration between mathematicians and mathematics education researchers. For each case we point to the theoretical and methodological orientation that informed the work and note that what is common in each case is not the particular orientation or approach, but rather the attention to both theory and practice.

In the first case from Australia, the initial situation is characterised by a divide between mathematicians and mathematics educators, who are located in different faculties (Science and Education) and do not need to share many teaching or research projects. Some collaboration has recently been emerging slowly, especially through national funded projects on the transition from school to university mathematics and the low retention rate in university mathematical sciences.

In the case of the USA, such a divide does not exist in the university structure. Mathematics departments often provide mathematics education positions and offer one third of the doctoral programs in mathematics education. The institutional conditions are therefore very different. An example of collaboration is provided by the project "Characteristics of Successful Programs in College Calculus," a national study of calculus programs developed by leaders of the Mathematical Association of America and several didacticians (Bressoud, Mesa, & Rasmussen, 2015). Let us just mention, as an output of the project, the participation of mathematicians and didacticians in the identification and illustration of programmatic features at both the classroom and department level that support student success in calculus.

The Mexican case shows an example of collaboration during a four-year project on teaching linear algebra using models. The project involves three research mathematicians and two didacticians who worked in the same university institution. The results show some conditions for a collaboration to be effective, such as the need for mathematicians to be open to experience for themselves innovative instructional methods, and some difficulties related to the diversity of approaches in mathematics education.

What we can see in the three cases is the growing importance of undergraduate mathematics education within mathematics and some joint projects to deal with the students' difficulties and "drop-outs" in scientific disciplines. Significant obstacles still remain. Some are due to the lack of recognition of educational research by mathematicians, who still seem to undervalue it relative to mathematics research. Others can be related to the consequent reluctance of mathematicians to get involved in educational projects that are demanding in time and effort but might not bring them enough return in terms of scientific prestige or publications.

Current interactions between mathematicians and didacticians in Australia

In Australia, as in many other countries, research in mathematics education is mainly carried out at faculties of education, while research in mathematics mainly takes place at "School of Science" departments. This separation is not only a matter of space. The evolution of the relationship between mathematicians and didacticians can be described as a long history of a big divide – and not only geographical… However, as is also rather common, there has always been some "cross-teaching" and, more recently, interactions and collaborations between the two communities are emerging slowly. Within the field of mathematics education, mathematicians are more present in large local conferences, like the Mathematics Education Research Group Australasia (MERGA). Also, UME dedicated conferences are growing, gaining visibility and prestige. We must mention in this context the biennial southern hemisphere symposia DELTA. Within the community of mathematicians, mathematics education is now formally a sub-group of the Australian Mathematics Society (AustMS) and it has become the second largest sub-group represented in the latest national AustMS conference.

In spite of this recent evolution, the institutional divide remains as an obstacle in many aspects. Teaching and education does not have a lot of recognition for mathematicians within their institution, where workload and financial reward is mainly based on publications. Therefore, the implementation of UME research in mathematics classrooms is sadly perceived as "on top of all other things", like research, teaching, administrative work, etc. We can identify a vicious circle with heavy consequences in terms of collaboration. It is generally hard to formally include mathematicians as principal or co-investigators in grants applications for mathematics education, since there is no track record in the field. They thus miss the corresponding financial gain for their department, which increases their reluctance to get involved.

There are however some success stories, such as the one we are here presenting. It constitutes a three-year project aiming to identify potential barriers that may contribute to the low retention rates in mathematical sciences subjects at university. This project involved four didacticians and five mathematicians from different universities and from the Australian Mathematical Science Institute. The main goal of the project was to investigate how mathematical expressions are "dealt with" (taught, understood and used) in first year university mathematical sciences subjects in comparison to students' experience at upper secondary levels.

Secondary and university mathematics: do they speak the same language?

Greater access to university education, the broadening of secondary school mathematics curricula, and the societal need for more expertise in Science, Technology, Engineering & Mathematics (STEM) all contribute to the challenge of teaching undergraduate mathematics students from diverse mathematical backgrounds. Tertiary teachers commonly lament the gap between students' apparent mathematical skills and knowledge and the ideal starting point for the curriculum they have planned. The project here presented starts from the assumption that, while mathematical thinking will develop throughout the school level, the bridge to mathematical thinking at university will be built to a great extent with symbols. At university, not only does mathematics become much more symbolic, but we argue that its writing is subtler and requires increased flexibility from the reader. One cannot take for granted that a symbol supposedly mastered at the secondary level will continue to be so at university because its meaning may not remain the same. The domain(s) with which a given symbol was prominently associated at secondary level often turn(s) out to be importantly extended at university level. The project proposes a framework for examining symbolic expressions and uses it to identify the elements that may cause students' bridge to tertiary mathematics to be weak or strong, and their passage over that bridge easy or hard.

It is commonly agreed that students' understanding and correct usage of mathematical notation plays a key role in their understanding of and fluency in mathematics. Much research has been undertaken at the secondary level to investigate students' understanding of symbols and what constitutes symbol sense (Arcavi, 1994; 2005), algebraic insight (Pierce & Stacey, 2001) and structure sense (Hoch & Dreyfus, 2006). Research at that level has shown that the very same features that make symbols a key element from which mathematics derives its power, namely conciseness and abstraction, can be a barrier to learning (Pierce, Stacey, & Bardini, 2010; MacGregor & Stacey, 1997). This issue is likely to be all the more crucial when it comes to mathematics at university. At this level, previous research showed that first-year university physics students' performance is highly correlated to their understanding of symbols (Torigoe & Gladding, 2007). We anticipated that similar outcomes might apply to other mathematical sciences.

Taking all of the above into account, the project developed tools to assist in examining the shift in the status of mathematical symbols from secondary to tertiary level and how this in turn may affect students' *symbolic literacy* (Bardini, Pierce, & Vincent, 2015) as an important element of their fluency in mathematics. Under the notion of symbolic literacy lies the notion of "symbol sense" described by Arcavi (1994; 2005), which includes among other components the ability to manipulate and "read through" symbolic expressions. The construct of the tools for analysing students' symbolic literacy developed in this project relied on: (i) the notion of "symbol template" introduced by Sherin (2001) and (ii) the epistemological approach specific to mathematical notations developed by Serfati (2005). In his work, Serfati both embraces the syntactical aspect of a symbol and investigates the underpinning mathematical concept(s) conveyed, by distinguishing three features of a symbol: its materiality, syntax and meaning. Sherin uses the term "symbol template" to refer to combinations of symbols that users recognise at a glance, rather than focusing on the symbol's syntax (in the sense of Serfati).

Over the three-year period of the project, first-year university students enrolled in mathematical subjects were surveyed fortnightly. The surveys included demographic questions and mathematical items designed to probe students' use and understanding of symbols and mathematical expressions across different topics they were studying at the time. This project proposed a framework with the potential to help analyse what students may be perceiving when manipulating – that is, reading, writing or making use of – mathematical symbols. In particular, it investigated what may be new to students and what they may have met before throughout their initial journey at the university, by taking the above-mentioned triple lens of materiality, syntax and meaning of symbols and, more generally, syntax templates.

The example we provide here has been chosen to illustrate the application of the framework, from its simplest form to a more refined setting, hopefully indicating the potentials of its use when considering students' symbolic literacy. The following question was responded to by 152 first-year university students studying a standard first-year mathematics subject at a highly ranked Australian university: "Explain the meaning of '-1' in each of the following: a) 'x^{-1}'; b) '$\sin^{-1} x$'; c) '$f^{-1}(x)$' where '$f(x)=3x + 1$'." The results (especially responses to (b)) indicated that a change in meaning of a familiar syntax template is problematic for many students, who respond to the syntax as they did when they first encountered it in junior secondary school, with one iconic example for (a) "x^{-1} is x is being times by itself negative one times" (applying blindly a template syntax disregarding the context or meaning). Results highlighted the subtle differences between each syntax template and the importance of carefully considering the various meanings that are signalled not only by the choice and order of pro-numerals[1] and operators but also by recognition of the mathematical context.

After analysing students' responses and interviews with their lecturers, one recommendation was the need for educators at all levels to explicitly draw students' attention to symbolic changes and not assume that they have somehow absorbed the subtle changes themselves, thus advocating the importance of explicit teaching. Noting new domains for familiar syntax templates and new, restricted or expanded contexts for symbols students have met before is not time-consuming, and can potentially have a great impact in students' increased flexibility at reading and writing mathematical expressions, thus progressively becoming more confident and at ease as they progress in their mathematical journey.

What lessons can we draw from the experience of the project? We can postulate that one of the positive conditions that contributed to the fruitful collaboration was that the project addressed problems mathematics departments were facing and struggling with, especially the issue of the retention rate and the consequences it might have. Moreover, the project was organised with a constant real collaboration between mathematicians and didacticians, the latter giving feedback from data collection promptly, with joint seminars to inform of the progress and difficulties met at the different levels (teaching and research). Participation in the project received institutional recognition, especially through its "advisory panel" and also allocated budget to help tutors with all practical details (e.g. data collection). It also relied on the dedication of certain mathematicians conscious that their students' learning difficulties could not be solved easily.

Collaborations between mathematicians and didacticians in the United States

The start of university mathematics education (UME) as a field in the US can be traced back to 1991 when three mathematicians, Ed Dubinsky, Jim Kaput, and Alan Schoenfeld organised the first special session on research in undergraduate mathematics education (RUME) at the annual meeting convened jointly by the two main mathematical professional societies, the Mathematical Association of America (MAA) and the American Mathematical Society (AMS). This was followed in 1992 by the AMS and MAA establishing the Committee on Research in Undergraduate Mathematics, which went on to publish several edited volumes on UME.

In 1996, the first RUME conference was held, bringing together mathematicians and the nascent cadre of researchers who received their doctorate in mathematics education with a focus on university mathematics education. This first conference had approximately 70 participants, but nowadays this annual event has over 350 participants. The growing interest in RUME then led in 2000 to the establishment of the first Special Interest Group of the MAA (SIGMAA), with a sole focus on RUME. Currently there are 13 SIGMAAs, several of which have strong ties to educational research. The latest SIGMAA to be created is one focusing on Inquiry Based Learning, primarily at the undergraduate level.

As this brief history attests, the mathematical professional societies in the US have been leaders and catalysts for RUME. All of this has led to what the former President of the MAA, the late Lynn Steen, saw as a cultural shift in the US. In 2011 he wrote the following:

> Professional meetings of university mathematicians, which in the mid-1980s were predominantly devoted to mathematical research and applications, are today a nearly equal mix of mathematics and mathematics education. For a community steeped in a tradition that focused only on research and exposition of mathematics, the very visible emphasis on teaching and learning is a major change in the culture.
>
> *(Steen, 2011, p.5)*

The growth of positive cooperation between mathematicians and mathematics education researchers (or didacticians) is reflected in what is now a decade-long collaboration between leaders at the MAA and researchers in UME, the majority of whom have positions in mathematics departments. The start of this collaboration was the *Characteristics of Successful Programs in College Calculus* project, which was a five-year grant from the US National Science Foundation (Bressoud, Mesa, & Rasmussen, 2015). The goals of this project were to improve our understanding of the demographics of students who enrol in calculus, to measure the impact of the various characteristics of calculus classes that are believed to influence student success, and to conduct explanatory case study analysis of exemplary programs in order to identify why and how these programs succeed. One of the findings from this project was the identification of seven common features across all five of the participating mathematics departments that offer a doctorate degree in mathematics and which have a relatively more successful calculus program. These features were identified through qualitative analysis of explanatory case studies of calculus programs that were selected as having more successful programs than comparable universities. The more successful programs were selected based on a national survey where success was operationalised in terms of passing rate, persistence, and confidence, interest, and enjoyment of mathematics. More details on these case studies and methodology can be found in Rasmussen, Ellis, Zazkis, & Bressoud (2015).

- *Coordination.* Calculus had a course coordinator who fostered a community of practice (Wenger, 1998) where instructors mutually engaged in the joint enterprise of calculus instruction, building common values, goals, and interpretations of student difficulties. Their shared repertoire, the elements that focused their collaborative efforts, included course pacing and coverage, homework assignments, and midterm and final exams.
- *Attending to local data.* There was someone in the department who routinely collected and analysed data in order to inform and assess program changes. Departments did this work themselves and did not rely on the university to do so. Data collected and analysed included pass rates, grade distributions, persistence, placement accuracy, and success in Calculus II.
- *Graduate teaching assistant (GTA) training.* The more successful calculus programme had substantive and well thought out GTA training programmes. These ranged from a week-long training prior to the semester together with follow up work during the semester to a semester course taken prior to teaching. The course included a significant amount of mentoring, practice teaching, and observing classes. GTAs were mentored in the use of active learning strategies in their recitation sections. The standard model of GTAs solving homework problems at the board was not the norm. The more successful calculus programmes were moving toward more interactive and student-centred recitation sections.
- *Active learning.* Calculus instructors were encouraged to use and experiment with active learning strategies. In some cases, the department chair sent out regular emails with links to articles or other information about teaching. One institution even had biweekly teaching seminars led by the math faculty or invited experts. Particular instructional approaches were not prescribed or required for faculty at any of the institutions.

- *Rigorous courses.* The more successful calculus programs tended to challenge students mathematically. They used textbooks and selected problems that required students to delve into concepts, work on modelling-type problems, or even proof-type problems. Techniques and skills were still highly valued. In some cases, these were assessed separately and a satisfactory score on this assessment was a requirement for passing the course.
- *Student support services.* Students were provided with out-of-class resources. Almost every institution had a well-run and well-utilised tutoring centre. In some cases, this was a calculus-only tutoring centre and, in other cases, the tutoring centre served linear algebra and differential equations. Tutoring labs had a director and tutors received training.
- *Effective placement.* Programmes tended to have more than one way to determine student readiness for calculus. This included placement exams (which were monitored to see if they were doing the job intended), gateway tests two weeks into the semester, and different calculus formats (e.g., more time) for students with lower algebra skills.

The body of research stemming from this project continues to be a source of inspiration and guidance for mathematics departments in the US to embark on research-based efforts to improve their introductory mathematics courses (e.g., Rasmussen et al., 2019). We now turn to a particular case of one such effort that involved the strong collaboration between mathematicians and mathematics education researchers, all of whom are in the same department.

A case study of "Reimagining the introductory mathematics program"

In the spring of 2014, spurred on by faculty discontent with students' knowledge and skills of prerequisite material, especially related to precalculus and calculus, unacceptably low passing rates in these courses, some dissatisfaction from engineering, and pressure from administration to do something about student success rates, the mathematics department at San Diego State University (SDSU) decided to embark on a major initiative to improve the learning and teaching of precalculus and calculus. Turning to the mathematics education research literature, they began by reflecting on how the department fared on the previously described seven characteristics of successful calculus programs and came to the hard realisation that their introductory mathematics program was seriously lacking on all seven characteristics. Rather than opting to make small, incremental changes, they took a systems approach (Austin, 2011) and decided on a comprehensive improvement plan that targeted all seven characteristics. As detailed in Apkarian, Bowers, O'Sullivan, and Rasmussen (2018), they conceptualised the roll-out of their improvement efforts in terms of the following four levels: administration, department, course, and individual. The administration level involved collection and use of local data and new student support services; the department level involved a new coordination system and a new placement system; the course level involved ensuring challenging, engaging courses and the use of active learning strategies; and the individual level involved professional graduate students. This vision to enact changes and the individual efforts to do so involved a leadership team comprised of both research mathematicians and researchers in undergraduate mathematics education.

For example, the department chair formed a calculus task force comprised of both mathematicians and didacticians who laid out a road map for the change efforts and who oversaw the implementation of the plan. Here we mention three aspects of the change efforts and roles that the two groups of researchers played. First, and central to the change efforts, was the creation of a coordination system for the precalculus and calculus courses. Prior to the changes, there was zero coordination across instructors, which often resulted in different instructors using different textbooks, even though they were teaching the same course in the same semester. Faculty at SDSU realised that this was not good for students and that it made it more challenging to build a community of practice where instructors supported each other in innovation. The department established three new coordinator positions, two of which were taken up by mathematicians and one of which was taken up by a mathematics education researcher. These coordinators worked together to improve their respective courses and to create better vertical alignment across courses.

Second, the department created an entirely new GTA professional development program aimed at supporting GTAs to implement active learning where students worked collaboratively on in-house developed challenging and engaging problems. The new professional development program consisted of a three-day workshop prior to the start of the semester, the creation of a new seminar during the academic year to continue to support GTAs as they developed and honed their practice, and a new lead GTA model where more experienced GTAs observed and supported their peers (Ellis, 2015). The GTA professional development program was largely designed by the mathematics education researchers, with mathematicians leading several of the sessions during the three-day workshop. A mathematics education researcher oversees the academic year seminar and regularly participates in weekly meetings convened by the course coordinators.

Third, the department was able to create a new Mathematics Learning Centre (MLC) to support students in precalculus and calculus. Prior to the change efforts there was a small, cramped room in the mathematics department run by GTAs with little to no oversight. By capitalising on the university's strategic plan, the mathematics department was able to obtain a much larger space in the library and the chair convinced upper administration that one of the mathematics education researchers in the department was the right person to be the director of the new MLC because she brought knowledge of the mathematics education literature as well as a collegial relationship with coordinators and instructors. This afforded considerable cross-talk and collaboration as everyone worked together to offer services to support and promote student success.

In conclusion, we see the case of SDSU as a prime example of the productive ways that faculty with different research foci but who are in the same department can productively work together to develop a shared vision for change, draw on respective expertise, and successfully roll out significant changes to the teaching and learning of introductory mathematics courses. While there were some bumps along the way in the enactment of the extensive change efforts, these bumps were not due to conflicts or disagreements between the mathematicians and mathematics education researchers, but rather due to the complexity of the innovation. The case of SDSU is an example of how positive collaborations between faculty in the same department, but with different research interests, can lead to positive changes in departmental culture and improved student success (Apkarian, 2018).

Building bridges between communities in Mexico

Mathematics education as a field of research has existed in Mexico for many years. The first institution to host a mathematics education department was a research institute that is part of the National Polytechnic Institute: CINVESTAV. The *Matemática Educativa* department was created with the goal of doing research that could provide a solid basis for elementary school teachers' training programs. Over the years, this goal was expanded to research in mathematics education in general, including research on university mathematics education.

Very soon new programmes were created in other universities and research centres around the country. Mathematics educators work mostly in education departments and research centres where research focuses on different aspects of mathematics education at different school levels including university, teacher training, and social or theoretical issues related to the discipline. Those who work in mathematics departments at universities are mainly focused on university mathematics education, but not only. It is important to clarify that in Mexico students who want to be teachers are not trained at university but in specific institutions called *Escuelas normales*; and that there are some teacher training programs at the university but no official training programs for teachers at the university level.

Since the early 90s, the Mexican Mathematical Society has included an interest section on mathematics education in its annual conference. Originally, mainly workshops designed for elementary and high school teachers were offered. But after a few years, talks on research on mathematics education and on mathematical teaching practices concerning all school levels were included as well. Presentations about university mathematics teaching are privileged at these conferences, but they are not concerned with research on mathematics education at this level. There are not many opportunities for interaction between the community of research mathematicians and that of mathematics educators.

An important achievement for the Mexican mathematics education community is that both mathematics researchers and mathematics education researchers have been involved in the development of the elementary and secondary school curriculum, and, at those levels, research results have been taken into consideration in new curricula. However, this is not the case for the university level. There, mathematicians are in charge of curricula both for mathematics degrees and for service courses imparted to other disciplines' students; thus mathematics educators are involved neither in decisions taken about ways of teaching nor on curricular changes.

Nowadays the science and technology government agency (CONACyT) is putting more effort into supporting research projects on mathematics education in general. This has fostered the interest of the mathematics community at the university level, which has been accompanied by a growing number of students in those master and doctoral programs where mathematics education at the university level is considered.

Regarding the teaching of mathematics at the university, most university courses are lecture based. There are some university teachers who have adopted other teaching methods, particularly with the use of technology, where students have opportunities to participate. Lecturing is not questioned by the mathematics community in spite of students' poor results and a high rate of student failure in many such courses. Research on mathematics education is tolerated in some mathematics departments but there is not support either for those researchers who work in this area, or for research mathematicians who are interested in collaborating on mathematics education projects; their research publications in this field are not necessarily considered for promotion or do not have the same value as research on mathematics.

An example of collaboration and interaction

In order to foster a closer collaboration between researchers in mathematics education and research mathematicians, a four-year research project on the teaching of linear algebra using models was developed at a Mexican university. Its aim was to study how interaction among mathematicians and mathematics educators fostered mathematicians' interest in changing their teaching methods. It also aims at testing the role of modelling activities in students' learning of linear algebra. The project was funded by the Mexican government and was based on two theoretical frameworks, Models and Modelling perspective (Lesh & Doerr, 2003) and APOS theory, developed by Ed Dubinsky and the RUMEC group (Dubinsky & McDonald, 2001), an international group of researchers that worked collaboratively in the 1990s.

At the first stage of the project, the aim was the development of modelling situations and the development of both modelling and conceptual activities for students to accompany each modelling situation to be used in the classroom with students enrolled in different programs: mathematics, engineering, economics and business. Research was conducted when each modelling situation was used in the classroom and, depending on the results obtained, they were or not modified. Once each situation was recognised to work as expected, a didactical sequence aimed at helping teachers who were not involved in the project to use the proposed teaching strategy in their teaching was developed.

Each week a two-hour seminar took place where all the project participants discussed ideas about possible real or realistic situations and the design of modelling situations. Some mathematics education theories were introduced and discussed, and activities based on APOS theory to foster students' construction of the concepts of interest were designed and discussed. The suggested teaching methodology involved students working in teams on the modelling situation and activities, whole group discussion at least twice per class and homework exercises for students.

The participants were three research mathematicians, three university mathematics teachers, and two mathematics education researchers. All of them volunteered to participate for the whole duration of the project. Each seminar was audio recorded with the agreement of the participants and notes were taken by two of them. The participants' development was then analysed. The resulting data were analysed by two mathematics education researchers in terms of participants' interactions. Lessons were observed and recorded by those participants not teaching that class and observations and recordings were analysed and discussed during the seminar. A study was conducted on participation in the seminar focusing on the transformation of mathematicians' ideas about students learning, mathematics education and teaching practices through participation in the seminar. It was found that the seminar was an interesting and enriching experience for all participants. Through active and respectful interaction, all participants were somehow transformed. The group as an entity also changed (Trigueros & Lozano, 2015).

Openness to innovation and learning about mathematics education research

A lot was learnt from the seminar. Participation transformed all the members. The seminar's dynamic was always rich in participation and openness. Mathematics researchers were interested in their students' learning and in mathematics applications. They had different levels of experience and they all used a lecture and some questioning approach in their lessons. In particular, they included some whole-group discussions when they introduced applications.. Their first interventions concerned the mathematics involved in some interesting applications, presenting applications that might be of interest to the students, and addressing difficulties students had with linear algebra. As the seminar progressed, they became interested in different teaching methodologies and some aspects of mathematics education theories. One of them started taking notes on his students' difficulties while working with the problems to be discussed in the seminar and, by the second year, he changed his teaching method. When using the modelling situations, he let the students work in small groups on their own and used whole-class discussion to compare students' ideas and strategies. He once commented "I have tried methods discussed here and in our readings. I find it is more interesting for students. I am enjoying it and I think it really works for students". "I am finding new ways to introduce the concepts, based on ideas discussed during the seminar". His participation, his questions, comments and interest in the seminar and in the classroom changed drastically.

Other participants also increased their participation in the analysis of the data obtained through class observation and changed their teaching strategies. Together with the above-mentioned participant, all considered important new aspects in their teaching and learning of mathematics, they valued the benefits of analysing mathematics and teaching from a mathematics education perspective, and they changed their teaching, so it became more collaborative and interactive. It was clear that involvement and interactions in the seminar and new experiences in their classroom helped them value new methodologies and mathematics education research (Trigueros & Lozano, 2018). Studies about different modelling situations were published collaboratively with them (Possani, Trigueros, Preciado, & Lozano, 2010). Two of them still participate in mathematics education studies and new collaborators have joined the group (Figueroa, Possani, & Tigueros, 2018; Salgado & Trigueros, 2015).

Other cases were not as successful, one of the mathematics researchers insisted on discussing those issues that were directly related to the mathematics involved. He is considered a good teacher by students and contributed to the project and the development of the group through sharing his knowledge and questioning the development of the project materials. He became more interested in education but not in mathematics education or new teaching methodologies. He considered the lecture method of teaching as consistent with the nature of the field itself, which accords with the views of many university mathematics teachers in contrast with those of mathematics education researchers who are convinced that participation and involvement in the solution of all types of mathematics situations help students learning and their appreciation of mathematical concepts.

He did not change his beliefs about students learning: "We all learned by being introduced first to definitions and theorems, then we used them to solve problems, and this worked fine", and close to the end of the project still showed doubts about the use of open modelling situations and activities: "Presenting the problem like that will take a lot of class time. … It would be better to have extra sessions to work on the models, once you have taught them what you have to teach … or give the problem as a project to do at home". He was overwhelmed by the complexity of mathematics education theories, and the changes in the class structure. However, although he was not fully convinced, he was willing to learn something about some of the theories that he considered were useful in the design of activities.

Some results

In retrospect, the project created a place where it was possible to develop positive interactions between, on the one hand, mathematics researchers and teachers at the university and, on the other, mathematics education researchers. It seems that mathematicians could develop confidence in mathematics educators' mathematical knowledge since this factor contributed to openness in discussion. It is also fundamental to convince mathematicians that research on university mathematics education needs the contribution of mathematicians. They are very helpful in designing complex situations and can critically contribute to the analysis of data while they can be convinced that different class methodologies may incite students' reflection and questions and whole-class discussions that can help them reach a deep understanding of the mathematical ideas involved in the class.

Another factor that may contribute to make more collaborations possible can be mathematicians' interest in students' learning, but convincing them to become open to experience innovative teaching methods is still not an easy endeavour. However, even if there is no such willingness, mathematicians can be helpful in reviewing and discussing activities and materials from the mathematical point of view. The project experience made us aware that sometimes we needed help and mathematicians were always willing to participate in such discussions and reviews.

Different approaches to mathematics education can be a limitation factor to mathematicians. They do not see in results the unity they would like to see; they are overwhelmed by differences in points of view, and this may distance them from being interested in knowing what mathematics educators have found. Diffusion of mathematics education approaches, methods and results with a focus on mathematicians' interests are widely needed. Some mathematicians that have been in touch with the results of the aforementioned project have shown interest in using the materials in their courses and some mathematics students involved in the projects' classes have joined with us to help developing new activities or to learn more about mathematics education.

The importance of collaboration between mathematicians and mathematics educators, through means where both groups have something to say and share, reinforces the need to look for spaces such as the project described here where interactions and dialogue conditions are made possible. From our experience, the development of projects and seminars with participation of mathematics researchers and teachers can contribute to convince them of the value of research in mathematics education. It seems that young mathematicians are more willing to participate in mathematics education projects, but one of the studies referred to in this experience involved senior mathematicians and some of them showed changes their perception of mathematics education and on students learning as well.

Conclusions and open questions

The three case studies presented in the chapter do not pretend any kind of generality. On the contrary, they correspond to isolated cases of good experiences and it is well-known that success stories do not depict reality, especially because bad experiences are always harder to present. The positive side is the existence of different forms of collaboration projects led by researchers in UME and involving mathematicians as teachers and designers. This variety is what this chapter illustrates.

Let us first highlight the influence of the institutional environment in the way didacticians and mathematicians can cooperate. The Australian case shows the difficulties of interaction when researchers belong to different departments and are therefore seen as belonging to different research communities. Only efforts from the outside or the consciousness of an important social issue – difficulties in the transition from secondary to university level, low rate of retention, etc. – can sow the seed for effective collaboration. To what extent this collaboration will be only temporary or will take a more regular shape remains however an open question. The US case presents another dimension from this perspective. There, didacticians and mathematicians are colleagues in the same department, share conferences and professional association. The starting point was also a crucial issue – students' failure in calculus courses – and the support of a national project. The interesting aspect of this case is the scope of the research, moving from a specific source of learning difficulties to the whole organisation of a subject's instruction by an entire department. Finally, the Mexican case shows a more local research project, also promoted by an external financing research institution and involving mathematicians and didacticians working in the same department. The interesting point refers to the difficulties encountered in the collaboration between educational researchers and mathematicians acting as teachers, due to the complexity of the research tools used by the former and the difficulties of the latter to accept it, a symptom that mathematics education is still not considered as a real field of research, with its own elaborations, methodologies and technicalities.

The three cases illustrate the possibility and also limitations of organising collaborations between mathematicians and researchers in mathematics education. In considering the interaction between the two communities, we must acknowledge the role and efforts made by the international associations ICM and ICMI in creating links and increasing the visibility of UME among mathematicians. The fact that some mathematicians evolve towards mathematics education or are involved in teacher education programmes might also possibly help bring closer the two communities. However, it is also crucial for UME to be recognised as an educational discipline among the others, to ensure the impact of research results at a more general level. Finding the optimal middle point between the specificity of the domain and the research community to exchange with does not seem easy and it definitely affects the conditions for UME research and its reception.

The three cases considered in this chapter do not address another important issue: the collaboration between didacticians and mathematicians teaching in programmes where mathematics is taught as a service subject (engineers, physicists, biologists, economists, etc.). Furthermore, the collaboration between disciplines might also be productive to disseminate some of the investigations and methodologies developed in UME to the teaching and learning of disciplines where mathematics plays an important role. What forms all these diverse collaborations might take and what conditions might enhance them remains an open question. Research at this institutional level is still incipient and deserves to be developed from an open perspective, to avoid unproductive corporate debates.

Always at the institutional level, developing international comparisons might be an interesting line to pursue. Much is to be learnt about national forms of professional development of university teacher education, the (sometimes almost nonexistent) participation of UME research in the preparation of university teachers, and its rivalry with more general educational disciplines. Finally, it is not only the type of department, school or faculty didacticians and mathematicians belong to that can make the difference. An important factor is also the balance between research and teaching in the promotion of university faculty. In many countries, the secondary role attributed to teaching with respect to research does not bode well for major changes in the short term.

From the perspective of research in UME, the analysis of different forms of cooperation between researchers and teachers raises several questions about the kind of background needed by both mathematicians and didacticians to ensure collaboration. For instance, it is clear that research in UME cannot take the mathematicians' vision of their own discipline for granted, either as a research or as a teaching domain. Therefore, often the collaboration needs to make this vision and the underlying assumptions explicit, which usually requires the mobilisation of complex research tools and methodologies. As we have seen in the Mexican case, mathematicians can be reluctant to accept the difficulties in adopting a given theoretical framework, as if research in UME could be done only based on common sense and some supplementary effort. The dissemination of UME approaches, methodologies and results to the mathematical community remains an open question. In the case of primary and secondary school levels, pre-service and in-service teacher education programmes are one of the main dissemination channels. The situation at the university seems much more complex. Maybe the progressive incorporation of some kind of university teacher training or support can be a productive path in this direction, as seems to be the funding of educational research projects involving mathematicians and didacticians (Bardini, Bosch, Hochmuth, Rasmussen, & Trigueros, 2016, p. 31). Studying the conditions to foster cooperation together with the epistemological and pedagogical obstacles that may appear opens new lines of research in the field of UME and should be approached as such.

Acknowledgments

We would like to thank Robyn Pierce (University of Melbourne, Melbourne Graduate School of Education) for his collaboration and the Instituto Tecnológico Autónomo de México for its support. The investigations reported in the chapter were supported by the Australian Research Council under the Discovery Project DP140100646; by CONACyT (Proy. 62375) in Mexico; by the Spanish project RTI2018–101153-B-C21 (MCIU/AEI/FEDER, UE); and by the US National Science Foundation (NSF) under Grant 1430540. The opinions expressed do not necessarily reflect the views of NSF

Note

1 A pro-numeral is a symbol, usually a letter,that mathematicians use to stand in place of a number.

References

Apkarian, N. (2018). *Transforming Precalculus to Calculus 2: A longitudinal study of social and structural change in a university mathematics department.* Unpublished Doctoral Dissertation, University of California San Diego and San Diego State University.

Apkarian, N., Bowers, J., O'Sullivan, M., & Rasmussen, C. (2018). A case study of change in the teaching and learning of Precalculus to Calculus 2: What we're doing with what we have. *PRIMUS*, 28(6), 528–549.

Arcavi, A. (1994). Symbol sense: Informal sense-making in formal mathematics. *For the Learning of Mathematics*, 14(3), 24–35.

Arcavi, A. (2005). Developing and Using Symbol Sense in Mathematics. *For the Learning of Mathematics*, 25(2), 42–48.

Austin, A. E. (2011). *Promoting evidence-based change in undergraduate science education.* National Academies, Board on Science Education. https://sites.nationalacademics.org/cs/groups/dbassesite/documents/webpage/dbasse_072578.pdf. Accessed 28 Jan 2019.

Bardini, C., Bosch, M., Hochmuth, R., Rasmussen, C., & Trigueros, M. (2016). Current interactions between mathematicians and researchers in mathematics education. In E. Nardi, C. Winsløw, & T. Hausberger (Eds.) *Proceedings of INDRUM 2016 First conference of the International Network for Didactic Research in University Mathematics* (pp. 29–31), Montpellier, France. https://hal.archives-ouvertes.fr/INDRUM2016.

Bardini, C., & Pierce, R. (2015). Assumed mathematics knowledge: the challenge of symbols. *International Journal of Innovation in Science and Mathematics Education*, 23(1), 1–9.

Bardini, C., Pierce, R., & Vincent, J. (2015). Contemplating symbolic literacy of first year mathematics students. In M. Marshman, V. Geiger, & A. Bennison (Eds.), *Mathematics education in the margins: Proceedings of the 38th annual conference of the Mathematics Education Research Group of Australasia* (pp. 77–84). Sunshine Coast: MERGA.

Bressoud, D. M., Carlson, M., Mesa, V., & Rasmussen, C. L. (2013). The calculus student: Insights from the MAA national study. *International Journal of Mathematical Education in Science and Technology*, 44(5), 685–698.

Bressoud, D., Mesa, V., & Rasmussen, C. (Eds.) (2015). *Insights and recommendations from the MAA national study of college calculus.* Washington, DC: The Mathematical Association of America.

Dubinsky, E., & McDonald, M. A. (2001). APOS: A Constructivist Theory of Learning in Undergraduate Mathematics Education Research. In: Holton, D. (Ed.) *The Teaching and Learning of Mathematics at University Level: An ICMI Study*, 276–282. Dordrecht: Kluwer.

Ellis, J. (2015). Three models of graduate student teaching preparation and development. In D. Bressoud, V. Mesa, & C. Rasmussen (Eds.), *Insights and recommendations from the MAA national study of college calculus* (pp. 117–122). Washington, DC: Mathematical Association of America.

Figueroa, A. P., Possani, E., & Trigueros, M. (2018). Matrix multiplication and transformations: an APOS approach. *The Journal of Mathematical Behavior*, 52, 77–91.

Hoch, M., & Dreyfus, T. (2006). Structure sense versus manipulation skills: an unexpected result. In J. Novotná, H. Moraová, M. Krátká & N. Stehlíková, (Eds.). *Proceedings 30th Conference of the International Group for the Psychology of Mathematics Education* (Vol.3, pp. 305–312). Prague: PME.

Lesh, R. & Doerr, H. (2003). *Beyond Constructivism, Models and Modeling Perspectives on Mathematics Problem Solving, Learning, and Teaching.* Mahwah, NJ: Lawrence Erlbaum Associates.

MacGregor, M., & Stacey, K. (1997). Students' understanding of algebraic notation: 11–15. *Educational Studies in Mathematics*, 33(1), 1–19.

Pierce, R., & Stacey, K. (2001). A framework for algebraic insight. In J. Bobis, B. Perry & M. Mitchelmore (Eds.), *Numeracy and Beyond: Proceedings of the 24th Annual Conference of the Mathematics Education Research Group of Australasia*, (pp. 418–425). Sydney: MERGA.

Pierce, R., Stacey, K., & Bardini, C. (2010). Linear functions: teaching strategies and students' conceptions associated with y = mx + c. *Pedagogies: An International Journal Special Issue: The teaching of Algebra*, 5(3), 202–205.

Possani, E., Trigueros, M., Preciado, G., & Lozano, M. D. (2010). Use of models in the teaching of linear algebra. In D. Kressner, *et al.* (Eds.), *Linear Algebra and its Applications (Special Issue devoted to the 15th ILAS Conference)*, 432(8), 2125–2140.

Rasmussen, C., Apkarian, N., Ellis, H., Johnson, E., Larsen, S., Bressoud, D., & the Progress through Calculus team (2019). Characteristics of Precalculus through Calculus 2 programs: Insights from a national census survey. *Journal for Research in Mathematics Education* 50(1), 24–37.

Rasmussen, C., Ellis, J., Zazkis, D., & Bressoud, D. (2014). Features of successful calculus programs at five doctoral degree granting institutions. In *Proceedings of the joint meeting of PME38 and PME-NA36* (Vol. 5, pp. 33–40).

Salgado, H., & Trigueros, M. (2015). Teaching eigenvalues and eigenvectors using models and APOS Theory. *The Journal of Mathematical Behavior*, 39, 100–120.

Serfati, M. (2005). *La Révolution Symbolique: La Constitution de l'Ecriture Symbolique Mathématique*. Paris: Éditions Petra.

Sherin, B. L. (2001). How Students Understand Physics Equations. *Cognition and Instruction*, 19(4), 479–541.

Steen, L. (2011). Challenges and transitions: Undergraduate mathematics 1990–2010. *Project Kaleidoscope 20th Anniversary Essay Collection*, 1–6. http://www.aacu.org/pkal/documents/MAA.pdf.

Torigoe, E., & Gladding, G. (2007). Symbols: Weapons of Math Destruction. *AIP Conference Proceedings*, 951(1), 200–203.

Trigueros, M., & Lozano, M. D. (2015). Teacher change: ideas emerging from a project for the teaching of university mathematics. *Teaching in Higher Education*, 20(7), 699–710.

Trigueros, M., & Lozano, M. D. (2018). Collaboration entre enseignants à l'université et émergence d'idées. In M. Abboud (Ed.), *Mathématiques en scène des ponts entre les disciplines. Actes du Colloque EMF2018* (pp. 1615–1623). Paris: Éditions de l'IREM de Paris. https://emf2018.sciencesconf.org/data/pages/SPE_4_Pre_actes_EMF2018.pdf.

Wenger, E. (1998). *Communities of practice: Learning, meaning, and identity*. Cambridge, UK: Cambridge University Press.

4

EDUCATION AND PROFESSIONAL DEVELOPMENT OF UNIVERSITY MATHEMATICS TEACHERS

Carl Winsløw, Rolf Biehler, Barbara Jaworski, Frode Rønning and Megan Wawro

Introduction

The title of this chapter may surprise some readers, as university teachers of mathematics are perhaps considered in no need of education: they typically hold a PhD in mathematics or some adjacent field, and in many places some on-site pedagogical training is also expected. However, as research has clearly evidenced, university mathematics teaching faces solid and perhaps increasing challenges. This is due to factors such as the varied and inhomogeneous student populations who attend mathematics courses at university level, and the didactic potential of computers, which in other contexts are increasingly involved in mathematical work. Is it time to think more systematically about the professional preparation and development of university mathematics teachers? In this chapter, we summarise some of the issues and the knowledge and experience that exist, particularly in the European setting and in research presented at the first two INDRUM conferences.

Let us meet Bob, a newly hired assistant professor of mathematics. With a recent PhD based on a thesis in abstract algebra, Bob was selected for the position among dozens of applicants from around the world because his research is excellent and gets published in outlets such as *Journal of Algebra*. During his first term at the prestigious university, several senior colleagues are on research leave, and it turns out to be necessary that he takes on the service course given to undergraduate biology students. The mathematics is simple for Bob (one-variable calculus, ordinary differential equations) but he has not thought about that material since his own time as a first-year student, eight years ago. His task is to lecture twice a week for a semester, in front of 250 biology undergraduates. What are his options?

Much as if he were to teach in an upper-secondary school, Bob could get the textbook and any teaching plans which may be available from last year's teacher and go by the book. Well, at least one thing is different: unlike the upper-secondary teacher, Bob is unlikely to have engaged in formal courses on teaching mathematics, or to have been exposed to research-based resources for teaching the contents and courses he has to undertake. Certainly, Bob has a lot of experience of doing classic blackboard lectures, but his audiences have so far been colleagues in abstract algebra.

Of course, the above vignette could have many variations. Rather than focusing on research in his new position, maybe Bob has some teaching assistant experience with Calculus and has been hired in a technical college where he will face a heavy teaching load. In this case, the main change for our young fellow may be transitioning from teaching small groups (problem sessions) to lecturing, while he is familiar with the details of the subject, students' difficulties and so on. Certainly, the situations of faculty members vary among higher education institutions. If some formal preparation of the university teacher is offered or required, it most certainly takes place within the school (organised at the university, in parallel with teaching – not at a separate, external institution before assuming regularly teaching); also it is likely to have a main focus on general aspects of teaching.

How do we know? What education is actually taken and offered? Does it matter? In fact, the literature so far offers a number of local, small-scale studies when it comes to describing the current state of university mathematics teachers' preparation for teaching (e.g. Speer & Hald, 2008), or the effects their various backgrounds may have on how they teach (e.g. González-Martín & Gomes, 2016; Mathieu-Soucy, Corriveau & Hardy, 2018). Other qualitative case-studies have been carried out to try to characterise the specific knowledge used or needed in situations of university mathematics teaching (e.g. Musgrave & Carlson, 2017; Speer & Wagner, 2009). Most of these studies explicitly relate to established frameworks for the study of mathematics teacher knowledge and education at the primary and secondary levels, while pointing out that they may at least have to be adapted to the special conditions found at the tertiary level. For research on mathematics teacher education at secondary and primary levels, not only sophisticated theoretical frameworks exist, but we also have a large number of studies of teacher education practices and their effects. These include large-scale international studies and contributions that date back at least half a century. No similar research effort has been devoted to the tertiary level, despite its undeniable challenges and importance.

Based on the work done before and during the plenary panel at INDRUM 2018, this chapter presents a number of reflections and experiences concerning the *preparation and in-service development of academics for the profession of teaching mathematics at university*. We refer to this activity field as University Mathematics Teacher Education (UMTE). We shall consider both how UMTE relates to, and goes beyond, the generic training which is (in most universities we know of) made available for university teachers at large. As the research literature bearing specifically on UMTE is, as outlined above, very scarce, we will present these cases on the background of closely related research areas – most notably,

- research into the education, knowledge, and development of mathematics teachers at other levels;
- research on university mathematics education (UME) at large.

The broad research questions, which this chapter will mainly address:

RQ1. What is the current, typical preparation of University Mathematics Teachers?
RQ2. What in-service opportunities for teacher development exist?
RQ3. Do the current preparation and opportunities for development meet the demands that exist or can be foreseen?
RQ4. Could the preparation and development opportunities be improved, for instance by giving university teachers (more) access to selected parts of current research on UME, and possibly also the opportunity to participate in research and development projects?
RQ5. What initiatives exist, and which could be imagined as beneficial – both to increase the impact and quality of research on UME, and of UME itself?

Naturally, answers to any of these questions are likely to vary between countries, and more locally between individual institutions and departments.

Here, we discuss specific cases and ideas from the authors' countries and institutions, as well as from the (sparse) related research literature. The aims are to build a first (and, naturally, partial) picture of how the professional knowledge of university mathematics teachers is currently established and sustained, and to identify what research is needed to obtain more systematic answers to the five research questions. Reflecting these aims, the chapter has three main parts: "Current UMTE practices", "Development UMTE" and "Future research on UMTE".

Current UMTE practices

In some countries, national legislation specifies requirements for university teachers' formal preparation to teach. This is the case, for instance, in Denmark, where all higher education is offered and regulated by the state. Since 1993, it has been a legal requirement that assistant professors at universities "in relation to the teaching tasks (…) receive systematic supervision and guidance, which concludes with a written assessment of the assistant professors' teaching qualifications" (Christiansen, 2016, p. 33). Tenure (typically hiring as associate professor) is based on such statements, and it is a formal requirement to undertake preparation for teaching within the first year of tenure, in case the new faculty member has not done so in advance (for instance, if hired from abroad). The detailed implementation of this policy is left to individual institutions, while it is monitored by the Ministry of Education.

In other countries, such as the United States, the institutional landscape is much more varied, as there are national, regional and private institutions of higher education. There is no nationally-required professional development for university mathematics professors related to the professional activity of teaching, although institutions and departments may, and often do, organise some form of formal training. It is a daunting task to provide an overview of these practices, as there are about 1500 mathematical sciences departments in the country (American Mathematical Society, n.d.). In this section, we present other special cases and experiences from the UK, Germany, Norway and the US.

UMTE practices in the UK

In the UK, university courses for (mathematics) teacher education vary from one institution to another and are largely general in nature; that is, they do not focus particularly on any disciplinary area. Despite the variations, certain national conditions motivate universities to focus on the quality of teaching, including the preparation of new teachers. These conditions include the competition to attract the best qualified students, and the fact that state funding depends strongly on quantity and quality in education. An example of how quality is assessed is the *Teaching Excellence Framework* (TEF), an exercise introduced by the government in England to assess higher education providers' levels of excellence in terms of teaching and ensuring students achieve good outcomes, whether with regard to graduate-level employment or further study (Office for Students, 2018) The TEF has awards, Gold, Silver and Bronze, which measure teaching excellence in three key areas:

Teaching quality: the extent to which teaching stimulates and challenges students, and maximises their engagement with their studies.
Learning environment: the effectiveness of resources and activities (such as libraries, laboratories and work experience) which support learning and improve retention, progression and attainment.
Student outcomes and learning gain: the extent to which all students achieve their educational and professional goals, in particular students from disadvantaged backgrounds.

One example of a university awarded Gold is Loughborough University (in 2017). The university is currently within the top 10 in England in all league tables that compare university education. When a new lecturer is appointed at this institution, she will have three years of probation, during which time she will have a reduced teaching load compared with more experienced colleagues. She will be required to follow a three-year general education programme, run by the University's Centre for Academic Practice (CAP) and requiring a portfolio for assessment each year. Also, each year, she will take on slightly more teaching until, after three years, she is teaching a complete load; providing her assessments are satisfactory, the probation will end and her contract will become permanent.

In some places, specific and optional professional development opportunities for mathematics faculty also exist. For instance, at Loughborough, until a few years ago, the Mathematics Education Centre (MEC: a Department in the School of Science alongside Mathematical Sciences, Physics, Chemistry and Computer Sciences) ran a series of seminars called the "How we Teach" seminars. At each seminar, one teacher of mathematics gave a talk (~30 mins) focusing on some aspect of her teaching. This was followed by discussion in the seminar focusing on issues in teaching. Teachers and mathematics educators engaged with and learned from each other. The first 20 seminars were video recorded, for those who had not been able to be present to view later. Jaworski and Matthews (2011) analysed 10 of these videos against research questions regarding content and issues discussed. Participation in the seminars and a reflective review became an option in the New Lecturer's Course at that time.

Currently, the university encourages a number of options in its various schools to support teaching. One of these, introduced and operating in the School of Science, is the TeStED (Teaching and Student Experience Development) programme, designed to provide workload-allocated time and funds to enable colleagues to innovate, develop and disseminate new strategically-focused initiatives in teaching and student-experience across the School and more widely. Although not overtly designed to educate teachers, this nevertheless encourages thoughtful activity in the design and evaluation of teaching.

UMTE practices in Germany

As in the UK, German UMTE varies from one university to another, and is mostly common to all subjects. Most universities offer professional development courses or certified programmes for postdocs who are not yet professors with tenure track. Some universities offer programmes for fresh tenure-track professors. Often centres for "university didactics" or "teaching support centres" are offering these courses. Most of these centres are neither research based nor subject-specific.

At Paderborn University (PU), the professional development course is part of the certification programme of the federal state (North Rhine-Westphalia). Some of those in charge of the programme have a research orientation and adopt the approach of "Scholarship of Teaching and Learning" as it is interpreted in Germany (cf. Huber et al., 2014). The programme consists of three modules, each with 70 hours workload. The topics of the first module are Activation/Motivation of students, Results of research on learning, Criteria for good teaching, Planning a course (structure and formats, learning and teaching goals, variation of methods), Role and self-image of university teachers, Attitudes towards teaching and students, Media use and e-learning, and Dealing with difficult teaching-learning situations.

The second module has some obligatory topics for everybody such as Testing and rating students (constructive alignment, competence-oriented testing, oral and written examinations), Coaching and consulting students, supervising students' work, and Evaluation of courses and programmes. For the rest of the module, participants can select from specific workshops on topics such as Supporting students' work with texts, Guiding teaching-assistants, Flipped classroom, etc.

The third, advanced module consists of project work. Based on the first two modules, the participants can design their own teaching experiment, which is partly supervised, on which they receive feedback, and on which they have to write a written reflection. In Paderborn, the third module is interpreted as designing and performing a research project in university teaching (Kordts-Freudinger et al., 2018). The participants have to define a research question regarding their teaching and related student learning processes. The theoretical material from the first modules has to be applied towards phrasing and addressing this question, and the participants get to experience the use of qualitative, quantitative, or mixed-methods research methods. The results can be published in a journal for university didactics. By these means, one may obtain a new perspective on one's own teaching. This experience is similar to what is described by Dewar et al. (2015, 2018) but is not subject specific. Here are some examples of questions from the projects: What is the impact of communication-supporting teaching method on the quality of knowledge? How can we improve the quality of student-teacher consultation in student support centres by means of new guidelines for interaction? What are student views on "good tutorial sessions" and how can these be used for developing guidelines for them? (Kordts-Freudinger et al., 2018, p. 6).

To enable participants to do such a project, general support is offered (continuous consulting, workshops on research methods, feedback on intermediate reports, etc.). But participants also have to find a mentor from their own discipline who can serve as a subject-specific consultant, and as a reviewer of the project plan and the final publication. In many cases, participants find it difficult to find qualified and willing mentors. Thus, the programme aims to take specificities of the respective subject into account, using Middendorf and Pace's (2004) approach to decode the disciplines and identify bottlenecks of understanding. These authors refer to the "Decoding the disciplines model" that was reportedly used with success by the Indiana University Faculty Learning Community (IUFLC). The seven steps model provides a guideline for identifying bottlenecks in learning a discipline, finding out how experts in the discipline think about these problems, making an a priori analysis of students' thinking, designing learning environments and assessments for students, analysing students' thinking and discussing how the resulting knowledge about learning can be shared. However, the growing literature on university mathematics teaching is not taken into account and the participants have to adapt the general approach to their subject themselves.

To include mathematics specific components in the first two modules and in the design of the research project in the third module, one would of course have to involve people with a background in mathematics education. This option is not available in all universities. Also, the number of mathematics participants at a given university is too small to justify separate module components. To establish mathematics specific components and supervision, nation-wide or at least state-wide initiatives would be needed.

UMTE practices in Norway

In Norway, the standard, compulsory, pedagogical courses for university teachers are also general in nature. However, a mathematics-specific course has been developed by MatRIC, the national Centre for Research, Innovation and Coordination of Mathematics Teaching, based at the University of Agder. This centre specialises in mathematics at university level and one of its initiatives is a national course directed towards the teaching and learning of mathematics in higher education. The course is offered, without charge, to recently appointed teachers of mathematics in higher education institutions, including postdocs and PhD candidates with teaching duties, as an optional addition to the generic university pedagogy courses. The course seems to fill a real need, as the mandatory courses are often criticised for being too general, whereas this course is designed to address many of the specific challenges mathematics teachers face.

The course was run for the first time in the academic year 2015–2016 and for the second time in 2017–2018. Participants are expected to participate in four sessions, 1–2 days each time, and to undertake course-related work between sessions. It is required that participants have teaching duties while taking the course, since the course-related work concerns their own teaching directly. Each participant prepares a portfolio composed of material and reports from course-related activity. Below is a list of topics that were covered in the course in 2017–2018:

- making mathematics as a service subject relevant to user programmes;
- task design – goals for tasks, creating tasks, managing activity;
- teaching large groups;
- researching mathematics teaching and learning in universities;
- assessing mathematical knowledge, understanding and competences;
- mathematics support;
- digital approaches to teaching and learning mathematics.

The course contains a small part on learning theories and general principles for teaching and supervision, leaving space for the main focus to be on practical approaches and on issues specific to teaching mathematics, in general and to particular groups of students such as engineering students or economics students.

Each participant is free to design his or her developmental project to carry out as part of the course. Ideas for the project are discussed at the first session. In the last session, each participant is expected to give a presentation of his/her project. A written report on the project will be part of the portfolio. To support the participants, some examples of possible themes are presented, such as:

- developing computer-aided assessment tasks in mathematics;
- developing examination questions or other course assessment tasks;
- developing lecturing competence – may involve an experienced colleague to observe and give comments;
- using digital technology (simulation/visualisation) to support teaching and learning of mathematics.

One of the participants in the 2017–2018 course taught a first course in Calculus at the time and wanted to increase student activity and participation in this course. He chose *comparative judgement* (Pollitt, 2012) as the topic for his project. Comparative judgement is based on the idea that the responses to a given task are compared in pairs and the aim of the comparison is to decide which response of the two is the "better one". In this case, the students did the comparisons. When a fairly large number of comparisons (judgements) are made, this method, implemented as peer assessment, has been shown to give valid and reliable results (Jones & Sirl, 2017). Indeed, the method is considered to be particularly well suited for assessing conceptual understanding. Bisson, Gilmore, Inglis and Jones (2016) conducted a study where the task to be evaluated was: "Explain what a *derivative* is to someone who hasn't encountered it before. Use diagrams, examples and writing to include everything you know about derivatives" (p. 148). This task was also used in this particular project in the MatRIC course. This is an example of how course projects thus offer an opportunity to make use of current research on university mathematics education in the preparation and early development of university mathematics teachers.

UMTE practices in the USA

To go beyond formal pre- and in-service courses for university mathematics teachers, and begin to explore our last three research questions, we will now turn to outline two well established, national, and research-based professional development opportunities that are thriving in the United States: Project NExT and the Academy for Inquiry-Based Learning.

Project NExT (which stands for "New Experiences in Teaching": Mathematical Association of America, n.d.) is a professional development programme organised by the Mathematical Association of America (MAA) for new or recent PhDs in the mathematical sciences. After an application process, fellows are accepted into yearly cohorts. Each cohort participates in various workshops (three days in the first year of being a fellow and one day in the second), as well as various special sessions at national mathematics conferences during their first and second years. In addition, fellows become part of an electronic network that links them with one another and with seasoned mathematicians and award-winning teachers. Since 1994, there have been over 1700 Project NExT Fellows.

Project NExT aims to provide training and support for new mathematics faculty members across all aspects of an academic career, namely: engaging in research and scholarship, finding exciting and interesting opportunities to engage in outreach and community services, participating in professional activities, and improving the teaching and learning of mathematics. In particular, Project NExT workshops run sessions related to aspects that are part of new mathematics professors' professional lives, such as: integrating innovative approaches to a variety of introductory and advanced courses; finding strategies for engaging students in specific mathematics courses; and supporting students from historically underserved groups.

The *Academy of Inquiry-Based Learning* (AIBL), is a federally-funded project in the United States that aims to create a supportive community for mathematics faculty who are interested in Inquiry-Based Learning (IBL) pedagogical approaches. The "twin pillars" that characterise most implementations of IBL are deep engagement in rich mathematics and opportunities to collaborate (Ernst, Hodge, & Yoshinobu, 2017); a suggested third pillar is instructor interest in and use of student thinking (Rasmussen, Marrongelle, Kwon, & Hodge, 2017). The IBL teaching method promotes collaborative group work, prioritises student engagement with authentic mathematical activity, and decentralises classroom authority.

A main goal of AIBL is "to improve undergraduate math education via increasing the use of IBL methods across a wide range of courses and institutional environments," to be accomplished by "hosting workshops and a supportive AIBL community committed to helping instructors adapt IBL methods to suit the specific local needs of their students and institution." The most common approach to becoming an AIBL participant, of which there have been over 350 since 2006, is to attend an IBL Workshop and then be supported by the larger community afterwards. In addition to the workshops, AIBL maintains various blogs, shared course notes, and the YouTube channel "AcademyIBL". It has at times offered small grants and travel scholarships to visit faculty mentors. It also maintains a strong social networking presence to help IBL adopters stay connected in community, and it is also beginning to create US-regional IBL consortia.

Developing UMTE practice

We now move from existing UMTE efforts to potential new ones, of which we have identified four main types: innovative professional development, identifying and rewarding quality teaching, development of new resources to support teachers, and collaboration with researchers specialising in UME.

New ideas for professional development

The majority of UMTE programmes seem not to focus on specific whole courses in mathematics such as calculus or linear algebra, but to address more general levels such as inquiry-based learning in mathematics. However, research in UME – for example on calculus – has been very substantial and could, in the future, be used for designing specific courses on "teaching calculus" for UMTE that take this knowledge into account. One example is the project "Effect and success criteria of support measures" (WiGeMath) at the universities of Paderborn and Hannover, in which specific knowledge elements for designing courses and support measures in the transition phase from secondary education to university are developed (Biehler et al., 2018; Kuklinski et al., 2018; Liebendörfer et al., 2017). The project is concerned with pre-university bridging courses, including first semester courses with titles such as "Introduction into mathematical working and thinking" (including transition to proof, problem solving, or mathematical reading and writing), with support centres for students, and with accompanying measures for standard courses (e-learning, electronic voting systems, and special tutorials). The project has developed a set of evaluation instruments that are adaptable to various course goals and can be used by teachers of such courses. These instruments contain pre- and post-tests for assessing mathematical knowledge and competences as well as mathematical learning strategies, attitudes and beliefs. Moreover, course evaluation instruments have been developed. Professional development workshops explain the theoretical background of these instruments to the teachers responsible for the courses and provide guidelines for the design of evaluations and data analysis with the instruments. The project has also developed a theoretical framework (Liebendörfer et al., 2017), a coordinate system in which concrete courses can be theoretically embedded and compared with each other with regard to goals, content and didactical elements. This framework aims to enable the participating teachers to reflect theoretically on their courses and support them in redesigning these courses. The project offers professional development workshops for these teachers.

Identify and reward quality teaching

Two of the questions that came up at the panel discussion during INDRUM 2018 were: "Are university instructors willing to invest time and effort to improve teaching? Does good teaching actually affect promotion in your country?" These are indeed crucial questions, and in a situation where the competition for appointments to a position is strong and the requirements for promotion are high, it is natural that faculty members will prioritise and invest their time and effort where it pays off the most. Traditionally, this means to prioritise research, as accomplishments in research are most important for getting ahead of the competition. Although university teachers often spend at least half of their working time on teaching, research is often given a far more prominent position than teaching in hiring and promotion processes.

In recent years, the quality of education at universities has attracted more attention, due to changing conditions of higher education. For example, in Norway, the number of students enrolled in higher education increased to more than 60% in the period 1995 to 2015 (Grepperud, 2016, p. 17). For the subject of mathematics, a particular challenge is that more and more subject fields are being "mathematised" so that students in subjects such as economics and biology have to learn mathematics at a higher level than before. In a 2017 White Paper (Kunnskapsdepartementet, 2017, p. 79), the Norwegian government announced that it will require Norwegian higher education institutions to establish a merit system in order to further the recognition and professional status of quality teaching.

In 2017, a system for awarding the title *excellent teaching practitioner* was launched at the Norwegian University of Science and Technology (NTNU) and at the University of Tromsø (UiT). Also, other universities have established, or are in the process of establishing, similar systems. Inspiration for development in Norway may be found in the UK system Career Framework for University Teaching (Royal

Academy of Engineering, 2018). This framework describes four levels, and for each level *promotion criteria* and *spheres of impact* are listed.

The main criteria for being recognised as an excellent teaching practitioner (at NTNU) are formulated in three parts. An excellent teaching practitioner

- has developed his/her teaching qualitatively over time;
- has a scientific approach to teaching and learning; and,
- is an engaged and skilful pedagogical leader.

These criteria are comparable to the criteria for the second level in the Royal Academy of Engineering system in the UK, denoted "skilled and collegial teacher". These criteria are "Skills: Employs evidence-based teaching to support student learning and engagement" and "Collaboration: Supports a collegial and collaborative learning environment" (2018, p. 39). To become an excellent teaching practitioner can be seen as akin to a promotion, meaning that an excellent teaching practitioner has documented qualifications regarding teaching that are well above the minimum requirements.

The system described is enacted, both at NTNU and UiT, across the whole university and therefore there is nothing in it which is particular to mathematics. However, it reflects the growing emphasis on educational quality that can be observed internationally. Also, it reflects the fact that research on teaching and learning in higher education is a field in rapid development, with mathematics at the forefront compared to other subjects.

In Norway, in particular in science and technology faculties, the requirements for pedagogical qualifications for full professors have recently been sharpened and made more explicit. And, at the Faculty of Information Technology and Electrical Engineering at NTNU, where mathematics is situated, a 2.5 ECTS module in Basic University Didactics for PhD students was scheduled for the first time in the autumn of 2019. These two examples show that pedagogical qualifications are now taken more seriously than before.

Creating and supporting inquiry-oriented task aequences for UME

A main obstacle for both junior as well as experienced instructors to implement a more inquiry-oriented approach to teaching may be not only a lack of experience, but also a potential lack of curricular materials such as textbooks and in-class activities designed for the subject matter they teach. Three related and US-based examples of such inquiry-oriented innovative curricular materials exist for differential equations (Rasmussen, Keene, Dunmyre, & Fortune, 2018), abstract algebra (Larsen, Johnson, & Weber, 2013), and linear algebra (Wawro, Zandieh, Rasmussen, & Andrews-Larson, 2013). All are grounded in the instructional design theory known as Realistic Mathematics Education (RME) (Freudenthal, 1991). Briefly stated, task sequences are based on experientially real starting points; classroom activity should support student development of *models-of* their mathematical activity that can be used as *models-for* subsequent mathematical activity; and student activity, with instructor guidance, should evolve towards the reinvention of formal notions and ways of reasoning.

Each of the three curricular materials contain a collection of task sequences that include instructor support materials for faculty interested in incorporating inquiry into their pedagogical approach. The instructor support materials vary across the curricula; for instance, the linear algebra one has three main foci: Learning Goals and Rationale, Student Thinking, and Implementation. In terms of professional development, this approach allows faculty interested in student-centred teaching approaches to have access to research-based materials, regardless of their location or local resources.

In order to support the uptake of inquiry-oriented instruction in mathematics departments in the United States, the Teaching Inquiry-Oriented Mathematics: Establishing Supports (TIMES) instructional support model was developed to support instructors in learning to teach in inquiry-oriented ways for each of the three aforementioned curricula in abstract algebra, linear algebra, and differential equations (Johnson, Keene, & Andrews-Larson, n.d.). The TIMES model was designed to build on current research into professional learning and instructional change, and it consists of three instructional supports: access to the instructional materials created by the course content developers, a three-day summer workshop, and online workgroups held for one hour per week during the semester in which instructors implement the materials. The TIMES model focuses on four components of inquiry-oriented instruction: generating student reasoning, building on student reasoning, fostering shared understanding, and formalising language and notation (Kuster, Johnson, Keene, & Andrews-Larson, 2017). Ongoing research into the implementation of these three curricula through the TIMES model is investigating aspects such as student learning gains in inquiry-oriented versus traditional classrooms, the levels of effectiveness of the various instructional supports, other factors that influence pedagogical decisions in UME, and the impact of inquiry-oriented instruction on student persistence in science, mathematics, technology, and engineering.

Engage UM-teachers and UME-researchers in shared inquiry

It is well known from international studies of mathematics teachers at other levels, that frameworks which allow teachers to maintain and share a reflective and experimental relationship to what and how they teach, constitute a crucial condition for developing teaching quality (see e.g. Stigler & Hiebert, 1999; Zack, Mousley, & Breen, 1997), and this perspective on teaching – which is less about "educating" individuals than about forming communities and shared resources with inquiry-based approaches to developing teaching – is found even in relatively early works in the field (such as Alsina, 2001).

More recently, this idea is related to *inquiry* as a *research paradigm*, used by mathematics education teachers to explore aspects of practice, often in and of themselves inquiry-based, and thus, to learn through insights gained. The research here is developmental research, resulting in new knowledge in practice for the practitioners involved and rigorous research findings to contribute to knowledge more widely (Jaworski, 2008, 2014; Rasmussen & Wawro, 2017).

A prime example is DATUM, a developmental research project for the *Development and Analysis of Teaching in Undergraduate Mathematics*, taking place in Auckland, New Zealand and involving professional development for mathematics lecturers: a group met regularly to discuss video excerpts of themselves lecturing, along with written pre- and post-lecture statements of their goal-oriented decision making (Schoenfeld, 2011). Researchers found evidence of improved teaching performance, identified key aspects of teaching practice in undergraduate mathematics, and developed further theoretical insight into lecturer behaviour in mathematics (Barton, Oates, Paterson, & Thomas, 2014).

Similarly to the team in Auckland, the Mathematics Education Centre (MEC) at Loughborough University has engaged over the years in developmental research projects, which both study the developmental process and contribute to that development. One of these projects, *Engineering Students Understanding Mathematics* (ESUM) involved the design of an innovation in a mathematics module for first-year materials engineering students. Three experienced teacher-researchers designed the innovation, one being the lecturer in the module, and a research associate gathered data and assisted in its analysis. The innovation was four-fold: inquiry-based mathematics questions, a computer-based exploratory medium, small group activity in tutorials and an assessed small-group project (Jaworski, 2014; Jaworski & Matthews, 2012; Jaworski, Robinson, Matthews, & Croft, 2012). Learning experiences for the teachers involved:

- Growth of awareness of students' perspectives and focuses in participation in the innovation, contrasting vividly with those of the teaching team;
- Experience of managing small-group activity with computers in a computer lab within university procedures, regulations and expected practices;
- Knowledge of design of inquiry-based questions and tasks and the ways in which students worked with and responded to them.

Another project, *Second-Year Mathematics Beyond Lectures* (SYMBoL: Duah, 2017; Duah, Croft, & Inglis, 2014), involved the employment of four interns, second-year mathematics undergraduates, who had participated in two mathematics modules during their second year. The modules, complex analysis and vector spaces, had achieved extremely low student scores over several years and the aim of the project was to develop resources to support student learning. The interns worked with mathematics lecturers for six weeks during the summer. Each afternoon, tea was offered and lecturers and interns talked together about mathematics and the design of resources. The resources designed by the interns were used in the following year's presentation of the modules. In parallel, a peer support programme was designed, based on the interns' work and involving 20 "peer-leaders" from third-year mathematics undergraduates. Peer leaders were trained by mathematics education lecturers to work in collaborative, inquiry-based ways with the second-year students studying the two modules.

Learning experiences for the involved university mathematics teachers involved:

- learning to talk about mathematics with students in a relaxed environment (i.e., not the formal lecture setting);
- gaining insights into new forms of resources that could support student understanding;
- experiencing the creation of a peer-learning programme of activity which supported students beyond the more formal lectures and tutoring;
- supporting students in their transition to a first year Engineering degree.

In one UK study (within TeStED, introduced above), mathematics education researchers wanted better to understand and support students during their initial transition into university engineering mathematics as part of an engineering degree. They surveyed first-year students from three engineering degree programmes and collected and analysed data from a mathematics diagnostic test and questionnaires relating to students' study habits, mathematics anxiety and dyslexia. Since they were, themselves, practitioners teaching first-year engineering students they were keen to develop a resource to help students with their mathematical study at university (including students with an additional need such as dyslexia). Over the two years allocated for the project, they developed an online revision tool that offered a comprehensive approach to revising the calculus topic of integration. The main features of the resource were a colour-coded, dynamic revision tool that resembled a mind-map. It allowed a student to select an area for revision and practice answering example questions while always keeping an overview of the various methods of integration on screen. It is intended that the resource will be made available for all students to use as preparation or revision and will be expanded to include other mathematics topics such as differentiation, matrices and complex numbers. Learning experiences for the teachers involved

- knowledge of engineering students' levels of achievement and study habits;
- growth of awareness of students with additional learning needs and how to cater for these needs;
- design of resources to support students in their mathematics revision.

From the examples above, it can be seen that developmental research conducted by mathematics teachers, possibly with associated experienced researchers, can be a powerful tool for professional and pedagogic development of the teachers who are involved. Such research feeds into the practice of those conducting the research, but also contributes to the professional environment involving teachers committed to exploring and improving teaching.

Issues for future research on UMTE

As seen in the preceding sections, there exists a wealth of (often small-scale) initiatives which contribute, either formally or informally, to the professional development of university teachers of mathematics. The impact and recognition of such projects often remain national; indeed, the sample of initiatives and experiences mentioned here mostly originate in the authors' home countries or even home universities. To achieve a broader overview of how university mathematics teachers build and develop their professional knowledge as teachers, a more systematic, international collection of evidence is needed, since both current problems and possible solutions are likely to be quite similar across a wide range of countries, even taking into account the variation in cultural and institutional conditions. It is possible that the details we have compiled here could motivate and inform an *international survey* with this aim.

When mathematics educators engage in teacher education at the primary and secondary level, they do so as researchers whose academic expertise and research basis can be regarded as well established. University mathematics teachers are often also researchers (in pure or applied mathematics, or some other field) and their needs and expectations for teacher education may differ substantially – for instance, they will typically feel very confident when it comes to content knowledge (here, knowledge of mathematics). On the other hand, their knowledge about (and frequently also interest in) didactical and pedagogical innovations may be quite limited. As researchers, they can also be expected to have substantial interest in evidence and justifications of methods and ideas proposed.

One of the questions presented at the panel discussion was: "What do we know about how different teaching practices in the different learning situations improve student learning?" This question may be typical of the kinds of general questions that university teachers may ask and expect mathematics education researchers to answer. This is highly relevant and at the same time very difficult to answer, since learning situations may produce different results with students of different backgrounds, and also because there is no obvious way to measure learning outcomes. For the university teacher, comparing exam grades is perhaps one important or obvious way. Even if we may question how good this measure is, we would expect that it gives some indication of learning outcomes. Another way to gain information about learning outcomes is to ask students to what extent they learn from taking part in different activities (as is commonly done in student evaluations, for instance at the University of Copenhagen). This is a more indirect measure, since it can only say something about perceived learning outcome. However, also this measure could be expected to be of some value.

In general, to orient the design and contents of university mathematics teacher education, we need to know more about what motivates faculty to prefer or to maintain specific instructional practices. And, by contrast, what could be supports and impetuses for instructional change? At the same time, needs could vary even between different groups of university teachers, depending on the students, subjects and levels they teach.

A well-known approach to identifying "key needs" for future or current teachers is the *job-analysis approach* (Bass & Ball 2004; Prediger, Quasthoff, Vogler & Heller, 2015), where the usual everyday practice of teachers is analysed with regard to situational requirements and demands. Of course, normative elements go into the analysis of job requirements. For university mathematics teaching, such research on key needs could conceivably lead to identify tasks such as

1. designing the content of a lecture;
2. giving a lecture;
3. designing tasks for weekly assignments for students;
4. collaboration with the lecture's teaching team;
5. setting the framework for student tutors' work (teaching assistants) in the weekly tutorial sessions;
6. designing final examinations;
7. participation in the faculty teaching team (programme design and re-design etc.);
8. evaluation and critical reflection of the course, consideration of its re-designing;
9. collaboration with educational researchers (general and researchers in UME).

One may further suspect that the last three items are less evident jobs, maybe merely wishful thinking from the perspective of mathematics education; many university mathematics teachers may not consider this as jobs they "have to do". At the same time, one can hypothesise that many of them think that jobs 1 to 6 are learned by doing, with no professional development needed.

We pick out four of these aspects for further scrutiny, in order to identify deficiencies in our present knowledge and resources for preparing for these jobs.

Designing the content of a lecture. A language would be needed for expressing different learning goals, competences and content foci of a lecture. Material that identifies fundamental or central ideas, provides insight into learning difficulties or obstacles for the students and that shows possible remedies. Such content-specific mathematics education knowledge (didactics) is available for teaching at school level, for instance to know about different ways to approach and organise the teaching of derivatives or integrals (cf. Greefrath, Oldenburg, Siller, Ulm, & Weigand, 2016). Similar expositions are inaccessible or unavailable when it comes to more advanced subjects (e.g. linear algebra) and their teaching at university level.

Giving a lecture. Here, missing resources include guidelines for better combining official blackboard writing with oral and written informal communication. A theoretical framework and concrete examples for *activating gestures*, such as posing oral questions, initiating small group tasks, or giving suitable questions for electronic voting systems ("clickers") – all within a specific subject – would be valuable, as would be guidelines for how to deal with students' questions during or after the lecture (see for instance Paoletti, Krupnik, Papadopoulos, Olsen, Fukawa-Connelly, & Weber. 2018).

Designing tasks for weekly assignments for students. End-of-chapter exercises are seldom constructed to nourish students' mathematical inquiry, or to engage them in other forms of more advanced mathematical work. For the teacher, then, missing resources include collections of alternative mathematical tasks for a given mathematical domain (for examples in linear algebra and vector analysis, see Wawro et al., 2013 and Gravesen et al., 2017). It would also be useful for them to know a theoretical framework for analysing and varying the cognitive and mathematical potential and demands of tasks for students.

Setting the framework for teaching assistants' work as they teach tutorial sessions (exercise classes, etc.). Research (e.g. Püschl, 2018; Kung & Speer, 2008) could be used for orientation towards appreciating and supporting the multiple roles of teaching assistants, which include: student-centred discussion of homework, moderating small group work (with minimal help), providing informative feedback when correcting homework, being a linkage between lecture goals and weekly tasks. Lecturers could need guidelines on how to conduct weekly meetings with teaching assistants, or engage in other forms of support and professional development of teaching assistants.

Conclusion

In this chapter, we have summarised some of the material that was shared during the panel at INDRUM 2018, including both projects and experiences by the authors and further reflections and ideas generated by the questions to the panel. The cases and examples presented do not in themselves suffice to give comprehensive answers to the research questions formulated in the introduction; in fact, we already noted the potential interest of an international survey on initial UMTE, as a means to go beyond very local accounts of practices and challenges.

We end this chapter by pointing out how the somewhat scattered and partly personal elements which we have presented can be summarised as a number of hypotheses, ideas and directions for further research into the questions.

As regards the current (initial) preparation of university mathematics teachers for their function as teachers, we have observed that generic courses on university pedagogy exist in several European countries as well as in North America, with considerable variation from one university to another – such courses seem to be offered to large groups of university teachers (across departments or even faculties) and to be mandatory for newly hired faculty in many universities. Courses or other forms of preparation that is more *specific* to mathematics seem to be much rarer, and to be optional. Nevertheless, the experiences described suggest that such courses may cater for perceived needs of early-career faculty members, and help them engage in innovation and development work related to their teaching; under further conditions, such work may be informed by, akin to, or even part of research on university mathematics education. We certainly need more research into the actual and potential links between generic university pedagogy courses on the one hand, and mathematics-specific preparation on the other, and we presently also know very little about how they affect participants' teaching.

It appears that in-service development opportunities for university mathematics teachers are almost always optional, and are often offered by institutional or national centres, which may (or may not) involve mathematics education researchers. Again, despite the frequent prevalence of subject-independent offers, a number of mathematics-specific initiatives have operated successfully over longer periods of time, often based on seminar or workshop style interaction among teachers and mathematics educators, where the discussion of teachers' concrete questions and experiences are central. In some cases, these meetings can lead to – or be part of – teachers engaging more directly in research, typically in collaboration with colleagues whose main expertise is in mathematics education. While all of this may be welcome, we suspect these cases touch a very small minority of university mathematics teachers. Also, as is often the case with optional professional development, it is likely that those who actively engage in it are not the ones who need it most. Moreover, as in the case of pre-university teaching, the examples illustrate how one can consider the development of university mathematics teaching as more than an individual task – both because needed change often requires the collaboration of a larger team (e.g., the mathematics department), and because of the value and impact of sharing professional experiences and insights with colleagues.

It is difficult to gauge to what extent the successful initiatives, which are optional for participants, would be equally successful if provided as mandatory preparation and development opportunities. We have noted that many universities have increased their focus on teaching quality, and to the extent that this is seriously reflected in the regulation of career opportunities, one can expect that the difference between mandatory and optional diminishes in this context. The question of the impact of different forms of UMTE on student achievement, which we have briefly touched upon, then becomes very important and researchable. It would also affect research on university mathematics education deeply, as this would be led to focus more closely on *questions raised by the profession*. This, in particular, would strengthen our focus on issues of relevance to UMTE; at the same time, results would be more readily and widely validated.

Focusing more closely on the questions and needs of university mathematics teachers involves, in particular, more focus on developing resources and materials and experimenting with innovative designs of tasks, texts and other resources for the university mathematics teacher. Most of these resources will necessarily be germane to specific mathematical topics, while others may apply to broader issues such as managing and supervising teaching assistants. In particular, we expect that more *content-specific research* will be needed if research on university mathematics education should achieve a role and impact on UMTE that is similar to what mathematics education research enjoys in pre- and in-service teacher education for the primary and secondary levels. As a result, Bob (the assistant professor we met in the introduction) would not only find relevant support in UMTE, but might even at some point choose to specialise in research and development of UME himself.

References

Alsina, C. (2001). Why the professor must be a stimulating teacher: Towards a new paradigm of teaching mathematics at university level. In D. Holton (Ed.), *The teaching and learning of mathematics at university level* (pp. 3–12). Dordrecht: Kluwer Academic Publishers.

American Mathematical Society. (n.d.). *Annual survey of the mathematical sciences: Departmental groupings*. Retrieved from http://www.ams.org/profession/data/annual-survey/groupings.

Barton, B., Oates, G., Paterson, J., & Thomas, M. (2014). A marriage of continuance: Professional development for mathematics lecturers. *Mathematics Education Research Journal*, 27(2), 147–164.

Bass, H., & Ball, D. L. (2004). A practice-based theory of mathematical knowledge for teaching: The case of mathematical reasoning. In J. Wang & B. Xu. J. Wang (Eds.), *Trends and challenges in mathematics education* (pp. 107–123). Shanghai: East China Normal University Press.

Biehler, R., Lankeit, E., Neuhaus, S., Hochmuth, R., Kuklinski, C., Leis, E., Liebendörfer, M. *et al.* (2018). Different goals for pre-university mathematical bridging courses – Comparative evaluations, instruments and selected results. In V. Durand-Guerrier, R. Hochmuth, S. Goodchild, & N. M. Hogstad (Eds.), *Proceedings of INDRUM 2018* (pp. 467–476). Kristiansand, Norway: University of Agder and INDRUM.

Bisson, M.-J., Gilmore, C., Inglis, M., & Jones, I. (2016). Measuring conceptual understanding using comparative judgement. *International Journal of Research in Undergraduate Mathematics Education*, 2(2), 141–164.

Christiansen, F. (2016). Stillingsstrukturens betydning for samspillet mellem forskning og undervisning. *Dansk Universitetspædagogisk Tidsskrift*, 21, 31–43.

Dewar, J. M., & Bennett, C. D. (Eds.). (2015). *Doing the scholarship of teaching and learning in mathematics*. Washington, DC: The Mathematical Association of America.

Dewar, J. M., Bennett, C. D., & Fisher, M. A. (2018). *The scholarship of teaching and learning – A guide for scientists, engineers, and mathematicians*. Oxford: Oxford University Press.

Duah, F. (2017). *Students as partners and students as change agents in the context of university mathematics*. Unpublished PhD thesis, Loughborough University, UK.

Duah, F., Croft, T., & Inglis M. (2014). Can peer assisted learning be effective in undergraduate mathematics? *International Journal of Mathematical Education in Science and Technology*, 45(4), 552–565.

Ernst, D. C., Hodge, A., & Yoshinobu, S. (2017). What is inquiry-based learning? *Notices of the American Mathematical Society*, 64(6), 570–574.

Freudenthal, H. (1991). *Revisiting mathematics education*. Dordrecht: Kluwer Academic Publishers.

Hernandes-Gomes, G., & González-Martín, A. (2016). Teaching Calculus in engineering courses. Different backgrounds, different personal relationships? In E. Nardi, C. Winsløw & T. Hausberger (Eds.), *Proceedings of INDRUM 2016* (pp. 201–210). Montpellier: University of Montpellier.

Gravesen, K. F., Grønbæk, N., & Winsløw, C. (2017). Task design for students' work with basic theory in analysis: The cases of multidimensional differentiability and curve integrals. *International Journal of Research in Undergraduate Mathematics Education*, 3(1), 9–33.

Greefrath, G., Oldenburg, R., Siller, H.-S., Ulm, V., & Weigand, H.-G. (2016). Aspects and "Grundvorstellungen" of the concepts of derivative and integral. *Journal für Mathematik-Didaktik*, 37(1), 99–129. doi:10.1007/s13138-016-0100-x.

Grepperud, G. (2016). *Innsats for kvalitet. Forslag til et meritteringssystem for undervisning ved UiT Norges arktiske universitet og NTNU*. Trondheim/Tromsø: NTNU and UiT.

Huber, L., Pilniok, A., Sethe, R., Szczyrba, B., & Vogel, M. (Eds.). (2014). *Forschendes Lehren im eigenen Fach. Scholarship of Teaching and Learning in Beispielen*. Bielefeld: Bertelsmann.

Jaworski, B. (2008). Building and sustaining inquiry communities in mathematics teaching development. Teachers and didacticians in collaboration. In K. Krainer (Volume Ed.) & T. Wood (Series Ed.), *International handbook of mathematics teacher education: Vol. 3. Participants in mathematics teacher education: Individuals, teams, communities and networks* (pp. 309–330). Rotterdam: Sense Publishers.

Jaworski, B. (2014). Unifying complexity in mathematics teaching-learning development: A theory-practice dialectic. In Y. Li, E. A. Silver, & S. Li (Eds.), *Transforming mathematics instruction: Multiple approaches and practices* (pp. 439–458). New York, NY: Springer.

Jaworski, B., & Matthews, J. (2011). How we teach mathematics: Discourses on/in university teaching. In M. Pytlak, E. Swoboda, & T. Rowland (Eds.), *Proceedings of the Seventh Congress of the European Society for Research in Mathematics Education* (pp. 2022–2032). Rzeszow: University of Rzeszow and ERME.

Jaworski, B., & Matthews, J. (2012). Developing teaching of mathematics to first year engineering students. *Teaching Mathematics and its Applications*, 30(4),178–185.

Johnson, E., Keene, K., & Andrews-Larson, C. (n.d.). *Teaching Inquiry-oriented Mathematics: Establishing Supports (TIMES)*. Retrieved from http://times.math.vt.edu.

Jones, I., & Sirl, D. (2017). Peer assessment of mathematical understanding using comparative judgement. *Nordic Studies in Mathematics Education*, 22(4), 147–164.

Kordts-Freudinger, R., Bücker, D., Braukmann, J., Schulte, R., & Velibeyoglu, N. (2018). *Scholarship of Teaching and Learning (SoTL) als Bestandteil hochschuldidaktischer Zertifikatsprogramme am Beispiel der Universität Paderborn*. Retrieved from https://www.nhhl-bibliothek.de/de/handbuch/gliederung/#/Beitragsdetailansicht/243/1180/Scholarship-of-Teaching-and-Learning-(SoTL)-als-Bestandteil-hochschuldidaktischer-Zertifikatsprogramme-am-Beispiel-der-Universitaet-Paderborn.

Kuklinski, C., Leis, E., Liebendörfer, M., Hochmuth, R., Biehler, R., Silke, N. *et al.* (2018). Evaluating innovative measures in university mathematics – The case of affective outcomes in a lecture focused on problem-solving. In V. Durand-Guerrier, R. Hochmuth, S. Goodchild, & N. M. Hogstad (Eds.), *Proceedings of INDRUM 2018* (pp. 527–536). Kristiansand, Norway: University of Agder and INDRUM.

Kunnskapsdepartementet. (2017). *St. meld. nr. 16 (2016–2017). Kultur for kvalitet i høyere utdanning*. Oslo: Author.

Kung, D., & Speer, N. (2008). Mathematics teaching assistants learning to teach: Recasting early teaching experiences as rich learning opportunities. *Studies in Graduate and Professional Student Development*, 12, 133–152.

Kuster, G., Johnson, E., Keene, K., & Andrews-Larson, C. (2017). Inquiry-oriented instruction: A conceptualization of the instructional components and practices. *PRIMUS*, 28(1), 13–30.

Larsen, S., Johnson, E., & Weber, K. (Eds.). (2013). The Teaching Abstract Algebra for Understanding project: Designing and scaling up a curriculum innovation [special issue]. *Journal of Mathematical Behavior*, 32(4).

Liebendörfer, M., Hochmuth, R., Biehler, R., Schaper, N., Kuklinski, C., Khellaf, S., Colberg, C., Schürmann, M., Rothe, L. (2017). A framework for goal dimensions of mathematics learning support in universities. In T. Dooley & G. Gueudet (Eds.), *Proceedings of the Tenth Congress of the European Society for Research in Mathematics Education* (pp. 2177–2184). Dublin: DCU Institute of Education and ERME.

Mathematical Association of America. (n.d.). *Project NExT*. Retrieved from https://www.maa.org/programs-and-communities/professional-development/project-next.

Mathieu-Soucy, C., Corriveau, C, & Hardy, N. (2018). Exploration of new post-secondary mathematics teachers' experience: preliminary results of a narrative inquiry. In V. Durand-Guerrier, R. Hochmuth, S. Goodchild, & N. M. Hogstad (Eds.), *Proceedings of INDRUM 2018* (pp. 527–536). Kristiansand, Norway: University of Agder and INDRUM.

Middendorf, J., & Pace, D. (2004). Decoding the disciplines: A model for helping students learn disciplinary ways of thinking. *New Directions for Teaching and Learning*, 98, 1–12.

Musgrave, S., & Carlson, M. (2017). Understanding and advancing graduate teaching assistants' mathematical knowledge for teaching. *Journal of Mathematical Behavior*, 45, 137–145.

Office for Students. (n.d.). *The Teaching Excellence and Student Outcome Framework (TEF). A short guide to the awards*. Retrieved from https://www.officeforstudents.org.uk/media/0c6bd23e-57b8-4f22-a236-fb27346cde6e/tef_short_guide_-june_2019_final.pdf.

Paoletti, T., Krupnik, V., Papadopoulos, D., Olsen, J., Fukawa-Connelly, T., & Weber, K. (2018). Teacher questioning and invitations to participate in advanced mathematics lectures. *Educational Studies in Mathematics*, 98, 1–17. doi:10.1007/s10649-018-9807-6.

Pollitt, A. (2012). The method of adaptive comparative judgment. *Assessment in Education: Principles, Policy & Practice*, 19, 281–300.

Prediger, S., Quasthoff, U., Vogler, A.-M., & Heller, V. (2015). How to elaborate what teachers should learn? Five steps for content specification of professional development programs, Exemplified by "Moves Supporting Participation in Classroom Discussions". *Journal für Mathematik-Didaktik*, 36(2), 233–257. doi:10.1007/s13138-015-0075-z.

Püschl, J. (2018). *Kriterien guter Mathematikübungen – Potentiale und Grenzen in der Aus- und Weiterbildung von studentischen TutorInnen*. Unpublished PhD thesis, University of Paderborn, Germany.

Rasmussen, C., Marrongelle, K., Kwon, O. N., & Hodge, A. (2017). Four goals for instructors using inquiry-based learning. *Notices of the American Mathematical Society*, 64(11), 1308–1310.

Rasmussen, C., Keene, K. A., Dunmyre, J., & Fortune, N. (2018). *Inquiry oriented differential equations: Course materials*. Retrieved from https://iode.wordpress.ncsu.edu.

Rasmussen C., & Wawro, M. (2017). Post-calculus research in undergraduate mathematics education. In J. Cai (Ed.), *Compendium for research in mathematics education* (pp. 551–581). Reston, VA: National Council of Teachers of Mathematics.

Royal Academy of Engineering. (2018). *The career framework for university teaching: Background and overview*. London: Author.

Schoenfeld, A. (2011). *How We Think: A Theory of Goal-Oriented Decision Making and Its Educational Applications*. New York, NY: Routledge.

Speer, N., & Hald, O. (2008). How do mathematicians learn to teach? Implications from research on teachers and teaching for graduate student professional development. In M. Carlson & C. Rasmussen (Eds.), *Making the connection: Research to practice in undergraduate mathematics education* (pp. 305–318). Washington, DC: Mathematical Association of America.

Speer, N., & Wagner, J. (2009). Knowledge needed by a teacher to provide analytic scaffolding during undergraduate mathematics classroom discussions. *Journal for Research in Mathematics Education*, 40(5), 530–562.

Stigler, J., & Hiebert, J. (1999). *The teaching gap: Best ideas from the world's teachers for improving education in the classroom*. New York, NY: The Free Press.

The Academy of Inquiry Based Learning. (n.d.). *Supporting instructors, empowering students, transforming mathematics learning*. Retrieved from http://www.inquirybasedlearning.org.

Wawro, M., Rasmussen, C., Zandieh, M., & Larson, C. (2013). Design research within undergraduate mathematics education: An example from introductory linear algebra. In T. Plomp & N. Nieveen (Eds.), *Educational design research – Part B: Illustrative cases* (pp. 905–925). Enschede: SLO.

Wawro, M., Zandieh, M., Rasmussen, C., & Andrews-Larson, C. (2013). *Inquiry oriented linear algebra: Course materials*. Retrieved from http://iola.math.vt.edu.

Zack, V., Mousley, J., & Breen, C. (Eds.). (1997). *Developing practice: Teachers' inquiry and educational change*. Geelong, Australia: Centre for Studies in Mathematics, Science and Environmental Education, Deakin University.

PART II

Teaching and learning of specific topics in university mathematics

5

CHALLENGING ISSUES IN THE TEACHING AND LEARNING OF CALCULUS AND ANALYSIS

María Trigueros, Stéphanie Bridoux, Ann O'Shea and Laura Branchetti

The importance of Analysis and the nuances involved in its teaching and learning continue to be a fruitful topic of interest for the mathematics education community at the university level. This chapter shows the variety and depth of research conducted recently, as reflected in the INDRUM meetings in 2016 and 2018. Transition from secondary school to university is a crucial issue in mathematics education, and within this the teaching and learning of Calculus and introductory Analysis is one of the most investigated topics. It might seem that once students overcome this initial transition, they would have less difficulty when studying other courses related to Analysis. This seems not to be the case. Participants at the INDRUM conferences have considered further transition points: for example, when students are introduced to new kinds of functions such as multivariable functions; when they are introduced to Analysis on the real line; and when they study Topology or Complex Analysis. All of these transitions have their own characteristics and we explore them in this chapter.

Calculus in the transition from secondary school to university

We note that in many countries no distinction is made between the fields of Calculus and Analysis. However, in some countries (for example the UK and USA) Calculus courses usually focus on differentiation and integration of functions of one or several variables where the emphasis is more on calculation rather than proof. In Analysis courses, students encounter a more rigorous version of this material and primary importance is given to formal definitions (for example the epsilon-delta definition of limits) and proof. Calculus and Introduction to Analysis are usually courses that students attend at the beginning of their university programmes and so the issue of transition is especially pertinent in this area.

In a report for ICME12, Thomas et al. (2015) carried out a survey of the research literature on the secondary-tertiary transition and were able to identify students' conceptions of functions, limits and infinity, as well as proofs and institutional factors as major obstacles in the area of Calculus and Analysis. For example, the concept of function has been shown to cause problems for students in the transition to studying Calculus at university (Tall & Vinner, 1981). These problems are often multi-faceted and include difficulties with the formal definition, but also with representations (Vandebrouck, 2011), and with the ability to view functions as objects as well as processes (Sfard, 1991). The link between graphical visualization, formalization and proof is also problematic. In particular, the use of definitions instead of, or at least together with, graphical representations is still a stumbling block for some students (Törner, Potari, & Zachariades, 2014).

The crucial role of real numbers

Artigue (2001) highlighted that at the beginning of undergraduate Calculus courses it may be necessary to reconstruct objects which are familiar to students from mathematics at school level (the typical example is that of real numbers) since many students will hold problematic conceptions of these objects.

In INDRUM 2016, some studies addressed the issue of conceptualization of real numbers. Kidron (2016) studied the conceptions of irrational numbers of students who were about to enter university and found several difficulties: students' conceptions of irrational numbers based on their representation as non-repeating infinite decimals were very weak, and most students identified the set of all decimals (finite and infinite) with the set of rational numbers. Vergnac and Durand-Guerrier (2014) considered students' knowledge about real numbers at the end of secondary school in France; they showed that most of the participants in their study had a conception of real numbers based only on decimal expressions. When asked to define real numbers, students often proposed representations of the set, like the line and the interval $(-\infty, +\infty)$, or referred to the set of real numbers as "the set of all numbers".

Most of these representations are not adequate to allow the distinction between dense and complete sets and to prove theorems needed in Analysis. In a study carried out by Bergé (2010), many university students showed a naive conception of the completeness of real numbers, which is often reformulated by students informally: for example "Completeness is a line without gaps, an infinity density of points" (p. 225). The author pointed out that such answers show that the representation of real numbers on the line does not necessarily help students to understand completeness. Also, students are not usually able to define real numbers in a formal manner. Durand-Guerrier and Vivier (2016) stressed the problem of visualization of dense sets using graphical representations on the number line in the first year of scientific university in France and showed that almost one third of the participants (35) were not aware of the fact that density does not imply completeness.

Ghedamsi (2016) considered whether numerical investigations consisting of progressive approximations of real numbers and sequences could have an impact on university students' learning of basic concepts of Analysis such as limits, functions, and sequences. A first result was that the students made a conceptual shift in their conceptions of numbers, which led them to overcome the problem of accessibility of real numbers and to provide real numbers, step by step, with the status of a mathematical object through meta-mathematical reflections.

Representations, visualization and perspectives in Calculus

The issue of representation in mathematics education has been widely investigated (Radford, Schubring, & Seeger, 2008). A successful transition requires different interpretations of familiar representations and the use of unfamiliar ones. In secondary school, students are usually introduced to mathematical objects such as functions, derivatives, and integrals as tools for algebraic calculation and not as mathematical concepts in their own right (Bloch & Gibel, 2016). Consequently, undergraduate students may continue to employ methods and conceptions used in secondary school and this may hinder their transition to mathematics at university. To study this problem, Bloch and Gibel (2016) developed a model to analyse students' work using the theoretical framework of the Theory of Didactical Situations (TDS). They found that students encountered problems due to the objects, signs and reasoning processes they had to cope with during the resolution of calculus problems. They showed that "*signs are multiform*" (p. 44) (e.g., different representations of derivatives are used) and, relying on Peirce's semiotics, they studied the evolution of signs throughout the reasoning processes and in proofs. The conclusion of the study is that students very rarely show the full competence necessary for reaching a level of understanding needed in formalization and proofs. The authors found that students usually do not look at a formula as a whole, but try to interpret each element separately, even though this may be meaningless in some tasks.

Vandebrouck and Leidwanger (2016) stressed another semiotic issue in the transition from secondary to tertiary level in relation to visualization: students are often expected to work in different ways on the same representations. In particular they may be expected to develop a sort of "big eye" that allows them to make inferences about the global behaviour of a function, while in other tasks they are only asked to manipulate expressions by applying algebraic rules. One major difficulty for students is the need to work with proofs. Inspired by Duval's (1999) decomposition of geometrical figures and of iconic and non-iconic visualization, Vandebrouck and Leidwanger (2016) introduced the notion of the decomposition of representations with perspective (DWP) of a formula. Since, according to Duval (1999), visualization can be produced in any register of representation, they focused on the non-iconic visualization of formulas and on the construct of perspectives (Vandebrouck, 2011) to describe and exemplify the large variety of skills that experts have developed to deal with formulas in Calculus. These perspectives are: decomposition with a global perspective (e.g., to see that a function is increasing); a local perspective (e.g., to identify different components of formulas to calculate a limit); or a point-wise perspective. They pointed out that, without sufficient work involving perspectives on functions, students may not properly understand the technical rules they are asked to remember and apply (such as the algebra of limits). They also observe that there is a lack of institutionalization of such different perspectives, which may cause students to seem to act chaotically; they suggest that the institutionalizing of these practices in secondary school could help prevent some problems in the transition to university.

The concept of tangent line to a curve is a crucial concept of Calculus and its conceptualization involves the transversal issues of representations, visualization, and perspectives. In secondary school, students are first presented the definition and properties of tangents to curves in Euclidean geometry (usually the tangent line to a circle), later tangents to curves are introduced. Vivier (2013) found that algebra is often absent from students' conceptions of this concept while this domain could allow an interesting transition from Geometry to Analysis. Páez Murillo, Montoya Delgadillo and Vivier (2016) analysed the problem of identifying the tangent line to a curve using the lens of *conception,* in the sense of Duroux (1983). They were able to categorize students' spontaneous conceptions of tangent into two blocks: the first concerning the geometrical domain, in particular the tangent to a circle, while the second concerned the domain of Analysis, linked directly to derivatives. In the first block of conceptions, Páez Murillo et al. included: the conception of the tangent as the line that has only one point of intersection with a curve; the conception of a tangent as a line that locally (but not necessarily globally) intersects the curve at a single point; and the conception of tangent as perpendicular to the radius of a circle. In the Analysis block, they found the following conceptions: a straight line with the same slope of the curve at that point; a linear function approximating the function at a point; a horizontal line at a maximum or a minimum. They realized that the notion of conception limited their interpretation of some data, so the authors included perspectives in their framework (Vandebrouck, 2011). They asserted that the point-wise, local and global perspectives helped in understanding the students' work. The authors observed that the didactical use of tangents to introduce differentiation seems to reduce such a rich mathematical concept solely to the point of view proposed by the derivative, and may not give the students opportunities for rich and interesting mathematical work on their spontaneous conceptions.

From the literature review and the new results we present in this section, some open problems emerge that require further research about the transition from secondary school to university Calculus. We stress in particular the need of a better understanding of students' difficulties due to the change of praxeologies and propose the trialling of good practices in didactical contexts to link secondary school and university, both concerning specific topics, like real numbers, tangents, functions, along with transversal issues, like iconic and not-iconic visualization, the use of representations, and the coordination of different perspectives.

Challenging issues in teaching and learning of limits and convergence

In many countries, convergence of numerical sequences and limits of functions are first taught in secondary school. Later, they are studied at university where a deeper comprehension of notions is required. In secondary school, mathematical tasks in Calculus and Analysis mainly involve the algebraic register (Vandebrouck, 2011). The emphasis is on the use of calculation rules, while argumentation and proof are often only weakly developed. The formal definition is sometimes taught but students are usually not required to work with it. At university, however, the formal definition is introduced and used to prove results. The manipulation of the definition requires knowledge of logic (including quantifiers, negation, and contraposition). The notion of limit (and in particular its definition) is also crucial to study other notions of Analysis such as derivatives and integrals and so the transition from secondary school to university brings significant changes in the mathematical work of students.

The study of limits of sequences and functions has been a subject of research for 40 years. Thus these notions have long been recognized as a source of difficulty for students; some of these difficulties have been linked to the complex logical structure of definitions and misconceptions about the meaning of definitions. There are several reviews of work on the teaching of the limits of sequences and functions (for example Oktaç & Vivier, 2016). Here we present research conducted on the difficulties of high school students and university students with limits and give examples of more recent research from INDRUM 2016 and 2018. These papers show very few changes over the years concerning student learning difficulties. We also show what new theoretical approaches are taken into account in research, and what perspectives might improve the conceptualization of limits.

On the learning of convergence of numerical sequences and limits of functions

Since the 1980s, research has focused on the study of students' difficulties with the notions of limits of sequences and functions. An important result concerns the intuitive representations that students develop about the notion of limit. Robert (1983) highlighted representations frequently observed among students at the beginning of university, such as the dynamic model where the convergence is described as "to get closer to". The static model is more formalized; here convergence is described in natural language such as "a sequence converges to l if any interval centred at l contains all the terms of the sequence from a certain rank". The monotone model is associated with the notion that only monotone and bounded sequences are convergent. Robert showed that these representations have consequences for the performance of students on tasks. Students with static conceptions have better results in exercises involving the notion of limit.

The complex logical structure of the notion of limit has also been the subject of much research (Durand-Guerrier & Arsac, 2005; Mamona-Downs, 2001). The use of the formal limit definition requires the manipulation of quantifiers or the ability to negate a sentence. However, many studies on secondary-university transition (for example Thomas et al., 2015) have revealed significant difficulties for students when working with notions of logic and the formalism contained in the definition of limit. The complex structure of the definition of convergence led Chorlay (2018) to study its conceptualization with a focus on the availability of logical prerequisites like manipulation of quantifiers. His results show a positive impact of the explicit teaching of logic before the teaching of limits. For example, after the intervention university students were able to better manipulate quantified statements and their negation.

The notion of limit has the characteristics of formalizing, unifying and generalizing notions (FUG notions) (Robert, 1983). These notions unify and generalize previous (even intuitive) notions taught in high school with a new formalism. The distance between new knowledge and old knowledge is often large, hence the difficulty in introducing FUG notions with a didactical situation in the sense of Brousseau (1997). Moreover, the formal nature of these notions makes it difficult for students to understand their meaning. This leads researchers to construct situations for FUG notions in which individual research phases are alternated with phases of institutionalization where the teacher gives specific comments on the formalism used or on changes in frames or registers. For example, the didactic engineering developed by Robert (1983) uses a representation of convergence in terms of strips that contain all the terms of the sequence from a certain rank on. The sequence involves changes of frames between drawings and formal writings. Bridoux (2016) further clarified the role of the teacher at each stage of this sequence to help students to move from a dynamic representation of convergence to a formal definition. Her experiment shows the robustness of Robert's didactic engineering.

The characteristics of the notion of limit raise the didactic question of how to strengthen the links between concept images, often developed in secondary school, and concept definitions (Tall & Vinner, 1981). In Robert's didactic engineering (1983), the formal definition of convergence of a sequence is constructed by students with specific help from the teacher to prove assertions such as "if a sequence converges to a strictly positive real number then the terms of the sequence eventually become positive". Bloch developed a didactic engineering to study area and perimeter of the Von Koch flake (Bloch & Gibel, 2011). Here, convergence is introduced in the geometrical register and with the idea of finding an approximation to a given order. Rogalski (2016) also relies on this idea to propose situations that highlight the raison d'être of the notion of limit. According to him, the introduction of the definition would thus become necessary and natural for students. He proposes situations that are based on the following three dimensions which aim to improve the conceptualization of the notion of limit: approximation, formalization and the role of "meta approaches" around convergence at the beginning of Analysis courses at university.

Limits of functions

Many studies have highlighted inherent learning difficulties with the notion of the limit of a function (e.g. Bloch, 2000; Cottril et al., 1996; Sierpinska, 1985). As Oktaç and Vivier (2016) explain, many results obtained on the learning of limits of sequences can be transposed to this setting. Here too dynamic conceptions are very resistant (Williams, 1991), and thus an important teaching goal is to enrich the dynamic representations of students with static conceptions. While the difficulties of students have not changed over the years, recent research has explored the teaching of limits with new theoretical approaches and research questions. We give some examples from INDRUM 2016 and 2018.

The fact that dynamic conceptions are sufficient in secondary education to compute limits leads students not to see the need for a formal definition. Using the notion of rationality, Lecorre (2016) addresses the question of the need to formalize in order to define a notion. He defines three types of rationalities:

> Pragmatic rationality consists in closely examining specific cases. ... The empirical rationality is used when one wants to have a general law. One uses facts to deduce generalizations. Theoretical rationality begins with theory (theorems, properties, definitions, axioms ...) to establish new properties and theorems.
>
> *(p. 84)*

Lecorre studied whether the concept of limits of functions can be an opportunity to develop a theoretical rationality. He developed and implemented a didactic engineering with high school students organized in the form of a scientific debate (Legrand, 2001), to bring out the need to define the notion of limit. This kind of practice structures the class in the form of a mini-community of researchers where the role of the teacher is to manage exchanges between the students. Lecorre offers different conjectures and students must decide if they are true or false. Here is an example: "If the limit of f is strictly smaller than the limit of g at $+\infty$ then a real A exists so that for every real x larger than A, $f(x)$ is strictly smaller than $g(x)$". Many students think that this conjecture is true and try to prove it. To structure the debate, the teacher suggests the study of the "monster"- a function f that remains below g (g is eventually constant and equal to 3) except for rare but regular peaks where f is greater than g on very small intervals. The teacher asks if "the monster" is an example, a counter-example or a "hors sujet" for the conjecture ("hors sujet" means that it does not fit the hypothesis). The analysis of the experiment shows that the students approach the question from a pragmatic point of view, starting with examples, and then taking a more theoretical point of view to produce a proof. The practice of scientific debate forces them to convince each other, moving from one rationality to another.

Many studies show the impact of static representations to describe infinite processes graphically and numerically. For example, the following statement is often given by students: "the closer x is to ∞, the closer $f(x)$ is to L". Swinyard and

Larsen (2012) developed a model to understand the formal definition of limit. As Ghedamsi and Lecorre (2018) explain,

> This model provides consistent arguments of how students reason about two infinite processes: 1) *f(x)* gets closer to L, and 2) the process of validating the limitation of the formal limit and encapsulation in the last step through the formalization of the infinite closeness notion (gets closer to) via the concept of arbitrary closeness.
>
> *(p. 34)*

Ghedamsi and Lecorre use "the monster" and they focus on the formalization of two infinite processes in the Archimedean continuum by using formal ε-statements like "there exists $A > 0$, for all $\varepsilon > 0$ such that for all $x > A$, $L - \varepsilon < f(x) < L + \varepsilon$" and "for all $\varepsilon > 0$, there exists $A > 0$ such that for all $x > A$, $L - \varepsilon < f(x) < L + \varepsilon$". The results highlight how ε-statements may assist students to reconsider their informal understanding of limit.

Vandebrouck and Leidwanger (2016) studied whether the knowledge acquired by students in secondary school enables them to work with representations of the notion of function by using global and local perspectives. They proposed a test for university students that requires the computation of limits, and conducted interviews with high school students. The students' answers revealed weak algebraic abilities, in particular, when they have to work with indeterminate forms. These conceptions may cause problems as the students make the transition from Calculus to Analysis.

Sierpinska (1985) describes students' difficulties with representations in terms of epistemological obstacles. For example, continuity is often described by students using a curve that is smooth and has no corners, and some students believe that only functions defined by a single algebraic formula are continuous, or that a continuous function is a function whose graph can be drawn without lifting the pen from the paper (Hanke, 2018).

Béchir and Vandebrouck (2018) developed software to introduce the formal definition of continuity and to lead to correct representation of the notion. To do this, they rely in particular on the review of works by Hanke and Shäfer (2017) which presented a classification (p. 2083) of the eight possible mental images that are reported in the literature:

- appearance of the graph of the function: "A graph of a continuous function must be connected";
- limits and approximation: "The left and right hand limits at each point must be equal";
- controlled wiggling: "If you wiggle x a bit, the function values will only wiggle a little bit too";
- connection to differentiability: "Each continuous function is differentiable";
- general properties of functions: "A continuous function is given by one formula and not defined piecewise";

- everyday language: "The function continues at each point and does not stop";
- reference to a formal definition: "I have to check whether the definition of continuity applies at each point";
- miscellaneous comments.

Béchir and Vandebrouck's (2018) tool (called "ICT-Analysis") has two matching display windows: one shows the cinematic dynamic point of view while in the second window the approximation-static point of view is highlighted. Three registers of representations of functions (numerical, graphical and algebraic) are coordinated to promote students' activities about conversions between registers. Students' activities and the teacher's discourse are analysed in terms of proximities (discourse elements that can foster students' understanding) (Bridoux, Grenier-Boley, Hache, & Robert, 2016). The main result is that the teacher gives many procedural mediations to introduce the tasks, to ensure an adequate progression of the students, and to take care of the instrumental process. This study highlights the importance of mediations in the conceptualization of such complex concepts.

The studies described in this section show that the learning difficulties related to the notion of limit (of sequences and functions) have been largely resistant to change over the last 40 years. Therefore, it is crucial to continue research in this domain to improve students' understanding.

Specific challenges with multivariable functions

University teachers usually dedicate no more than two or three sessions to the introduction of two-variable functions. This may be due to an expectation that once students overcome the difficulties involved in the transition from secondary school to university and complete the one-real variable function Calculus courses, they will be able to extend and generalize their knowledge to different types of functions. Research has found, however, that changing the nature of variables and functions demands a reconstruction of knowledge and involves a new level of abstraction.

In this section we discuss research studies that address students' needs when taking a multivariable Calculus course. We describe some of the challenges students face when introduced to two-variable functions. We illustrate how research results can be used to inform teaching intended to help students addressing these challenges.

On the learning of two-variable functions

As outlined previously, the concept of a single-variable function is difficult for most students. The study of other kinds of functions had not been considered by researchers until recently. Most studies conducted so far have shown that, after successfully completing one-variable Calculus courses, students consider functions as a correspondence rule (e.g. Dorko & Weber, 2014). This conception of function is carried on to two-variable functions even for students who have completed a Multivariable Calculus course (Martínez-Planell & Trigueros Gaisman, 2012;

Weber & Thompson, 2014). It prevents students from recognizing the role played by the notions of domain, range and codomain in the definition of function. For example, students who are able to consider the uniqueness of functional value and the vertical line test, are not aware that uniqueness is also part of the definition of two-variable functions.

All studies about the learning of functions of two variables, independently of the theoretical approach used, underline the crucial role of the use of different registers of representation (Duval, 2006) in developing students' deep understanding of this concept. Martínez-Planell and Trigueros (2013) point out that it is necessary for students to construct a strong $\mathbf{R}^3$ schema starting from point-by-point evaluation actions on functions, using flexibly different representations of surfaces, fundamental planes (planes of the form $x = c$, $y = c$ and $z = c$) and the domain of functions on the XY-plane. Other findings include students' need to distinguish between transversal sections and their projection onto the corresponding coordinate plane, and how informal language used by teachers in the description of strategies or procedures acts as an obstacle in students' learning. Some studies have underlined the importance of the ability to visualize surfaces as the graphs of two-variable functions, and to construct and demonstrate arguments about their properties (Dorko & Weber, 2014). In a recent study (Martínez-Planell & Trigueros, 2016), three cycles of research about the learning and teaching of two-variable functions were developed using Actions, Processes, Objects, Schemas (APOS) Theory (Arnon et al., 2013) and theory of semiotic representations (Duval, 2006). Results suggested that the use of the designed activities improved students' understanding: they overcame most of the difficulties found in the literature, were able to visualize and anticipate dynamic imagery, and developed coherent arguments and proofs about the properties of functions.

Limits and continuity of two-variable functions have received less attention in the research literature. Through the analysis of conceptual knowledge of preservice mathematics teachers, Biber (2010) found that even those students who showed a good understanding of limits and continuity of one-variable functions, had difficulty in explaining and arguing about these concepts when situations involved two-variable functions. Mamona-Downs and Megalou (2013) focused only on students' understanding of the limiting behaviour of these functions. They found that some of the difficulties reported in the literature for one-variable functions, such as the use of quantifiers or the use of graphical representations, persist and are even magnified when students work with two-variable functions.

Derivative and differential of two-variable functions

There are few studies published relating to differential calculus in several variables. In an early study, Tall (1992) proposed the use of a geometric model to visualize differentials in three dimensions. Weber (2012) studied how the rate of change of two-variable functions helped students to build this notion in space. Thompson, Manogue, Roundy, and Mountcastle (2012) pointed out that in the context of a Thermodynamics course students had difficulty in interpreting partial derivatives geometrically.

A similar model to Tall's was used by McGee and Moore-Russo (2015) to explore the effect of a semiotic chain in students' understanding of partial and directional derivatives of functions of two variables. They concluded that the synergy of different representations helped students to understand rates of change in space, while Martínez-Planell, Trigueros, and McGee (2015) applied the same model together with APOS Theory to study different components of the differential calculus of two-variable functions and the relation of the tangent plane at a point to the directional derivative to overcome difficulties related to limits and to give meaning to direction in the definition of slope in 3D-space. Trigueros, Martínez-Planell, & McGee (2018) analysed students' understanding of the differential after they had finished a multivariable calculus course where activities based on APOS Theory were used. The authors compared results with those obtained by students in a control course where those activities were not used; they showed that the construction of a process conception of directional derivative was hard for students and, more importantly, that even students who appeared to have constructed the concept of slope in a two-dimensional context as a process or object, struggled to generalize it to slope in space. The few students who could use the definition of directional derivative had trouble when using it to interpret what they were doing or to visualize slopes in questions including graphs, even with help. Unexpectedly, most students had difficulties in locating a direction vector which would constrain their geometric interpretation of a directional derivative as a slope. Coordination of processes related to the graph of a function with the same processes on the tangent plane considered as a local approximation to the function proved to be hard for students; they needed help to draw the tangent line necessary to discuss the sign of a directional derivative. The authors considered that developing visualization skills would help students in constructing a schema for $\mathbf{R}^3$ and the development of understanding of concepts related to functions of two variables and their calculus. They refined and incorporated new activities to address the specific difficulties observed. For example, activities where a flexible use of representation registers is developed were used to give meaning to the analytical definitions of concepts and the formulas used to calculate directional derivatives, as well as activities to recognize the role of the tangent plane as a local approximation of the function in different representations.

Trigueros, Martínez-Planell and McGee (2018) analysed students' responses to questions about the differential at a point and its relation to the tangent plane. Only one student (out of 26) from the experimental section showed evidence of a process conception of this concept. Five students from the same section showed evidence of an action conception and the others could not remember anything about this concept. Most respondents seemed to depend entirely on an empty symbolic representation and interpreted it as "some kind of derivative", and did not consider the relation between differential and tangent plane. Results suggested change is needed in the genetic decomposition of the concept of differential and that teaching activities should include opportunities for students to construct the differential of a function of two variables at a point as a function of two independent variables, (dx, dy) representing horizontal change, and to coordinate the differential and the tangent plane. Making links to other Calculus concepts such as directional derivative and integration was found to be important since the notion of differential seems to be isolated from other concepts.

Riemann sums and integrals of two-variable functions

Few studies in the literature deal with students' difficulties and understanding of the integral in multivariable Calculus. McGee and Martínez-Planell (2014) showed that a lecture-based course did not promote students' understanding, while activities using semiotic chains and developing synergy among representations did. Trigueros and Martínez-Planell (2018) used APOS Theory to study the relation between double integrals and Riemann sums with students who had taken a lecture-based course in Multivariable Calculus. The results were rich and unexpected: many students had difficulties considering a given rectangle as the domain of a function or a sample point on the graph of the function. Most of them considered geometric and symbolic representations as unrelated. Only one of 10 students showed evidence that he had constructed Riemann sums and their relation to double integrals as processes. The authors also focused on finding constructions relating Riemann sums and double integrals. They found that most students showed an unexpected confusion between area and volume when interpreting graphs of functions of two variables in three-dimensional space; although all students were able to calculate the value of the product $f(0,1)\Delta x\Delta y$, they had trouble when explaining its geometrical meaning as the volume of a rectangular prism. The authors related this difficulty to the fact that students' construction of space did not allow them to visualize what teachers mean when talking about a surface in space (Trigueros & Martínez-Planell, 2010) and the double integral as the volume under a surface. These authors also found that many students considered functions only as a correspondence rule so they had difficulties when interpreting the domain of functions and the function in a graphical representation. Most students found it difficult to deal with and visualize Riemann sums and their relation to double integrals. Even students who interpreted Riemann sums in terms of adding the volume of prisms, showed not to have constructed their relation to the function in question.

Results of the experiences described in this section show that Multivariable Calculus should not be considered as a simple generalization of one-variable Calculus. The transition involves the need to take care of students' construction of the notions of function, tangent plane, direction vectors, partial and directional derivatives, differential, volume under a surface and the role of the Riemann sum. Some open questions remain. When students understand Calculus of two variables, will they be able to generalize their learning to multivariable functions? Would it be appropriate to introduce each topic by working on functions of two variables and then help students to generalize it? The answer to these questions is important so that students are well prepared to face new challenges in Analysis courses.

The relation with Further Analysis and Complex Analysis

Although most of the literature on the teaching and learning of Analysis focuses on the topics students meet in a first course, some studies have considered the issues that arise in upper level undergraduate courses. Clearly courses on Topology build on the conceptions that students have developed in earlier introductory courses in Analysis (Bergé, 2008) as well as their experiences of formal definitions and proof, as do modules on Complex Analysis and Functional Analysis. Furthermore, the understanding that students develop in these upper-level courses may help them to view elementary notions from Pre-calculus and Calculus in a new light. Thus the links between Calculus and Introductory Analysis courses, and more advanced courses are worthy of study here; we consider the influence of experience with Analysis on the real line on the development of topological understanding, as well as the opportunities which may arise towards the end of an undergraduate degree programme to deepen understanding of elementary topics.

Journey from Analysis on the real line to Analysis in topological spaces

Two papers at the INDRUM conferences (Hamza & O'Shea, 2016; Hanke, 2018) refer to ways in which students' conceptions of topological notions and experience of proof (often formed in the setting of the real line) can influence their thinking in later years.

Hanke (2018) examined student-generated proofs of the fact that a function from $\mathbf{R}$ to $\mathbf{R}$ is continuous if and only if its graph is a path-connected subset of $\mathbf{R}^2$. This result is familiar to Calculus students and may even be treated as the definition of continuity in that context. However, the proof requires some sophisticated topological ideas. Hanke (2018) studies students' use of concepts and mental images as they attempt to prove the result. In particular, he considers how students use previously-seen theorems to build a new proof. He finds that these types of actions are usually successful if the conditions of the theorem in question are satisfied, however sometimes students attempt to use results even when the conditions are not satisfied. In these cases, students seem to be unaware of the complexity of certain concepts (for example that continuous functions need not be monotonic). It may be that these students are influenced by "typical" examples of concepts. Alcock and Weber (2010) found that the use of examples in proof construction could cause problems if students' understanding is not closely connected to definitions. This is evidence that a well-developed concept image (Tall & Vinner, 1981) is crucial in order to build correct proofs.

Little work has been done on investigating the challenges which arise in advanced courses on Analysis or Topology, and in particular on students' concept images in this setting. Bridoux (2011) studied students' conceptions of topological concepts in Euclidean space and found that the complex logical structure of the definition of a closed set in terms of balls led students to define this kind of set by giving a definition that is satisfied by all sets. Hamza and O'Shea (2016) found five main categories of concept images of open sets in a metric space setting held by undergraduate students based on: the formal definition, the idea of the boundary of a set, open sets in Euclidean space, the union of open balls, and visualization using generic pictures of intervals or discs. Some of these are directly related to the teaching in the course (the formal definition, and the result that any open set in a metric space is the union of a collection of open balls), however others seem to be based on students' previous experience of the topology of $\mathbf{R}$ or $\mathbf{R}^n$. Students in this study often referred to properties such as connectedness or completeness when describing or working on tasks about open sets in a general metric space.

Since many of the concepts that students meet in courses on Topology are generalizations of properties of the real line, it is natural that students' previous experience in that context would influence the development of their understanding. This is reminiscent of McGowen and Tall's (2010) idea of a met-before, which is the notion that one's current mental structures are influenced by earlier experiences, and of the tacit models of Fischbein (1989). These tacit models or met-befores can help to build intuition and mental imagery, but also may cause learners to make incorrect assumptions, for example that all paths are injective (Hanke, 2018) or that all open sets are connected (Hamza & O'Shea, 2016). It makes sense therefore for teachers to gain knowledge of the likely progression of conceptual development based on previous experience. This is a point made also by Branchetti and Durand-Guerrier (2018): they identified students' difficulties in learning properties of $\mathbf{N}$ and $\mathbf{Q}$ concerning their cardinalities, their standard order relations, and the comparison between discreteness and density. Such difficulties were compared with a historical-epistemological analysis of the development of such notions; the authors identified conflicts inherent in the students' concept images that were at the basis of their difficulties and showed that they were due to students' previous experiences in secondary school with the images of the line and its properties. Durand-Guerrier and Vivier (2016) reported on a study of how students' conceptualization of the real numbers is related to learning in an Analysis course. They claim that the study of Analysis requires a strong foundation built on the understanding of the completeness of the real numbers. Oktaç and Vivier (2016) highlighted the role that the understanding of the topology of $\mathbf{R}$ plays in the transition to Analysis and called for more work on this topic. Completeness is one of the concepts that link Calculus, Analysis, and Topology courses. Durand-Guerrier (2016) studied first-year undergraduates' conceptions of the continuum; she identified difficulties with distinguishing between the properties of discreteness, density, and completeness when applied to subsets of $\mathbf{R}$. She proposed that they could be overcome by using her didactical design which aimed to foster understanding of these concepts by introducing the main elements of Dedekind's construction of the real numbers and related tasks. She was able to show that this approach was successful in addressing the distinctions mentioned above both with secondary students and with pre-service teachers. Bergé (2008) studied the treatment of the completeness property of the real numbers over a series of four Calculus and Analysis courses in an undergraduate programme. Her analysis showed that the topic is treated in progressively more formal and abstract manner as students move from Calculus courses to Analysis courses. She advocates that more thought is given by syllabus designers to the change of perspective required of learners. In a study of students' views on the usefulness of the supremum in the context of completeness Bergé (2016) claimed that the types of tasks that students engage with, and the situations in which they encounter the notion of supremum, can have a major effect on their appreciation of the concept of completeness.

Journey from Calculus to Advanced Analysis and back again

A theme evident from the INDRUM conferences is how experience gained in upper level Analysis courses can help students revisit concepts from Calculus or Pre-calculus courses and provide opportunities to deepen understanding of these concepts. Here the influence of Felix Klein's work is felt; he identified the "double discontinuity" (Klein, 1908/1932, p. 1) in mathematics teacher education where the connections between mathematics at school and university levels are not always apparent.

Hanke and Schäfer (2018) worked on the design of a course in Complex Analysis for future teachers. The students in this module followed a traditional syllabus for two-thirds of the course, while the final segment of the module concentrated on the identification of aspects of the material helpful when teaching at secondary school. Students were given the opportunity to design and trial tasks for secondary school pupils using dynamic geometry software. The aim was to investigate how students could combine specialized and pedagogical content knowledge (Ball, Thames, & Phelps, 2008) in their teaching. Courses in Complex Analysis are often taken by students towards the end of their undergraduate program, and may be seen as not being directly relevant for teachers. However, as Hanke and Schäfer (2018) point out, these courses could be used as places to deepen students' knowledge of concepts such as polynomials, trigonometric functions, and power series, and results such as the Fundamental Theorem of Algebra.

Despite the importance of Complex Analysis, very little research has been carried out on the teaching and learning that takes place in these courses (apart from studies of students' conceptions of complex numbers). Some work has been done on the use of dynamic geometry software to help students develop geometric reasoning about the derivative of a complex-valued function (Troup, 2018). In this study, students were able to develop insight concerning the complex derivative as a local property and of its geometric effect as a rotation and dilation. Troup (2018) advises instructors to present both the algebraic and geometric view of derivatives and the connections between them in order to help students build understanding. Soto-Johnson, Hancock, and Oehrtman (2016) investigated the interplay between formal and more intuitive notions when mathematicians explain continuity of complex-valued functions. They caution instructors to be explicit about the shortcomings of less formal conceptions (for example the idea that continuous functions are those that preserve closeness) while acknowledging the important role these have in building intuition and aiding reasoning.

Kondratieva and Winsløw (2018) (drawing on Winsløw, 2016) were also inspired by Klein's advice to present mathematics to students in a less compartmentalized way so that connections between different fields become evident. They explore two teaching situations where knowledge from courses in Analysis can be linked to more elementary notions; the first of these concerns the definition of angles and trigonometric functions using ideas from Vector Analysis, and the second uses skills from a Calculus course to motivate and prove special cases of a theorem in Fourier Analysis. The links explored in this study have the potential to help upper-level university students to build a deeper understanding of basic material as well as to utilise previous knowledge and skills to build understanding of concepts in advanced Analysis modules. Kondratieva and Winsløw call for more research on how mathematics programs are designed and on identifying connections between different elements of these programs. This involves careful consideration of the links between the theoretical material found in courses such as Analysis and the practical skills developed in Calculus modules.

We have seen here that the links between courses on Calculus, introductory Analysis, and further Analysis are very important and are worthy of further study. The relationships between these courses are complicated; we have seen that students' conceptions built during their early years as undergraduates can affect their understanding of topics encountered later in their degree programs. We have also seen that courses taken towards the end of a student's degree can be used to revisit earlier work and to make connections with school mathematics. More information about these bi-directional influences would be beneficial to instructors and program designers. Furthermore, we have identified the need for more research on the teaching and learning of topics in areas such as Complex Analysis and Topology.

Conclusion

This chapter shows the variety and depth of research on the teaching and learning of Calculus and Analysis conducted in the last few years using different and sound theoretical frameworks, as reflected in the INDRUM meetings. The studies discussed here show in particular the interest in reaching a deeper comprehension of the transition to studying mathematics at university. The analysis of the challenges involved in the transition between secondary school Calculus and university Calculus bring to the fore the necessity to carefully give meaning to the formalism involved in the definition of limit and its role in the definition of other important Calculus concepts. Topics discussed in this chapter also show the importance of introducing real numbers and their properties by involving students in the discussion and in active work that may give sense to their role in the study of Calculus and later in Analysis.

This chapter shows that there are other challenging issues that have not received much attention, moving beyond the study of Calculus of one-variable functions to dealing with other kinds of functions, such as two-variable functions. The analysis of the challenges involved in understanding three-dimensional space and the definition and representation of two-variable functions, the changes needed in the definition of limit and continuity and the introduction of all the Calculus concepts related to these functions reveals the need for further study in this area. It is important that research is carried out into conceptions of other types of functions and how an understanding of them can be interrelated to develop better teaching strategies and ultimately more coherent student understanding. The studies presented at INDRUM conferences related to a later transition from Calculus to Analysis courses underline the complex links between these courses. We see a need to carry out further research about the influence exerted by what students learn in their first courses and how the experience obtained from later courses can be better understood in terms of rigour, argumentation and proof, and then used.

Topics discussed in this chapter underline the role of rigour in the teaching of Calculus and Analysis. Several transversal axes can be identified through the sections of this chapter. The role of formalism and definition that has been addressed above has a ubiquitous presence in the chapter. All of the sections discuss the importance of students' development of flexibility in using different representations and their potential to deeply understand Calculus and Analysis concepts as well as the role of visualization in the interpretation of different significant Calculus and Analysis ideas.

Despite the presence of the use of technology in most high school curricula, this chapter shows that this topic remains an underdeveloped field of research. This may be a good perspective to develop for future INDRUM conferences.

References

Alcock, L., & Weber, K. (2010). Undergraduates' Example Use in Proof Construction: Purposes and Effectiveness. *Investigations in Mathematics Learning*, 3(1), 1–22.

Arnon, I., Cottrill, J., Dubinsky, E., Oktaç, A., Roa Fuentes, S., Trigueros, M., & Weller, K. (2013). *APOS Theory: A framework for research and curriculum development in mathematics education*. New York: Springer Verlag.

Artigue, M. (2001). What can we learn from educational research at the university level? In D. Holton (Ed.), *The teaching and learning of mathematics at university level* (pp. 207–220). Heidelberg: Springer Netherlands.

Ball, D. L., Thames, M. H., & Phelps, G. (2008). Content knowledge for teaching: What makes it special? *Journal of Teacher Education*, 59(5), 389–407.

Béchir, S., & Vandebrouck, F. (2018). Teaching and learning continuity with technologies. *INDRUM 2018 Proceedings*, pp. 74–83.

Bergé, A. (2008). The completeness property of the set of real numbers in the transition from calculus to analysis. *Educational Studies in Mathematics*, 67, 217–235.

Bergé, A. (2010). Students' perceptions of the completeness property of the set of real numbers. *International Journal of Mathematical Education in Science and Technology*, 41, 217–227.

Bergé, A. (2016). Le rôle de la borne supérieure (ou supremum) dans l'apprentissage du système des nombres réels. *INDRUM 2016 Proceedings*, pp. 33–42.

Biber, A. Ç. (2010). *Concept knowledge related to the limits and continuity concepts in one-variable functions and in two-variable functions of prospective mathematics teachers*. Unpublished doctoral dissertation, Gazi University, Ankara, Turkey.

Bloch, I. (2000). *L'enseignement de l'analyse à la charnière lycée / université – Savoirs, connaissances et conditions relatives à la validation*. Doctoral thesis, Université Bordeaux I, Bordeaux, France.

Bloch, I., & Gibel, P. (2011). Un modèle d'analyse des raisonnements dans les situations didactiques: étude des niveaux de preuves dans une situation d'enseignement de la notion de limite. *Recherches en Didactique des Mathématiques*, 31(2), 191–228.

Bloch, I., & Gibel, P. (2016). A model to analyse the complexity of calculus knowledge at the beginning of University course. Two examples: parametric curves and differential equations. *INDRUM 2016 Proceedings*, pp. 53–62.

Branchetti, L., & Durand-Guerrier, V. (2018). Are all denumerable sets of numbers order-isomorphic? *INDRUM 2018 Proceedings*, pp. 14–23.

Bridoux, S. (2016). Introduire la notion de convergence avec une ingénierie des années 1980: rêve ou réalité didactique pour l'enseignant?*INDRUM 2016 Proceedings*, pp. 53–62.

Bridoux, S. (2011). *Enseignement des premières notions de topologie à l'université - Une étude de cas*. Doctoral thesis, Université Paris-Diderot - Paris VII, Paris, France.

Bridoux, S., Grenier-Boley, N., Hache, C., Robert, A. (2016). Les moments d'exposition des connaissances: analyses et exemples. *Annales de didactique et de sciences cognitives*, 21, 187–233.

Brousseau, G. (1997). *Theory of Didactical Situations*. Dordrecht: Springer.

Chorlay, R. (2018). An empirical study of the understanding of formal propositions about sequences, with a focus on infinite limits. *INDRUM 2018 Proceedings*, pp. 24–33.

Cottrill, J., Dubinsky, E., Nichols, D., Schwinngendorf, K., Thomas, K., & Vidakovic, D. (1996). Understanding the limit concept: Beginning with a coordinated process schema. *Journal of Mathematical Behavior*, 15, 167–192.

Dorko, A., & Weber, E. (2014). Generalising calculus ideas from two dimensions to three: how multivariable calculus students think about domain and range. *Research in Mathematics Education*, 16(3), 269–287.

Durand-Guerrier, V. (2016). Conceptualization of the continuum, an educational challenge for undergraduate students. *International Journal of Research in Undergraduate Mathematics Education*, 2, 338–361.

Durand-Guerrier, V., & Arsac, G. (2005). An epistemological and didactic study of a specific calculus reasoning rule, *Educational Studies in Mathematics*, 60, 149–172.

Durand-Guerrier, V., & Vivier, L. (2016). Densité de D, Complétude de R et analyse réelle: Première approche. *INDRUM 2016 Proceedings*, pp. 143–152.

Duroux, Alain (1983). La valeur absolue. Difficultés majeures pour une notion mineure. *Petit x*, 3, 43–67.

Duval, R. (1999). Representation, vision and visualisation: cognitive functions in mathematical thinking. Basic issues for learning. In F. Hitt & M. Santos (Eds.) *Proceedings of the 21st Conference of the North American chapter of the international group for the Psychology of Mathematics Education*. 1, 3–26.

Duval, R. (2006). A cognitive analysis of problems of comprehension in a learning of mathematics. *Educational Studies in Mathematics*, 61(1), 103–131.

Fischbein, E. (1989). Tacit Models and Mathematical Reasoning. *For the Learning of Mathematics*, 9(2), 9–14.

Ghedamsi, I. (2016). Entrée des étudiants dans l'Analyse formelle de début d'université: Potentialité des méthodes numériques d'approximation. *INDRUM 2016 Proceedings*, pp. 63–72.

Ghedamsi, I., & Lecorre, T. (2018). Students' understanding of ε-statements involving equality and limit. *INDRUM 2018 Proceedings*, pp. 34–43.

Hamza, S., & O'Shea, A. (2016). Concept image in metric spaces. *INDRUM 2016 Proceedings*, pp. 113–122.

Hanke, E. (2018). "A function is continuous if and only if you can draw its graph without lifting the pen from the paper" – Concept usage in proofs by students in a topology course. *INDRUM 2018 Proceedings*, pp. 44–53.

Hanke, E., & Shäfer, I. (2017). *Students view of continuity: an empirical analysis of mental images and their usage*, Proceedings of the 10th Congress of the European Society for Research in Mathematics Education (CERME), February 1–5, 2017. Dublin, Ireland: DCU Institute of Education & ERME.

Hanke, E., & Schäfer, I. (2018). Learning complex analysis in different branches – Project Spotlight-Y for future teachers. *INDRUM 2018 Proceedings*, pp. 54–63.

Kidron, I. (2016). Understanding irrational numbers by means of their representations as non- repeating decimals. *INDRUM 2016 Proceedings*, pp. 73–82.

Klein, F. (1908/1932). *Elementary Mathematics from an advanced standpoint* (E. Hedrick and C. Noble, Trans.). London: MacMillan.

Kondratieva, M., & Winsløw, C. (2018). Klein's Plan B in the Early Teaching of Analysis: Two Theoretical Cases of Exploring Mathematical Links. *International Journal of Research in Undergraduate Mathematics Education*, 4, 119–138.

Lecorre, T. (2016). Rationality and concept of limit. *INDRUM 2016 Proceedings*, pp. 83–92.

Legrand, M. (2001). Scientific debate in mathematics courses. In D. Holton (Ed.), *The teaching and learning of mathematics at university level: An ICMI study* (pp. 127–135). Netherlands: Kluwer Academic.

Mamona-Downs, J. (2001). Letting the intuitive bear on the formal: A didactical approach for the understanding of the limit of a sequence. *Educational Studies in Mathematics*, 48, pp. 259–288.

Mamona-Downs, J. K., & Megalou, F. J. (2013). Students' understanding of limiting behavior at a point for functions from R^2 to R. *Journal of Mathematical Behavior*, 32, 53–68.

Martínez-Planell, R., & Trigueros, M. (2012). Students' understanding of the general notion of a function of two variables. *Educational Studies in Mathematics*, 81(3), 365–384.

Martínez-Planell, R., & Trigueros, M. (2013). Graphs of functions of two variables: Results from the design of instruction. *International Journal of Mathematical Education in Science and Technology*, 44(5), 663–672.

Martínez-Planell, R., Trigueros, M., & McGee, D., (2015). On students' understanding of the differential calculus of functions of two variables. *Journal of Mathematical Behavior*, 38, 57–86.

Martínez-Planell, R., & Trigueros, M. (2016). Cycles of APOS based research. The case of functions of two variables, *Proceedings of 40th Annual Conference of the International Annual Meeting of the international group for the Psychology of Mathematics Education*, Vol. 3, pp. 283–290. Szeged.

McGee, D., & Martínez-Planell, R. (2014). A study of semiotic registers in the development of the definite integral of functions of two and three variables. *International Journal of Science and Mathematics Education*, 12, 883–916.

McGee, D., & Moore-Russo, D. (2015). Impact of explicit presentation of slopes in three dimensions on students' understanding of derivatives in multivariable calculus. *International Journal of Science and Mathematics Education*, 13(2), 357–384.

McGowen, M., A., & Tall, D. O. (2010). Metaphor or met-before? The effects of previous experience on practice and theory of learning mathematics, *Journal of Mathematical Behavior*, 29(3), 169–179.

Oktac, A., & Vivier, L. (2016). Conversion, change, transition … in research about analysis. In B. R. Hodgson, A. Kuzniak, J. B. Lagrange (Eds.) *The Didactics of Mathematics: Approaches and Issues*. Switzerland: Springer International.

Páez Murillo, R. E., Montoya Delgadillo, E., & Vivier, L. (2016). Conceptions spontanées et perspectives de la notion de tangente pour des étudiants de début d'université. *INDRUM 2016 Proceedings*, pp. 103–112.

Radford, L., Schubring, G., & Seeger, F. (2008). *Semiotics in mathematics education: Epistemology, history, classroom, and culture*. Rotterdam: Sense Publishers.

Robert, A. (1983). L'enseignement de la convergence des suites numériques en DEUG, *Bulletin de l'APMEP*, 340, 431–449.

Rogalski, M. (2016). Revenir au sens de la notion de limite par certaines de ses raisons d'être: un chantier pour le début de l'analyse à l'université. *INDRUM 2016 Proceedings*, pp. 133–142.

Sfard, A. (1991). On the dual nature of mathematical conceptions: reflections on processes and objects as different sides of the same coin. *Educational Studies in Mathematics*, 22(1), 1–36.

Sierpinska, A. (1985). Obstacles épistémologiques relatifs à la notion de limite. *Recherches en Didactique des Mathématiques*, 6 (1), 5–67.

Soto-Johnson, H., Hancock, B., & Oehrtman, M. (2016). The Interplay Between Mathematicians' Conceptual and Ideational Mathematics about Continuity of Complex-Valued Functions. *International Journal of Research in Undergraduate Mathematics Education*, 2, 362–389.

Swinyard, C., & Larsen, S. (2012). What does it mean to understand the formal definition of limit? Insights gained from engaging students in reinvention. *Journal for Research in Mathematics Education*, 43(4), 465–493.

Tall, D. (1992) Visualizing differentials in two and three dimensions. *Teaching Mathematics and Its Applications*, 11(1), 1–7.

Tall, D., & Vinner, S. (1981). Concept image and concept definition in mathematics with particular reference to limits and continuity. *Educational Studies in Mathematics*, 12(2), 151–169.

Thomas M. O. J., de Freitas Druck, I, Huillet, D, Ju, M.-K., Nardi, E., Rassmussen, C., & Xie, J. (2015). *Key Mathematical Concepts in the Transition from Secondary School to University*. In S. Cho (Ed.) *The Proceedings of the 12th International Congress on Mathematical Education*. Cham: Springer.

Thompson, J., Manogue, C., Roundy, D., & Mountcastle, D. (2012). Representations of partial derivatives in thermodynamics. In C. Singh, N. Sanjay Rebello & P. V. Engelhardt (Eds.), *Proceedings of the 2011 Physics Education Research Conference*, pp. 85–88. Omaha, AIP.

Törner, G., Potari, D., & Zachariades, T. (2014) Calculus in European classrooms: curriculum and teaching in different educational and cultural contexts. *ZDM Mathematics Education*, 46, 549–580.

Trigueros, M., & Martínez-Planell, R. (2010). Geometrical representations in the learning of two-variable functions, *Educational Studies in Mathematics*, 73(1), 3–19.

Trigueros, M., & Martínez-Planell, R. (2018). On students' understanding of Riemann sums of integrals of functions of two variables. *INDRUM 2018 Proceedings*, pp. 64–73.

Trigueros, M., Martínez-Planell, R., & McGee, D. (2018). Student understanding of the relation between tangent plane and the total differential of two-variable functions. *International Journal of Research in Undergraduate Mathematics Education*, 4(1) 181–197.

Troup, J. (2018). Developing Reasoning about the Derivative of a Complex-Valued Function with the Aid of Geometer's Sketchpad. *International Journal of Research in Undergraduate Mathematics Education*, 5, 3–26.

Vandebrouck, F. (2011). Perspectives et domaines de travail pour l'étude des fonctions. *Annales de Didactique et de Sciences cognitives*, 16, 149–185.

Vandebrouck, F., & Leidwanger, S. (2016). Students' visualization of functions from secondary to tertiary level. *INDRUM 2016 Proceedings*, pp. 153–162.

Vergnac, M., & Durand-Guerrier, V. (2014). Le concept de nombre réel au lycée et en début d'université, un objet problématique, *Petit x*, 96, 7–28.

Vivier, L. (2013). Without derivatives or limits: from visual and geometrical points of view to algebraic methods for identifying tangent lines, *International Journal of Mathematic Education in Science and Technology*, 44(5), 711–717.

Weber E. D. (2012). *Students' ways of thinking about two-variable functions and rate of change in space*. PhD Dissertation. Arizona State University, Phoenix, USA.

Weber, E. D., & Thompson, P. W. (2014). Students' images of two-variable functions and their graphs. *Educational Studies in Mathematics*, 87(1), 67–85.

Williams, S. (1991). Models of limit held by college calculus students. *Journal for Research in Mathematics Education*, 22 (1991), 219–236.

Winsløw, C. (2016). Angles, trigonometric functions, and university level analysis, *INDRUM 2016 Proceedings*, pp. 163–172.

6

TASK DESIGN IN CALCULUS AND ANALYSIS

Fabrice Vandebrouck, Erik Hanke and Rafael Martínez-Planell

This chapter about task design in calculus and analysis is divided into three parts. In the first one, we consider different perspectives on task design in the INDRUM papers about calculus and analysis topics. In the second one, we consider the assessment of task design and students' task performance. In the last part, we enlarge our scope by considering papers suggesting the need for task design in calculus and analysis as well as task design for the purpose of research.

Several perspectives on task design

The aim of the first part is to show, through questions about the conceptualisation of the notion of limit, that task design can be looked at from different perspectives, from constructivist views based only on the task itself to more technological and socio-culturally oriented views which focus on the role of the teacher and tools used.

From constructivist to more socio-constructivist and cultural perspectives

In France, task design was introduced a long time ago in association with didactical engineering (Artigue, 2014; González-Martín, Bloch, Durand-Guerrier, & Maschietto, 2014). In this first sense, task design is associated with the Theory of Didactical Situations (TDS; Brousseau, 1997), mostly in a constructivist vision of learning, and this approach is still present in some research presented at the INDRUM conference. For instance, Ghedamsi (2016) has built two situations aiming at the introduction of the notion of limit via numerical approximation of real numbers by sequences of rational numbers. In this context, task design means the building of a "milieu" such that the students can reach the notion of limit through the designed task and the interactions (actions, retroactions) with the milieu by themselves. The situation evolves through action, formulation, validation and institutionalisation phases associated with the evolution of the milieu itself. The task design supposes strong epistemological and historical analyses of the notion at stake and strong a priori analyses of the designed tasks in order to analyse the findings in the a posteriori analysis and predict students' possible activities. In her study, Ghedamsi justifies her choices for the design with a historical perspective on understanding of real numbers with attention to approximations. She suggests two situations according to the values of some didactical variables that are crucial for the learning objectives: the real numbers which are approximated (rational or irrational), the nature of the sequence which approximates (explicit or recursive, growing or not, …), and the nature of the approximation and error. The two situations are the approximation of the square root of 2 by Euclid's method (antiphérèse) and the solution of $cos(x) = x$ by Newton's algorithm. Ghedamsi's conclusion highlights especially the success of the situations with respect to students' understanding of real numbers as mathematical objects together with a meta-mathematical reflection about their existence and the process of their approximation. However, the situation of the fixed point of the *cos* function allows to go deeper into the conceptualisation of real numbers because it gives access to real numbers which cannot be expressed in terms of elementary functions or symbols (like the square root of 2) and which oblige students to use formal procedures in order to deal with these numbers.

In the epistemological work done in order to build new tasks which aim at introducing the concepts of convergence and real numbers, Rogalski (2016) recalls several dimensions which can structure students' conceptualisations of the notion of limit and which can be some basis for task design: first of all, the need for approximations of real numbers as Ghedamsi (2016) argued, secondly the need for formalisation in order to enter in some proof paths, and thirdly the need for understanding about mathematics practices at a meta-level.

The focus of Bridoux (2016) is to respond to the second and third need. First, she argues that the Theory of Didactical Situations is not sufficient in itself to introduce notions such as limit and convergence because of their FUG aspects (formalisation, unification, generalisation) in the sense of Robert (Robert & Hache, 2013): The definition of limit formalises intuitive aspects of convergence already encountered by students, it unifies different pre-existing students' ideas of convergence, and it permits a generalisation which tries to overcome problems about convergence and limits that students can face. This means that the notion of limit is such a complex notion that a constructivist approach of task design is not sufficient for the students to understand it deeply. This implies the need of a meta-discourse of the teacher in order to justify the need for the formalisation and the need of the limit as a formalised concept to enter in the domain of analysis at the university level, as Rogalski (2016) argues. The designed task in Bridoux's paper comes from old research papers of Robert in the eighties (Robert, 1982; 1983). Her paper is about the classification of ten sequences, drawing graphical representations, discussions about students' classifications and then answering two questions which need the formalisation of limit: 1) Is a positive sequence which has 0 as limit decreasing after a certain rank? 2) If a sequence has a strictly positive limit, are all the terms strictly positive after a certain rank? There is an emphasis on the role of the teacher at each step of the engineering. Indeed, the situation alternates between adidactical phases in the same sense as in the TDS and interactive phases when the teacher gives procedural and constructive assistance to students using different kinds of didactical processes such as the interplay between different settings ("jeux de cadres" in the sense of Douady, 1986), a shift between several points of view on the tasks or the shift between several registers of representations which Duval (1993) highlighted as a cognitive benefit. Bridoux shows that this engineering is still robust with some adaptations she explains more closely.

The focus of Lecorre (2016) and Ghedamsi and Lecorre (2018) is also about the formalisation of the notion of limit. However, the papers deal with functions whereas the two previous ones from Ghedamsi (2016) and Bridoux (2016) were in the context of sequences. Cognitive analyses are made to supplement the epistemological ones. For instance, Lecorre mentions the cognitive shift which has to be done by students between dynamic conceptions of limits to static conceptions (Robert, 1982; Williams, 1991). The dynamic conceptions are often associated to a cinematic point of view in the graphical register where moving the independent variable x leads to moving of $f(x)$ towards the limit, possibly reaching it. The static point of view is associated to the approximation process where the chosen degree of approximation pulls the independent variable of the function and leads to the search of a real number A (in the case of limits of functions at infinity), or a so-called η – or δ in some papers – in the case of continuity. This cognitive characteristic remains strong in the domain of analysis and for the introduction of formal definitions such as continuity as we will see below. In Lecorre's engineering, there is a focus on the need of formalisation, which is a common point with Bridoux's paper. As a first question, the students are asked to discuss the conjecture "If the limit of f is strictly less than the limit of g at plus infinity then for every real x, $f(x)$ is strictly less than $g(x)$" (Lecorre, 2016, p. 85). Then Lecorre introduces a "monster function" to the students with which they should rethink their conjectures and feel the need to establish a definition of the limit of a function at plus infinity. The role of the monster function is also to enrich discussions between students and to destabilise their dynamic conceptions. Like Bridoux, Lecorre highlights the role and the discourse of the teacher. The task itself is not sufficient for the students to reach the notion of limit in its formal definition by themselves. Lecorre uses the concept of "scientific debate" to ensure the link between students' activities with the task and the institutional process.

Task design in calculus and analysis is renewed today in line with Chevallard's paradigm of questioning the world (Chevallard, 2015). The scope is supposed to be larger. The starting point is the global difficulty of students to relate the content of analysis courses at the university level (formal theory of limits and continuity, function spaces etc.) to the calculus they have encountered in secondary schools. As Winsløw (2016) notes, the abstract theory seems to constitute a universe in itself. The new paradigm of questioning the world is based on Klein's ideas about Plan A and Plan B (Klein, 2016). Plan A is a compartmentalised approach to mathematics whereas Plan B, "by contrast, involves a more holistic approach which emphasises and exploits connections between different sectors" (Winsløw, 2016, p. 164). In the paradigm of questioning the world, as Klein recommends, Plan B is a better strategy for presenting mathematics to students. Winsløw investigates the specific notions of angles, sine and cosine. Several plans A appear in secondary school with three distinct contexts: triangle trigonometry, analytic geometry in the circle, and functions through tables and periodic graphs. Winsløw develops ideas for some mathematical organisations of type B in university analysis highlighting the importance of curve length as a link between the contexts; this link has never been questioned in the sense of Chevallard (2015). However, as Winsløw concludes, the task design based on such a Plan B organisation can remain too ambitious (cf. also Kondratieva & Winsløw, 2018).

Task design with technologies

We now turn our attention to papers which deal with technological artefacts that can be used throughout a course to support student learning in calculus and analysis.

The task design by Sghaier and Vandebrouck (2018) builds on ideas from the previous subsection and creates tasks by integrating technological devices. They develop a digital tool aiming at introducing the concept of local continuity of a function together with its formal definition. Concept image and concept definition (Tall & Vinner, 1981) together with the notion of point of view (Bkouche, 1996) are addressed with dynamic and static conceptions as introduced by Robert (1982) and used by Lecorre (2016). The tool and the associated tasks are designed in order for students to make links between two aspects of continuity: the dynamic aspect related to graphical and cinematic approaches of continuity, and the static aspect associated to more formal approaches through the lens of approximation. The knowledge to be learned about continuity is referred to conceptualisation in the sense of Vergnaud (2009). Conceptualisation is equally the cognitive product and the developmental process which occur within students' actions over a class of situations characteristic of the concept involved, continuity in this paper. The tasks by Sghaier and Vandebrouck are analysed with didactical tools which have been developed in the French community dealing with Vergnaud's cognitive references: technical tasks which are direct applications of the concepts involved versus tasks with cognitive adaptations of knowledge such as shifts between several registers of representations, uses of several points of view on the tasks etc. There are adidactical aspects in the designed situation in which the task is inserted, in spirit of the TDS and the constructivist approach. However, as in several previous papers, the role of the teacher remains crucial. The students are only supposed to develop some mathematical cognitive activities while linking the two points of view of continuity which are highlighted in the tool. In this paper, the discourse of the teacher during the interactive phases is theorised in term of "proximities": teacher's discourse elements which are supposed to be developed within the students' zone of proximal development according to the mathematical activities that have been developed in the tool. In this respect, the task design is supposed to take into account explicitly some Vygotskian approaches occurring in the conceptualisation process (Vygotsky, 1986).

In another paper, Hogstad, Isabwe, and Vos (2016) use the software Sim2Bil which combines a simulation of two cars, velocity graphs, and an input for velocity functions. Their focus is on how engineering students use visualisations in their mathematical communication. This perspective is more socio-cultural. This means that the tool is no longer mostly a medium for students to construct new knowledge with well-designed tasks but it is rather a medium through which students can communicate knowledge in the social context, together with language, gesture, and mathematical representations. The students are supposed to observe the movement of two cars and some representations about distances and velocities in the tool according to parameters they can choose. The task is to find the velocities of the two cars (v_1 and v_2 respectively) so that v_2 is half of v_1 when the cars reach the finish line simultaneously at four seconds. The students are then asked if they can prove that their answer is correct. This initiates collaboration between the students, giving them room to follow their own approaches. As the authors were interested in visualisation, they observe three different media for visualisations: first Sim2Bil-based visualisations (the cars running in the simulation window and the area under the graphs of the functions in the graph window), second some paper-based visualisations (sketches and detailed drawings), and third gestures, for instance mimicking the kinematic movement of cars or indicating a parabolic graph. They conclude that although the task could have been posed without Sim2Bil (by symbolical calculations) the tool is central in the task by offering visualisations. Such a tool offers rich opportunities to visualise mathematics and to connect different mathematical representations and applications.

In his paper, Hoppenbrock (2016) analyses the role of clicker tasks in triggering productive discussion in small groups of students (peer instruction) and in supporting the construction of new mathematical knowledge in the group. The study uses the notion of epistemological triangles (Steinbring, 2005) to analyse the discussion of one of six audio-recorded groups of three students in a large lecture hall of an analysis course. The clicker task that is analysed is a multiple-choice question in which students are presented with three choices for the formalised statement "f has a local maximum at x_0". Two of the choices involve changing the order of quantifiers. This question is expected to be challenging for students since, as discussed by Dubinsky and Yiparaki (2000), students have difficulty distinguishing between AE (for all there exist) and EA (there exist for all) statements. The focus of Hoppenbrock's paper is finding characteristics of clicker tasks that will promote significant discussion resulting in new knowledge construction. Based on Dekker and Elshout-Mohr (1998), Hoppenbrock proposes that tasks should be meaningful, complex, require different student abilities, and aim at level raising. The example he develops supports this proposal. The author emphasises the need to further investigate the interplay between the clicker tasks and the skills, motivation, and knowledge of the students in the group as well as the group dynamics. Finally, the implementation of similar clicker-based tasks for the definition of convergence of sequences or continuity is suggested as a possible pedagogical tool.

The article by Gaspar Martins (2018) documents the effect of using a weekly online quiz system during an entire semester for two courses: calculus of one variable and multivariable calculus. The data include responses from surveys answered by a large percentage of students in both courses. Survey results suggest that most students regularly used the quiz system, quizzes fomented studying regularly, helped students discover what material they did not know enough about, and helped students rate their own learning level and learn new material. Further, a large percentage of survey respondents reported that they found quizzes useful and fair, and that they perceived that quizzes helped them attain a higher grade. Indeed, comparative data from five academic years shows that semesters in which the quiz system was used resulted in higher passing rates on both courses. Gaspar Martins suggests the potential of task design for online quiz systems to help students keep up to date in their studies and improve passing rates.

Tasks on topics for other disciplines and preservice teacher training

Now we deal with the use of tasks about understanding of concepts from calculus and analysis (including complex analysis) of students who do not major in pure mathematics and preservice teachers in a lecture on mathematics in their study programme. These studies have different focus. The research project by Feudel (2016, 2018) deals with the interpretation of the derivative in the context of economics, González-Martín and Hernandes-Gomes (2018) work on applications of multivariable integration in the context of engineering, and Hanke and Schäfer (2018) establish a link between mathematics and mathematics education in a course on complex analysis. What is common in all these pieces of research is that the studies address mathematics in different, more or less applied contexts and tasks are used to cross boundaries between different professional practices (Akkerman & Bakker, 2011).

Feudel (2018) investigates students' understanding of the economic interpretation of the derivative of a cost function. The literature suggested that economics students, but also students of other disciplines that apply calculus concepts, have difficulties in interpreting the derivative $C'(x)$ of a cost function C (x measures the unit of production) in terms of recognising $C'(x)$ as a rate of change instead of absolute change and in terms of the corresponding units. Feudel (2018) describes that in the economic context, $C'(x)$ is often interpreted as the additional cost of the next unit since $C(x + h) - C(x) \approx hC'(x)$ for $h \approx 0$, and $h = 1$ could be considered small in economics (Feudel, 2016). However, this does not resolve the problem between rate of change and absolute change. In interviews, Feudel (2018) asks economics students to determine the marginal cost and the unit for a cost function C at a given point and to determine the relationship between $C'(x)$ and the additional cost when x is increased by one unit. Moreover, he asks explicitly about the connections the students see between the mathematical concept of derivative and the economic context. Feudel (2018) points out that the interviewer could react to the answers of the students and provoke cognitive conflicts (Tall & Vinner, 1981), and claims that it might be helpful to thematise the relationship between the mathematical notion of derivative and the economic notion of marginal cost by explicitly referring them to each other, not only introducing the students to each of the notions separately. The study suggests the potential benefit of tasks designed to help economics students link the mathematical notions and their applications or interpretations more closely.

González-Martín and Hernandes-Gomes (2018) provide a textbook analysis on how the first moment of area is presented in textbooks for engineering students. The approach is similar to Feudel's studies (2016; 2018) since González-Martín and Hernandes-Gomes also investigate how a notion of mathematics is related to the use of this notion in an applied field. Using the Anthropological Theory of the Didactic (ATD; Chevallard, 1999; Bosch & Gascón, 2014), the authors analyse tasks in calculus and mechanics to identify praxeologies around the determination of the moment of area $\iint_A ydA$. They also refer to Castela (2016) who claims that praxeologies used in mathematics may undergo a transposition when the mathematical notions are regarded from an applied science as an effect of boundary crossing. González-Martín and Hernandes-Gomes come to the conclusion that techniques to work with integrals of the given type, which are part of mathematics courses on this topic, are not necessarily required to be used in the applied context the students face in their courses on mechanics of materials, e.g., in their textbooks. From the first analysis of the mechanics textbook, the authors recognise that the first moment of area is defined as an integral like $\iint_A ydA$, but in the actual computation of moments of areas the evaluation of integrals is most of the time not necessary. Rather, techniques from geometry or previously derived formulae are used to evaluate the moments of area, and thus implicitly the corresponding integrals. To sum up, in this study tasks are not designed but current task design is used to identify mathematical praxeologies in calculus and engineering about the first moment of area.

The design research project "Spotlight-Y" (Hanke & Schäfer, 2018) belongs to the discussion of the "double discontinuity" in mathematics teacher training: the fragmentation between the courses on mathematics and mathematics education many preservice teachers experience within their study programs (Hefendehl-Hebeker, 2013; Winsløw & Grønbæk, 2014), mathematics teacher knowledge (Ball, Thames, & Phelps, 2008), and the "higher standpoint" on school mathematics that Klein (2016) called for. The authors describe the format of a lecture in complex analysis in which the last third of the lecture is split into two branches, one for pure mathematics students and one for preservice teachers. Tasks appear at different points. Firstly, the preservice teachers get the task to create and implement learning arrangements for gifted pupils from secondary schools about a phenomenon of the lecture on complex analysis, and thus, secondly, the preservice teachers develop tasks and learning materials themselves. The intention is that the preservice teachers experience the relevance of the mathematics they learn at university for their future career as teachers. But the tasks have another function. They create an institutional opportunity for the students to link the mathematics of their study programme with their courses on mathematics education, in this case a seminar on task design, which aims to help the students to prepare tasks. Thus, one can say that the initial task acts as an initial boundary object and the tasks and materials the preservice teachers develop constitute an emerging boundary object between university mathematics and mathematics education (Hanke & Bikner-Ahsbahs, 2019). Hanke's and Schäfer's empirical findings about two learning arrangements, one on differentiation as linearisation and one on Taylor series expansion, which were developed by their students and on the planning of which these students reflected, show that it is a challenging task for the preservice teachers to work on the preparation of a mathematically demanding phenomenon for pupils, and linkages between these topics of university mathematics and mathematics education did not happen automatically.

The results of the first and third research projects described in this section show that many students need assistance when they need to bring together mathematical knowledge with knowledge from profession-specific domains. It is not enough to teach the different perspectives next to each other.

Assessment of task design and students' task performance

The assessment of solutions to calculus tasks is often associated to the task design itself and has also been addressed in several INDRUM papers. Each of them draws on elements of well-established theories but provides adjustments and refinements needed to assess students' solutions.

The change of methods used in schools to methods required in university mathematics, one part of the transition from school to university, is investigated by Bloch and Gibel (2016). Based on TDS, they draw on a model to analyse adidactical situations and use it to analyse reasoning processes in 14 students' solutions to tasks about parametric curves and ordinary differential equations in an exam. The authors "classify the objects, signs and reasoning processes they [students] have to cope with during the resolution of calculus problems" (p. 44). The model includes the consideration of functions of reasoning, levels of argumentation and signs, and the distinction between three levels of didactic milieus: heuristic/exploration level, formulation/validation level, and institutionalisation level. Bloch and Gibel arrive at the conclusion that their students remain at low level because they often get stuck in calculations or do not adapt their previous knowledge. Some students try to work at higher level but do not succeed because, again, they either fail in calculations or do not manage to interpret their calculations. The authors also recognise that a holistic view on formulas does not seem very present; instead, parts of formulas are interpreted separately. Finally, the authors hypothesise that the phenomenon that students remain at low levels of reasoning is not directly linked to teachers' didactical choices, rather it is a lack of association between syntactic and semantic methods.

In his case study, Hanke (2018) investigates students' written proofs of a theorem in a topology course in terms of a refinement of "object layers" (Oerter, 1982; Schäfer, 2010). The task was to show that a function $\mathbf{R} \to \mathbf{R}$ is continuous if and only if its graph is path-connected. Relying on the RBC-model of the Abstraction in Context methodology (Dreyfus & Kidron, 2014), Hanke reconstructs the traces of the epistemic actions of recognising and building-with which are performed on the mathematical objects introduced in the course of the written proof attempts by the students, and how implications in the course of proof are achieved. The notions of "singular", "contextual" and "formal" layer of objects (Oerter, 1982) reflect the levels of concreteness of objects used. Schäfer (2010) applied these layers in mathematics education in a combination with RBC, and Hanke introduces the specifically mathematical "propositional layer" which includes theorems (e.g., "If $F: X \rightarrow Y$ is any function between a path-connected topological space X and any topological space Y, then the graph of F is path-connected if F is continuous", p. 48). The propositional layer is useful because the application of a mathematical proposition is very common in proofs and relies on the fact that the requirements in the proposition have to be recognised in advance, thus the important properties of the objects appearing in the course of proof have to be abstracted, in order to justify a building-with action on the propositional layer (cf. the similar approach in terms of "structuralist praxeologies" by Hausberger, 2018). Hanke shows that recognising and building-with actions on different layers were often successful, but also that propositional building-with actions can be carried out well by the students even if the prerequisites in a theorem have falsely been recognised to be valid.

Whereas Bloch and Gibel (2016), and Hanke (2018) propose new theoretical constructs that may be used to supplement established theories (TDS and AiC respectively) in the assessment of student task performance, other papers use existing theoretical constructs to further expound on the assessment of tasks, this time from the institutional point of view afforded by the ATD. Brandes and Hardy (2018) use the ATD to assess the tasks and activities of students in a multivariable calculus course. Student success in the course they consider depends largely on their performance in a final exam. This is reflected in the knowledge to be taught (KT, mathematical knowledge listed in the sections and exercises in the course outline) and in the knowledge to be learned (KL, the subset of KT that students need to learn in order to provide solutions in final exams), which can be used to describe what is effectively done in the course. Essentially, Brandes and Hardy argue that the technological-theoretical component of the KT praxeology disappears in the KL praxeology which will now be restricted to a practical block which is a deformation of the practical block in the KT praxeology. This is in the sense that in the KL praxeology students are limited to tasks that promote reasoning based on identification of similarities (IS), as described by Lithner (2004). IS reasoning merely requires students to recognise a type of task and to apply a similar known technique, circumventing the need for technological-theoretical arguments in problem solving. Thus, the course loses its original purpose of bridging the gap between pre-university mathematics and more advanced courses as described by the transition types of Winsløw, Barquero, Vleeschouwer, and Hardy (2014), where students would first transition from praxeologies restricted mainly to practical blocks to praxeologies that begin to incorporate a theoretical block, and then transition to praxeologies where past theoretical blocks turn into practical blocks that include theorem proving. Citing these transition types, Broley and Hardy (2018) argue in a similar paper for the need to investigate students' transition from introductory calculus and multivariable calculus courses that may be modelled with praxeologies with a missing theoretical component, to the introductory analysis courses where students are expected to work with or on the theoretical component. They discuss the plan to construct a praxeological model of the knowledge to be learned in introductory mathematical analysis courses. Both papers use the ATD to discuss the types of tasks that students actually face in the multivariable calculus and analysis courses and implicitly uncover the courses' need of a deep re-structuring that will perhaps require the design of Research and Study Paths as advocated in the ATD, or similar such restructuring as described in other theories.

Research with or suggesting task design

In this part we consider papers where the results suggest the need to design or renew the design of tasks for the teaching of calculus and analysis. In contrast to the first part of this chapter, task design has still to reach classroom use and is mainly put forth by the reported research. Some of these papers are based on a cognitive theory and support the development of tasks to help students perform mental constructions in which they have shown difficulty, or of tasks to take advantage of unexpected constructions that students seem able to do. In some papers, tasks are components of the research itself and reveal specific difficulties that students encounter with calculus and analysis concepts. In contrast to many of the articles considered in the first part of this chapter, these tasks are not part of the design of Didactical Situations (TDS) or of Research and Study Paths (ATD), but rather directly address student behaviour when solving problems involving a specific mathematical notion. Of course, results of cognitive studies may be used in conjunction with other task design perspectives as shown, for example, in Salgado and Trigueros (2015) where modelling activities are interweaved with APOS-based tasks (Action-Process-Object-Schema; Arnon et al., 2014). Other papers in this section base the need of task design on epistemological studies, the analysis of institutional documents, or questionnaires. In these cases, the research results suggest that tasks be designed as part of didactical situations or couched in a modelling context.

Task design as one research component

In addition to derivatives, continuity and limits for real-valued functions, addressed in the majority of papers about education in calculus and analysis, some more topologically oriented topics have also been dealt with by the INDRUM participants. Durand-Guerrier and Vivier (2016) investigated students' usage of density and completeness of the real numbers, Branchetti and Durand-Guerrier (2018) looked at concept images for the ordering of subsets of real numbers, and Hamza and O'Shea (2016) studied students' concept images and definitions of open sets in metric spaces. These papers share that tasks were designed for their research purpose instead of a teaching purpose. Chorlay's (2018) questionnaire, on the other hand, was originally intended to be part of a larger project about the formalisation of limit from the perspective of didactical engineering.

Durand-Guerrier and Vivier (2016) describe the first stage of on-going research into students' learning of the notion of density and completeness of subsets of the real numbers, and how this impacts students' understanding of the main ideas of analysis. They analysed some questions in a questionnaire to 35 students of a first course on analysis which covered the notions of continuity of functions and sets (all proper cuts of which are Dedekind cuts), (order) density and discreteness taking into account the limits of visualisation in analysis. An important set in their tasks is the set **D** of decimal numbers which is dense. The students are asked to give examples of intervals which contain exactly one (or two, respectively) natural numbers, and the same for decimal numbers. Also, they should answer whether $f(x) = 2$ necessarily has a solution in **N**, **D**, **Q** or **R** when the function f was given by its graph. Or, in a task about approximation, it was asked whether one can deduce from a table with function values, a discrete graph of the table values, and an additional requirement for the function (such as piecewise monotonicity or continuity, respectively) whether the function has one fixed point, none or several. Durand-Guerrier and Vivier claim that an adequate conceptualisation of density is important for the introduction of the real numbers as the completion of the rational numbers, and the difference between being dense in itself and completeness is fundamental for the beginnings of analysis and the intermediate value theorem. Their tasks show that close to one third of the students did not observe the density of **D** (or its usefulness in the application to the tasks), possibly due to the misunderstanding of **D** as one of the sets $\{k/10^n : k \in \mathbf{N}\}$ for some natural number n, or not recognising that there are subsets of the real numbers which are neither discrete nor continuous.

Branchetti and Durand-Guerrier (2018) took a closer look at the interaction between university students' and teachers' concept images of ordered, order dense and denumerable sets. The authors argue that concept images of one of the notions of ordering, order density and discreteness can cause cognitive conflict when dealing with another, and they aim to identify the interplay between those concept images more closely. The tasks were (paraphrased): Why can **Q** be ordered even if **Q** is order dense (i.e. consecutive rational numbers do not exist), and why is there a bijection between **N** which is discrete and **Q** which is dense? The tasks Branchetti and Durand-Guerrier prepared are guided by historical and epistemological analyses of the concepts, and were given to master students and teachers. These questions were also embedded into a didactical scenario, i.e. the master students should indicate how they would explain the answers to the questions to pupils and the teachers how they would address these questions in class. As a result, teachers often relied on definitions, whereas some of the students had difficulties in relating the concepts to each other. The authors argue that the latter is due to their concept images and the focus on the definitions of the concepts will not be sufficient to overcome the compartmentalisation (Tall & Vinner, 1981) in students' concept images. Further, Branchetti and Durand-Guerrier claim that "this lack of awareness of the links between the concepts of density-in-itself and order of **Q** might prevent them [French master students] from designing appropriate learning situations, once they pass the selection procedure exam and become teachers" (p. 22).

Hamza and O'Shea (2016) start off the research on students' concept images and concept definitions of open sets in metric spaces. This research belongs to the transition from real analysis to advanced courses. The theoretical background of their research is formed by the idea of concept image and concept definition (Tall & Vinner, 1981) and their research is grounded by the claim that concept images are formed in early contact with the notions at hand, referencing "met-befores" by McGowen and Tall (2010) and "tacit models" by Fischbein (1989). Ten students from a course on metric spaces were interviewed. The design of the tasks in the interviews incorporates different activities (defining, explaining, solving exercises). The participants were asked to define open sets and how to explain the notion to a friend (this is similar to the didactical situation Branchetti and Durand-Guerrier (2018) offer to their participants). Further, among the interview tasks, the students had to argue, whether $B = \{m - 1, m, m + 1\}$ (for some integer m) is an open ball in $\mathbb{Z}$ with the standard metric, describe a certain metric on the set of all real sequences in their own words, and argue whether $\{(a_n): a_1 = 0\}$ is open in the resulting metric space. For instance, Hamza and O'Shea find that the students rely on the definition or a characterisation from the lecture, often paraphrased in their own words, but sought for explanations using non-mathematical language (e.g., not reaching the boundary of a set). In the solution processes of the tasks, many students used generic geometrical pictures like circles to visualise open balls, some did not identify B to be open because they got mixed up with open balls in the set of real numbers and properties like connectedness, and some students had difficulties to find a visualisation for the sets in question and relied on the definitions of metric and open ball. However, such tasks as those in Hamza's and O'Shea's study provide opportunities to make students aware of their concept images and enable them to express these.

The questionnaire by Chorlay (2018) asks to identify true or false statements about sequences of real numbers diverging to infinity, e.g., "If lim $u_n = +\infty$ then $\forall A \in R \exists n_A \in N$ such that $u_{n_A} \geq A$" (p. 28). The main objective is to get insight into students' recognition and understanding of quantifier usage in propositions given through infinite limits. The students were also asked to state the converses of the propositions and to justify why they considered which of these statements to be wrong. It turned out that several students did not validate the implication given in the tasks but rather tried to adapt the right-hand side of the implication to fit the definition of infinite limit. This was done often by grade 12 students and less so by university students who had been exposed to logic before. An interpretation might be that students who adapt the original task to a comparison with the definition of infinite limits in fact know the definition and use it as a "template against which other quantified formulae ought to be contrasted" (p. 28). In summary, Chorlay's research shows that tasks can not only be misinterpreted by the participants but that these misinterpretations may be caused by "the level of familiarity with predicate calculus" (p. 32), suggesting that other curricular issues not related to analysis are relevant for students' engagement in these tasks.

Research suggesting task design

In two APOS-based papers, a detailed conjecture (genetic decomposition) of mental constructions students may use to understand particular topics in the multivariable calculus is posed and tested with student interviews.

In one of these papers, Trigueros, Martínez-Planell, and McGee (2018) discuss how students' knowledge about the total differential seemed to be unrelated to their knowledge about the tangent plane. An initial genetic decomposition (a priori conjecture of possible mental constructions students may do to understand a mathematical notion) was proposed. Study results show that students have difficulty with some of the conjectured constructions and that they also make some unexpected constructions. In particular, most students did not relate vertical change on a tangent plane at a point (a,b) with the total differential at that point, df(a,b). Students did not relate processes on the function, like partial derivatives, vertical change, and directional derivatives, with the corresponding processes on the tangent plane. Further, students did not think of the total differential at a point as a function on the independent variables dx and dy. Overall, students showed much difficulty converting between different registers of representation (Duval, 2006). These results underscored the need to refine the genetic decomposition. The refined decomposition may be seen as suggesting corresponding task design to help students do the conjectured constructions. In a follow-up to their 2016 paper, Trigueros, Martínez-Planell, and McGee (2018) take advantage of the dialogue between APOS and the ATD (Bosch, Gascón, & Trigueros, 2017) to examine students' individual and institutional constructions of the total differential, thus exemplifying the possible interplay between these two very different theories.

In another APOS-based paper, Trigueros and Martínez-Planell (2018) explore students' understanding of the relation between integrals of functions of two variables defined on rectangles and corresponding Riemann sums. The authors only consider a partial genetic decomposition of students' possible constructions of the integral of a function of two variables. Their findings show that students that had finished a traditional multivariable calculus course failed to interpret geometrically even the most basic ideas relating Riemann sums to double integrals. The students showed difficulty visualising the relationship between the graph of the function and its rectangular domain, both when the function was given graphically and symbolically. They also showed difficulty interpreting geometrically a term $f(a,b)\Delta x\Delta y$ of a Riemann sum as the volume of a rectangular prism, and thus, as may be expected, could not distinguish in simple cases when such a term would be an underestimate, an overestimate, or produce the exact value of the double integral over the corresponding rectangle. Also, most students were not able to geometrically interpret a four-term Riemann sum, being limited to saying as a memorised phrase lacking geometric meaning that it was an approximation of the double integral. This paper underscores the need of task design to help students visualise the geometric meaning of ideas relating Riemann sums and double integrals.

Vandebrouck and Leidwanger (2016) examine students' computations of limits in a paper based on the notions of iconic and non-iconic visualisation (Duval, 1999), and deconstructions with a pointwise, local, or global perspective (Deconstruction With Perspectives, DWP; Montoya Delgadillo, Páez Murillo, Vandebrouck, & Vivier, 2018). In Vandebrouck's and Leidwanger's paper, students' work suggests that algebraic difficulties constrain their limit computations leading them to error and to the inability to treat some limits using results on indeterminate forms. In some of these cases, students were able to successfully do some DWP to compute limits. For example, some students would avoid the use of results pertaining to indeterminate forms and algebraic manipulations by arguing that since $\exp(x)$ grows much faster than $\ln(x)$, the limit at infinity of $\exp(-x)\ln(x)$ is 0, or that since x^2 grows faster than x, the infinite limit of $(x^2-1)/(x+3)$ is $+\infty$. Students also showed misconceptions when trying to apply DWP. For example, some students would erroneously argue that the limit at infinity of $\exp(x)(1 - \sqrt{x})$ is $+\infty$ since the exponential dominates, not taking into account that the exponential occurs in a product where the other factor will be negative. Another type of difficulty shown by students in their limit computations is the tendency to mix together the local and pointwise perspectives. So for example, a student might erroneously argue that the limit at 0 of $\sin(x)/x$ is 0 since, for the numerator, $\sin(0)$ is 0 (pointwise perspective) while the denominator x only tends to 0 (local perspective). Vandebrouck and Leidwanger observe that students' work on the limit problems suggest that tasks could be designed to take advantage of the knowledge of DWP that students bring from secondary school, so that they can use DWP with understanding and avoid mistakes, and so that they will be able to compute limits using DWP as an expert would.

Dealing with the same idea of perspectives while also considering the role of representation registers (Duval, 2006), Páez Murillo, Montoya Delgadillo, and Vivier (2016) discuss students' spontaneous conceptions of the notion of tangent line. They analysed the written responses given by first year university students in Chile, Mexico, and France to one question: For you, what is a tangent line? Based on previous research, the authors propose five types of conceptions of tangent line: unique point of intersection (UPI), perpendicular to a ray (PR), slope (S), process of derivation (PD), and horizontal tangent (HT). These are presented as two blocks, one coming from the domain of geometry (UPI, PR), and directly related to the tangent line to a circle at a given point; and the other coming from the domain of analysis and directly related to derivation (S, PD, HT). The authors then describe how the punctual, local, and global perspectives may appear and interrelate in each conception type while also considering registers of representation. They suggest to interpret the geometric and the analytic conception blocks as part of Kuzniak's (2011) Mathematical Working Space (Espace de Travail Mathématique, ETM), relating them to two of the paradigms of standard analysis (geometrical-arithmetic and calculating paradigms) as described by Kuzniak,

Montoya Delgadillo, Vandebrouck, and Vivier (2016) and Kuzniak, Tanguay, and Elia (2016). They argue that since ETMs take into account representations, visualisation, theoretical knowledge, and their evolving interrelationship, they may allow to better explain the relation between tangent line conceptions, perspectives, and representations. They finish with what may be taken as a call to design tasks that take into account students' spontaneous conceptions of tangent lines to allow students to refine their own ideas and to avoid limiting students' work to the calculating paradigm.

Kouki, Belhaj Amor, and Hachaichi (2016) study the teaching and learning of Taylor series expansions (développements limités). They start by presenting a historical-epistemological analysis based on Douady's (1986) interplay between different settings (jeux de cadre) and the tool-object dialectic. In their didactic analysis, Kouki et al. also take into account the semantic/syntactic dimensions (Kouki & Ghedamsi, 2012) as well as the semiotic dimension (Duval, 1993). Kouki et al. argue that in the historical course of the development of the notion of an infinite series expansion several phases can be distinguished. They go on to show that different techniques were developed by mathematicians in transiting across these phases which required the articulation of the syntactic and semantic dimensions while also considering the geometric, algebraic, numeric, graphical, and analytic registers. The epistemological analysis by Kouki et al. also shows how the tool-object dialectic is reflected in the evolution of the Taylor series concept. Then they go on to analyse institutional documents, like programmes, manuals, and course handouts. In contrast, here they find that explorations into the semantic dimension of Taylor series are lacking and that there are also limitations in the mobilisation of the graphical and numerical registers. This suggests the development of tasks to help better articulate the semantic and syntactic dimensions, particularly concerning numerical and graphical work. Task design would use modelling and a construction of the Taylor series notion guided by the tool-object dialectic.

Bergé (2016) studies the evolution of students' understanding of the role of the supremum as a tool in developing basic properties of analysis. The study is part of a larger project that uses the ATD as framework. Like Kouki et al. (2016), it makes reference to ideas of Douady (1986), particularly the tool-object dialectic, and starts with an epistemological analysis of the evolution of the notion of interest, in this case the completeness of the real numbers. Following that, the response to a questionnaire given to students who had been introduced to completeness through the notion of supremum in a four-course sequence of calculus and analysis were considered. The study concentrates on the analysis of one of the questions in which students were asked about the utility of the notion of supremum. Results show that most students in the second course of the sequence (multivariable calculus) either give erroneous answers, seem to be unaware of the utility of the supremum, or (the larger percentage) have a restricted understanding of supremum only as an upper bound. Students in the third course of the sequence have an increased awareness of the role of the supremum in the proof of some of the basic theorems of analysis and by the fourth course most of them are able to give specific examples of its use. These students had re-encountered the notion of supremum in the third and fourth courses as a tool to define or determine specific numbers. This suggests the need to find a situation that will integrate the use of completeness as a tool in the mathematical organisation used to introduce the real numbers, so that students will experience the reason of being of the supremum and completeness early in the sequence of calculus and analysis courses.

Kidron (2018) investigates 10th, 11th, and 12th year students' understanding of infinite non-repeating decimals. She used a questionnaire in which a total of 188 students were to comment on an imagined discussion between two students: one who argued that it was possible to form infinite non-repeating decimal numbers (by writing digits at random after the decimal point) and the other one who argued that any infinite decimal must be formed by dividing two whole numbers. Among other results, she finds that although around 80% of the students claimed that they had studied irrational numbers in class, 55% of the students believed that any decimal (finite or infinite) is the result of the division of two whole numbers, and only 19% of students believed that an infinite non-repeating decimal is not the result of the division of two whole numbers. Kidron proposes that students' reluctance to deal with irrational numbers is due to their intuition that the existence of numbers is beyond their control, their request to define irrational numbers by means of rational numbers alone, and the need to know the process that leads to the infinite decimal expansion. This explanation is consistent with and suggests the need of task design at university level dealing with the construction of the real numbers, such as that presented in the first part of this chapter.

Conclusion

The 2016 and 2018 INDRUM meetings show some patterns in the reported task design for analysis and calculus, and suggest possible future trends in task design. A large proportion of the research in the analysis and calculus working groups considered the problem of formalisation of basic notions, most notably the construction of the real numbers, supremum, convergence of sequences, and continuity. Given that many of these papers reported initial steps in this research direction, or implementation attempts that are not yet completed, one may expect that these topics will continue to receive as much attention in the near future.

Many of the papers attending to the basic aspects of analysis were based on the Theory of Didactical Situations, although some were based on the Anthropological Theory of the Didactic, or on Douady's (1986) tool-object dialectic. Many of the TDS and ATD papers appeared mostly in the form of epistemological analyses that precede and suggest the design of situations, or in the evaluation of implementations of such situations. Some papers were based on cognitive theories, mainly APOS and the notion of DWP (deconstructions with perspectives, Montoya Delgadillo et al., 2018), and their results suggest the need to design tasks to improve student understanding of some specific ideas. Other theoretical frameworks like Abstraction in Context (AiC) and the Mathematical Working Space model (Kuzniak,

2011) were also present. Semiotic challenges appeared in many of the reported research and this led to the frequent incorporation of Duval's ideas (1999; 2006) within other theoretical frameworks.

In what has been a growing tendency for some time (see for example Sfard and Cobb, 2014), research considering the role of socio-cultural aspects in the design and assessment of tasks was also widely represented, and this may be expected to continue as the role of teachers, student group work, and open class discussions are further explored.

The incorporation of different types of technology in the design and implementation of student-centred tasks may be predicted to be another direction of future research growth. While the number of papers that directly dealt with the use of technology was not large, it is only natural to expect an eventual shift in that direction as the everyday use of technological artefacts by teachers and students continues to increase and the potential and advantages of using technology for the teaching and learning of mathematics is further explored, documented, and discussed within the community.

In terms of the mathematical content of the research, much attention was given to foundational and basic aspects of analysis, and relatively less attention was given to a variety of other topics in the differential and integral calculus, multivariable calculus, differential equations, complex analysis, and topology.

Research reported in this chapter considered task designs for student use or suggested the need or improvement of such designs, as when incorporating technology, for example. We also considered research on theoretical issues dealing with task assessment and, in what may be yet another growing tendency, research on calculus and analysis for students of other disciplines, including future teachers.

REFERENCES

Akkerman, S. F., & Bakker, A. (2011). Boundary crossing and boundary objects. *Review of Educational Research*, 81(2), 132–169.

Arnon, I., Cottrill, J., Dubinsky, E., Oktaç, A., Roa Fuentes, S., Trigueros, M., & Weller, K. (2014). *APOS theory. A framework for research and curriculum development in mathematics education*. New York, NY: Springer.

Artigue, M. (2014). Perspectives on design research: The case of didactical engineering. In A. Bikner-Ahsbahs, C. Knipping, & N. Presmeg (Eds.), *Approaches to qualitative research in mathematics education. Examples of methodology and methods* (pp. 467–476). Dordrecht, Netherlands: Springer.

Ball, D., Thames, M. H., & Phelps, G. (2008). Content knowledge for teaching. What makes it special? *Journal of Teacher Education*, 59(5), 389–407.

Bergé, A. (2016). Le rôle de la borne supérieure (ou supremum) dans l'apprentissage du système des nombres réels. *INDRUM 2016 Proceedings*, pp. 33–42.

Bkouche, R. (1996). Points de vue sur l'enseignement de l'analyse: des limites et de la continuité dans l'enseignement. *Repères IREM*, 24, 67–76.

Bloch, I., & Gibel, F. (2016). A model to analyse the complexity of calculus knowledge at the beginning of university course. *INDRUM 2016 Proceedings*, pp. 43–52.

Bosch, M., & Gascón, J. (2014). Introduction to the Anthropological Theory of the Didactic (ATD). In A. Bikner-Ahsbahs & S. Prediger (Eds.), *Networking of theories as a research practice in mathematics education* (pp. 67–83). Cham, Switzerland: Springer.

Bosch, M., Gascón, J., & Trigueros, M. (2017). Dialogue between theories interpreted as research praxeologies: the case of APOS and the ATD. *Educational Studies in Mathematics*, 95(1), 39–52.

Branchetti, L., & Durand-Guerrier, V. (2018). Are all denumerable sets of numbers order-isomorphic?*INDRUM 2018 Proceedings*, pp. 12–21.

Brandes, H., & Hardy, N. (2018). From single to multi-variable calculus: a transition? *INDRUM 2018 Proceedings*, pp. 477–486.

Bridoux, S. (2016). Introduire la notion de convergence avec une ingénierie des années 1980: rêve ou réalité didactique pour l'enseignant?*INDRUM 2016 Proceedings*, pp. 53–62.

Broley, L., & Hardy, N. (2018). A study of transitions in an undergraduate mathematics program. *INDRUM 2018 Proceedings*, pp. 487–496.

Brousseau, G. (1997). *Theory of didactical situations*. Dordrecht, Netherlands: Kluwer.

Castela, C. (2016). When praxeologies move from an institution to another: an epistemological approach to boundary crossing. In R. Göller, R. Biehler, R. Hochmuth, & H.-G. Rück (Eds.) *Proceedings of the KHDM Conference: Didactics of Mathematics in Higher Education as a Scientific Discipline* (pp. 153–161). Kassel, Germany: Universitätsbibliothek Kassel.

Chevallard, Y. (1999). L'analyse des pratiques enseignantes en théorie anthropologique du didactique. *Recherches en Didactique des Mathématiques*, 19(2), 221–266.

Chevallard, Y. (2015). Teaching mathematics in tomorrow's society: A case for an oncoming counter paradigm. In S. J. Cho (Ed.), *Proceedings of the 12th International Congress on Mathematical Education* (pp. 173–188). Cham, Switzerland: Springer.

Chorlay, R. (2018). An empirical study of the understanding of formal propositions about sequences, with a focus on infinite limits. *INDRUM 2018 Proceedings*, pp. 24–33.

Dekker, R., & Elshout-Mohr, M. (1998). A process model for interaction and mathematical level raising. *Educational Studies in Mathematics*, 35(3), 303–314.

Douady, R. (1986). Jeux de cadres et dialectique outil-objet. *Recherches en Didactique des Mathématiques*, 7(2), 5–31.

Dreyfus, T., & Kidron, I. (2014). Introduction to abstraction in context (AiC). In A. Bikner-Ahsbahs & S. Prediger (Eds.), *Networking of theories as a research practice in mathematics education* (pp. 85–96). Cham, Switzerland: Springer.

Dubinsky, E. (1991). Reflective abstraction in advanced mathematical thinking. In D. Tall (Ed.), *Advanced mathematical thinking* (pp. 95–123). Dordrecht, Netherlands: Kluwer.

Dubinsky, E., & Yiparaki, O. (2000). On student understanding of AE and EA quantification. In E. Dubinsky, A. H. Schoenfeld, & J. Kaput (Eds.), *Research in collegiate mathematics education IV* (pp. 239–289). Providence, RI: American Mathematical Society.

Durand-Guerrier, V., & Vivier, L. (2016). Densité de D, complétude de R et analyse réelle. Première approche. *INDRUM 2016 Proceedings*, pp. 142–152.

Duval, R. (1993). Registres de représentation sémiotique et fonctionnement cognitif de la pensée. *Annales de Didactique et de Sciences Cognitives*, 5, 37–65.

Duval, R. (1999). Representation, vision and visualization: Cognitive functions in mathematics thinking. Basic issues for learning. In F. Hitt & M. Santos (Eds.), *Proceedings of the Twenty First Annual Meeting of the North American Chapter of the International Group for the Psychology of Mathematics Education. Vol. 1 (October 23–26, 1999)* (pp. 3–26). Columbus, OH: ERIC Clearinghouse for Science, Mathematics, and Environmental Education.

Duval, R. (2006). A cognitive analysis of problems of comprehension in a learning of mathematics. *Educational Studies in Mathematics*, 61(1), 103–131.

Feudel, F. (2016). Relevant knowledge concerning the derivative concept for students of economics – A normative point of view and students' perspectives. *INDRUM 2016 Proceedings*, pp. 181–190.

Feudel, F. (2018). C'(x) = C(x+1)-C(x)? – Students' connections between the derivative and its economic interpretation in the context of marginal cost. *INDRUM 2018 Proceedings*, pp. 95–104.

Fischbein, E. (1989). Tacit models and mathematical reasoning. *For the Learning of Mathematics*, 9(2), 9–14.

Gaspar Martins, S. (2018). Weekly homework quizzes as formative assessment for engineering students are a fair and effective strategy to increase learning?*INDRUM 2018 Proceedings*, pp. 105–114.

Ghedamsi, I. (2016). Entrée des étudiants dans l'analyse formelle de début d'université: Potentialité des méthodes numériques d'approximation. *INDRUM 2016 Proceedings*, pp. 63–72.

Ghedamsi, I., & Lecorre, T. (2018). Students' understanding of ε-statements involving equality and limit. *INDRUM 2018 Proceedings*, pp. 34–43.

González-Martín, A. S., Bloch, I., Durand-Guerrier, V., & Maschietto, M. (2014). Didactic situations and didactical engineering in university mathematics: cases from the study of calculus and proof. *Research in Mathematics Education*, 16(2), 117–134.

González-Martín, A. S., & Hernandes-Gomes, G. (2018). The use of integrals of Mechanics of Materials textbooks for engineering students: the case of the first moment of an area. *INDRUM 2018 Proceedings*, pp. 115–124.

Hamza, S., & O'Shea, A. (2016). Concept images of open sets in metric spaces. *INDRUM 2016 Proceedings*, pp. 113–122.

Hanke, E. (2018). "A function is continuous if and only if you can draw its graph without lifting the pen from the paper" – Concept usage in proofs by students in a topology course. *INDRUM 2018 Proceedings*, pp. 44–53.

Hanke, E., & Bikner-Ahsbahs, A. (2019). Boundary crossing by design(ing): a design principle for linking mathematics and mathematics education in preservice teacher training. In U. T. Jankvist, M. van den Heuvel-Panhuizen, & M. Veldhuis (Eds.), *Proceedings of the Eleventh Congress of the European Society for Research in Mathematics Education (CERME11, February 6–10, 2019)* (pp. 3046–3053). Utrecht, Netherlands: Freudenthal Group & Freudenthal Institute, Utrecht University and ERME.

Hanke, E., & Schäfer, I. (2018). Learning complex analysis in different branches – Project Spotlight-Y for future teachers. *INDRUM 2018 Proceedings*, pp. 54–63.

Hausberger, T. (2018). Structuralist praxeologies as a research program on the teaching and learning of abstract algebra. *International Journal of Research in Undergraduate Mathematics Education*, 4(1), 74–93.

Hefendehl-Hebeker, L. (2013). Doppelte Diskontinuität oder die Chance der Brückenschläge. In C. Ableitinger, J. Kramer, & S. Prediger (Eds.), *Zur doppelten Diskontinuität in der Gymnasiallehrerbildung. Ansätze zu Verknüpfungen der fachinhaltlichen Ausbildung mit schulischen Vorerfahrungen und Erfordernissen* (pp. 151–168). Wiesbaden, Germany: Springer.

Hogstad, N. M., Isabwe, G. M. N., & Vos, P. (2016). Engineering students' use of visualizations to communicate about representations and applications in a technological environment. *INDRUM 2016 Proceedings*, pp. 211–220.

Hoppenbrock, A. (2016). Multiple choice questions and peer instruction as pedagogical tools to learn the mathematical language. *INDRUM 2016 Proceedings*, pp. 454–463.

Klein, F. (2016). *Elementary mathematics from a higher standpoint* (Vol. 1). Berlin, Germany: Springer.

Kidron, I. (2018). Students' conceptions of irrational numbers. *International Journal of Research in Undergraduate Mathematics Education*, 4(1), 94–118.

Kondratieva, M., & Winsløw, C. (2018). Klein's plan B in the early teaching of analysis: Two theoretical cases of exploring mathematical links. *International Journal of Research in Undergraduate Mathematics Education*, 4(1), 119–138.

Kouki, R., & Ghedamsi, I. (2012) Limite des méthodes syntaxiques en algèbre du secondaire. In J.-L. Dorier & S. Gousseau Coutat (Eds.), *Enseignement des mathématique et contrat social: enjeux et défis pour le 21ème siècle. Actes du Colloque EMF 2012 (Genève, 3–7 février 2012)* (pp. 435–444). Geneva, Switzerland: Université de Genève.

Kouki, R., Belhaj Amor, F., & Hachaichi, Y. (2016). Comparaison entre l'évolution historique ayant mené aux développements limités et leur pratique d'enseignement au début de l'université: Entre syntaxe et sémantique. *INDRUM 2016 Proceedings*, pp. 123–132.

Kuzniak, A. (2011). L'espace de travail mathématique et ses genèses. *Annales de Didactique et de Sciences Cognitives*, 16, 9–24.

Kuzniak, A., Montoya Delgadillo, E., Vandebrouck, F., & Vivier, L. (2016). Le travail mathématique en analyse de la fin du secondaire au début du supérieur: identification et construction. In Y. Matheron *et al.* (Eds.), *Enjeux et débats en didactique des mathématiques. XVIIIe école d'été de didactique des mathématiques* (pp. 47–66). Grenoble, France: La Pensée Sauvage.

Kuzniak, A., Tanguay, D., & Elia, I. (2016). Mathematical working spaces in schooling: an introduction. *ZDM*, 48(6), 721–737.

Lecorre, T. (2016). Rationality and concept of limit. *INDRUM 2016 Proceedings*, pp. 83–92.

Lithner, J. (2004). Mathematical reasoning in calculus textbook exercises. *The Journal of Mathematical Behavior*, 23(4), 405–427.

McGowen, M. A., & Tall, D. (2010). Metaphor or met-before? The effects of previous experience on practice and theory of learning mathematics. *Journal of Mathematical Behaviour*, 29(3), 169–179.

Montoya Delgadillo, E., Páez Murillo R., Vandebrouck F., & Vivier L. (2018). Deconstruction with localization perspective in the learning of analysis. *International Journal of Research in Undergraduate Mathematics Education*, 4(1), 139–160.

Oerter, R. (1982). Interaktion als Individuum-Umwelt-Bezug. In E. D. Lantermann (Ed.), *Wechselwirkungen* (pp. 101–127). Göttingen, Germany: Verlag für Psychologie.

Páez Murillo, R. E., Montoya Delgadillo, E., & Vivier, L. (2016). Conceptions spontanées et perspectives de la notion de tangente pour des étudiants de début d'université. *INDRUM 2016 Proceedings*, pp. 103–112.

Robert, A. (1982). L'acquisition de la notion de convergence des suites dans l'enseignement supérieur. *Recherches en Didactique des Mathématiques*. 3(3), 307–341.

Robert, A. (1983). L'enseignement de la convergence des suites numériques en DEUG. *Bulletin de l'APMEP*, 340, 431–449.

Robert, A., & Hache, C. (2013). Why and how to understand what is at stake in a mathematics class. In F. Vandebrouck (Ed.), *Mathematics Classrooms: Students' activities and teachers' practices* (pp. 23–74). Rotterdam, Netherlands: Sense.

Rogalski, M. (2016). Revenir au sens de la notion de limite par certaines de ses raisons d'être: un chantier pour le début de l'analyse à l'université. *INDRUM 2016 Proceedings*, pp. 133–142.

Salgado, H., & Trigueros, M. (2015). Teaching eigenvalues and eigenvectors using models and APOS Theory. *The Journal of Mathematical Behavior*, 39, 100–120.

Schäfer, I. (2010). Towards a refinement of epistemic actions: A case study of processes of recognition on spirolaterals. In M. F. Pinto & T. F. Kawasaki (Eds.), *Proceedings of the 34th*

Conference of the International Group for the Psychology of Mathematics Education, Vol. 4 (pp. 161–168). Belo Horizonte, Brazil: PME.

Sfard, A., & Cobb, P. (2014). Research in Mathematics Education: What can it teach us about Human Learning? In R. Sawyer (Ed.), *The Cambridge Handbook of the Learning Sciences* (pp. 545–564). Cambridge, MA: Cambridge University Press.

Sghaier, B., & Vandebrouck, F. (2018). Teaching and learning continuity with technologies. *INDRUM 2018 Proceedings*, pp. 74–83.

Steinbring, H. (2005). *The construction of new mathematical knowledge in classroom interaction. An epistemological perspective* (Mathematics Education Library Vol. 38). Berlin/New York: Springer.

Tall, D., & Vinner, S. (1981). Concept image and concept definition in mathematics with particular reference to limits and continuity. *Educational Studies in Mathematics*, 12(2), 151–169.

Trigueros, M., & Martínez-Planell, R. (2018). On students' understanding of Riemann sums of integrals of functions of two variables. *INDRUM 2018 Proceedings*, pp. 64–73.

Trigueros, M., Martínez-Planell, R., & McGee, D. (2018). Student understanding of the relation between tangent plane and the total differential of two-variable functions. *International Journal of Research in Undergraduate Mathematics Education*, 4(1), 181–197.

Vandebrouck, F., & Leidwanger, S. (2016). Students' visualization of functions from secondary to tertiary level. *INDRUM 2018 Proceedings*, pp. 153–162.

Vergnaud, G. (2009). The theory of conceptual fields. *Human Development*, 52, 83–94.

Vygotsky, L. (1986). *Thought and language*. Cambridge, MA: MIT Press.

Williams, S. R. (1991). Models of limit held by college calculus students. *Journal for Research in Mathematics Education*, 22(3), 219–236.

Winsløw, C. (2016). Angles, trigonometric functions, and university level analysis. *INDRUM 2016 Proceedings*, pp. 163–172.

Winsløw, C., Barquero, B., Vleeschouwer, M. D., & Hardy, N. (2014). An institutional approach to university mathematics education: from dual vector spaces to questioning the world. *Research in Mathematics Education*, 16(2), 95–111.

Winsløw, C., & Grønbæk, N. (2014). Klein's double discontinuity revisited: Contemporary challenges for universities preparing teachers to teach calculus. *Recherches en Didactique des Mathématiques*, 34(1), 59–86.

7

DISCRETE MATHEMATICS, COMPUTER SCIENCE, LOGIC, PROOF[1], AND THEIR RELATIONSHIPS

Faïza Chellougui, Viviane Durand-Guerrier and Antoine Meyer

This chapter relies on original contributions to the first two INDRUM conferences, summarizing their research lines, the theoretical and methodological issues raised, their results and perspectives. It aims at providing a synthesis of the research field as it appeared through the presentations and discussions at INDRUM. It presents some of the challenges faced by university students when dealing with discrete mathematics, computer science, logic, and proof and proving in their mathematical activity.

The chapter is structured into three main parts, all concerned with university levels. The first part describes the issues of teaching and learning discrete mathematics, combinatorics and their interactions with computer science. The second part presents the place and role of logic in defining, reasoning, proof and proving. The third part is devoted to the need for logic in mathematics and computer science.

Discrete mathematics and links with computer science

In this part, we first present reasons for teaching discrete mathematics at university level. Then we provide three examples of situations for teaching discrete mathematics, two of which stand at the interface with computer science.

Reasons for teaching discrete mathematics at university level

As pointed out by Ouvrier-Buffet, Meyer and Modeste (2018), discrete mathematics is a recent field linked to computer science. The authors underline the fact that teaching and learning discrete mathematics involves mathematical skills and heuristics (e.g. different kinds of proof and reasoning techniques, several ways of modelling etc.) and also develops objects, concepts, methods and tools that are relevant for computer science. This, in turn, raises new types of questions for mathematics. More generally, several aspects of discrete mathematics are at the interface between mathematics and computer science. An important example is the concept of algorithm (Modeste, 2012, 2016). Durand-Guerrier, Meyer and Modeste (2019) elaborate on the links between algorithms, proofs, and programs, and argue that the question of proof in mathematics and computer science is a central didactical issue, in particular at undergraduate levels. Considering that several topics are often gathered under the term *discrete mathematics*, Ouvrier-Buffet et al. (2018) claim that a way to characterize this field is to emphasize the specific modes of reasoning and the discrete nature of the structures that are involved. Knuth (1974, p.329) pointed out that many parts of mathematics use methods referring to discrete mathematics (e.g. group theory, number theory, geometry, algebraic combinatorics, graph theory, or cryptography) and wondered whether the traditional curriculum – calculus courses, etc. – should be revised to include more of these discrete aspects. The crucial role of discrete mathematics for developing heuristics and reasoning skills is pointed out by Abdallah (2018). She notes that in the debates around proof and proving, many examples in discrete mathematics are provided. Grenier and Payan (1998) provide evidence that discrete mathematics is relevant for changing the students' view on proof: discrete objects and situations are in general easily accessible and favour the devolution of problems; it is relevant in that context to explore particular simple cases, to solve them and then to move on to more sophisticated ones, and further to generalization. Alcock (2009) considers that such examples enhance the semantic development of mathematical concepts and proving skills. Nowadays, in some countries, such as France, the place of discrete mathematics in curricula is still not well established (Ouvrier-Buffet et al., 2018), contrary to other countries with a strong tradition such as Hungary, or the USA where it is part of curricular recommendations (Epp 2016). Ouvrier-Buffet et al. (2018) advocate that "discrete mathematics courses are relevant to a wide variety of majors at university level, including mathematics, number theory, computer science, and engineering" (ibid, p. 256). They claim that for countries such as France where the teaching of discrete mathematics is not well established, a key research question is "How to design curricula and didactical engineering[2] for the first university years?" (ibid, p. 258). In the next section, we present examples of situations for teaching discrete mathematics that were presented and discussed during the conferences, as a first step for dealing with this question.

Examples of situations for teaching discrete mathematics at university level

The brief overview presented in the previous section provides evidence that the crucial contribution of discrete mathematics to foster proof and proving skills in mathematics and in computer science is acknowledged by a number of researchers. This was defended and discussed during the two INDRUM conferences, relying on concrete examples of activities both in mathematics and in computer science. In this section, we present some of these examples and activities, trying to make clear the dialectic between proof and conceptualization, in the sense that working on proofs fosters conceptualization (Durand-Guerrier & Tanguay, 2018), and conversely deepening the appropriation of concepts enhances the capability to engage in proof.

Addressing generalization in discrete mathematics

Generalization is a key aspect of mathematics that fresh university students are expected to be able to grasp. Indeed, an important part of university courses (e.g. Calculus) consists in conjecturing and proving universal statements (theorems), most often using proofs by generic element, that are then used in proofs through inferences. In addition, students are expected to be able to engage by themselves in a proving process. However, there is research-based evidence that university students face strong difficulties regarding this aspect of mathematical activity. One can hypothesize that even in Calculus, the mathematical objects at stake are too complex to allow fresh university students to explore them in a fruitful way. In other words, students will be more inclined to look for a relevant theorem in order to solve a general problem in one step, than to explore the properties of objects by examining particular cases or small examples of a given class in order to make conjectures on a possible generalization and then try to prove it. Lockwood and Reed (2018) support the idea that combinatorial enumeration problems are relevant to develop rich mathematical thinking. In their paper, they focus on the role of generalization in students' counting, relying on a model developed in Lockwood (2013) in a set-oriented perspective, with three main components: formulae/ expressions; counting processes; sets of outcomes, and their interrelation. They report on two experiments with students following a Calculus course, aiming to shed light on the following research question:

In what ways can specific examples, concrete outcomes, and particular contexts be leveraged to foster generalizing activity in a combinatorial setting? (ibid, p. 244).

In the first experiment, four volunteer students having not yet followed a course on discrete mathematics were involved. They were invited to solve tasks aiming to shed light on the fundamental role of "concrete examples, concrete outcomes and particular contexts in combinatorial reasoning and of its relevance for generalization". Lockwood and Reed (2018) report on the resolution by a student named Carlson of the "horse race problem": *There are 10 horses in a race. In how many different ways can the horses finish in first, second, and third place?* They noticed that Carlson had solved the problem in a non-standard way, by dividing the total number of outcomes (namely 10!) by the number of outcomes for the 7 horses that were not ranked (namely 7!). During the experiment, when confronted with various other arrangement problems that he explicitly recognized as similar to the horse race problem, Carlson repeatedly resorted to this resolution, using the same mode of reasoning (related to the counting process of arranging all objects and then dividing by the number of ways to arrange the leftover objects). Finally, he refers once more to this problem in order to provide the general formula for arranging a certain number of objects from a finite set of distinct objects. Thus, the horse race problem served for Carlson as a generic example (in the sense of Mason & Pimm, 1984) of this kind of combinatorial problem. In addition, this exemplifies the model developed by the author in Lockwood (2013), by shedding light on the interrelation between its three components.

In the second experiment, a Calculus student, Tyler, was interviewed by the author, while working on a password activity:

> In the interview Tyler was counting two kinds of passwords – those involving either upper case As and/or Bs, and those involving As, Bs, and the number 1 (with repetition allowed). Tyler had initially engaged in systematic listing activity to count the number of possible 4-character AB passwords.
>
> *(ibid, p. 251)*

Later in the interview, the student showed that he recognized, in a more complex password problem, the opportunity to reduce it to the prior problem, and engaged in generalization. Lockwood and Reed (2018) argue that: "reasoning carefully about the concrete examples and actually engaging in concrete listing activity may have helped to solidify a broader combinatorial process." (ibid, p. 252).

These two examples are developed more in depth in Lockwood and Reed (2018). The short synthesis above stresses the role of concrete examples and concrete outcomes for leveraging generalization in combinatorial setting. This is consonant with several aspects of the work developed by Gardes (2013) in the frame of number theory, in particular with the concepts of gesture adapted from the philosophy of mathematics (Cavaillès, 1981; Châtelet, 1993; Longo, 2005). The main features of this perspective are summarized in the second part of this chapter.

Two examples at the interface with computer science

In this section, we briefly present two situations in discrete mathematics developed in Ouvrier-Buffet et al. (2018), designed as examples

> to illustrate the potentialities of discrete mathematics to engage students in learning modelling, proving, and mathematical reasoning and also to underscore the validity and the interest of keeping in mind the algorithmic point of view and the connections with computer science.
>
> *(ibid, p. 259).*

The first example is a research situation for the classroom in the sense of Grenier and Payan (1998). The object at stake is the discrete straight line (the colouring of squares, or "pixels", on a regular rectangular grid, in order to give the best possible visual impression of a straight line). Ouvrier-Buffet (2006) had previously analysed the evolution potential of zero-definitions[3] of the concept of "discrete straight line". Professional researchers in discrete geometry (both mathematicians and computer scientists) use several definitions, but the proof of their equivalence remains worth considering. From a mathematical point of view, discrete geometrical objects, and more specifically discrete straight lines, can be approached in several ways, such as discrete differential analysis, algorithms involving combinatorial analysis, or through an arithmetic definition. The author underlines the fact that approaching the discretization of a real straight line by checking linearity conditions is directly related to number-theoretical issues in the approximation of real numbers by rational numbers. From a didactical point of view, discrete straight lines are accessible through their representations (e.g. perceptive and analytical aspects of geometry) and the (real) straight line can act as a referent. Their definitions and properties are not institutionalized (a concept is institutionalized if it has a place in an official curriculum). Ouvrier-Buffet (2006) reports on an experiment implemented with fresh university students, who were invited to elaborate a definition of a discrete straight line. She underlines several approaches dealing with this concept, which she refers to as "real straight line" (What is the "nearest" pixel to a real line? What kind of modelling should be used?), "regularity" (What are the properties of the sequence of stages –called the *chaincode string*?), and "axiomatization" (What about the existence of the intersection of two discrete straight lines? Is a discrete straight line unique?). Each point of view brings about several zero-definitions. To engage into an axiomatic perspective carries great difficulty, requiring to deal with both the perceptive aspect of a straight line and the axiomatic perspective, the latter requiring some implicit skills and knowledge in students (e.g. building a theory and choosing among competing definitions). The main results of this experiment underscore the ability of students to engage in a defining activity with this concept. Although students did not choose an axiomatic perspective, they were able to engage in reasoning involving methods of approximation close to those used for real straight lines (e.g. statistical linear approximation, op. cit. p.277) and also recursive arguments in order to characterize the sequence of stages of pixels (how can we modify a sequence to obtain a better regularity?). Ouvrier-Buffet et al. (2018) conclude that

> The ongoing mathematical problems in discrete geometry are intimately related to questions in other fields of mathematics and computer science. The construction and the manipulation of algorithms are important for this purpose.
>
> *(ibid, p. 260).*

This example supports the relevance of the introduction of research problems in discrete geometry in mathematics education (Grenier & Payan, 1998).

The second situation presented in Ouvrier-Buffet et al. (2018) deals with a classical "divide and conquer"-type algorithm computing the *n*th power (for some natural number *n*) of some floating-point number (or integer) *a*. This situation is formulated as a typical *algorithmic problem*: given *a* and *n*, compute a^n. As usual in this kind of setting, one expects an algorithm as an answer, and not simply the mathematical definition of the *n*th power of *a*, which is considered known or given. The algorithm provided for computing a^n is called "exponentiation by squaring". It relies on the observation that for all n, $a^n = (a^2)^{n/2}$ when n is even, and $a^n = (a^2)^{(n-1)/2}$ otherwise. It is presented as a recursive function *power* (a, n) written in the Python programming language. The main motivation for this particular way of computing a^n is of course that it performs far fewer operations than a more naive algorithm (such as multiplying a by itself $n - 1$ times). This activity includes considerations regarding its analysis. In particular, the typical questions of termination, correctness and algorithmic complexity are discussed, which mobilizes competencies related to infinite-descent arguments, proofs by contradiction, by recurrence or by induction, logic, recurrent sequences, counting, etc.

The authors' point of view is that tackling these questions mobilises specific mathematical objects and concepts, notably from the above-mentioned fields of arithmetic, discrete mathematics, combinatorics, etc., and provides an opportunity to illustrate, on a simple example, the types of reasoning and proofs which are involved in the analysis of simple algorithms and programs. The article only presents the case of a recursive function, but similar tools can also be used when working with iterative algorithms and programs.

Conclusion of the first part

There is evidence that some mathematical knowledge and skills are required in order to become proficient in computer science. Conversely one may postulate that reasoning about algorithms may provide an incentive towards mathematical activity. The authors therefore advocate the need for continued cooperation between mathematics and computer science teachers and between researchers on mathematics and computer science education in order to better understand the interactions between these disciplines when teaching and learning them.

Logic, defining, proof and proving at university level

In this second part, we consider the place and role of logic in defining, reasoning, proof and proving at university level in the teaching and learning of mathematics, relying mainly of the papers presented during the two INDRUM conferences. Here Logic refers to the domain dealing with reference and with validity of inference. As is generally the case in Mathematics education, we restrict ourselves to first-order logic, which encompasses the interplay between syntax and semantics that is crucial for mathematical activity (Durand-Guerrier & Dawkins, 2018). One hypothesis is that the role of logic in the teaching and learning of mathematics at university is often underestimated by university teachers. This could be a consequence of a rather common assumption that practising mathematics in higher secondary school should guarantee that the logical competences required for advanced mathematics are available to students entering university. However, there is research-based evidence that this is not the case. During the two INDRUM conferences, in line with Durand-Guerrier, Boero, Douek, Epp and Tanguay (2012), several contributions supported the relevance of teaching logic for undergraduates. In the first section, we report on two pieces of research presented in Chellougui (2016) and Brown (2018). The first aims to shed light on the logical complexity of the concept of supremum that is likely to hinder its appropriation by students, and hence advocates the need for addressing the logical aspects of its definition and of its use at university level; the second emphasizes the logical complexity of proof by contradiction. In the second section, we present two case studies aiming to analyse the links between logical skills and students' activity in proof and proving processes.

Logical complexity in mathematical activity at university level

In this section, we examine the relevance of and interest in taking into consideration the logical complexity of the definition of advanced mathematical concepts at university level. In several countries, university and college faculty teachers commonly complain that many students lack the logical competences, such as the use and interpretation of quantifiers (Dubinsky & Yiparaki, 2000; Selden & Selden, 1995), or access to indirect modes of reasoning (Antonini & Mariotti, 2008), to learn advanced mathematics. This contradicts the view that the knowledge and practice of mathematics gained during secondary education sufficiently develops the required logical abilities. In a didactical perspective, recurrent, resistant and widespread difficulties question both the mathematics at stake and the teaching choices. This means that didactic research on logical issues in the teaching and learning of mathematics and computer science at university should not focus only on students' difficulties, but also on intrinsic mathematical difficulties and on teachers' practices, in order to design activities focused on these difficulties. The first point is addressed in the first subsection, which relies on Chellougui (2016). The second point is partially addressed in the second subsection, which relies mainly on Brown (2018).

Logical complexity of the concept of supremum and related students' difficulties

Chellougui (2016) reported on empirical research experiments aiming to investigate the impact of the logical complexity of the concept of supremum on its appropriation by undergraduates. She first acknowledged the logical complexity of the definition by providing a logical analysis of the classical definition of the supremum given in natural language.

"Given a set E endowed with a generic order relation $\leq$, an element $\alpha \in \mathrm{E}$ is called a least upper bound (or supremum) of a non-empty bounded above set $A \subseteq E$ if and only if α is an upper bound for A and α is the smallest upper bound for A." (translated from Chellougui 2016, p. 269)

In this definition, many components are interwoven: objects (*elements of a set*); two structures (*the initial ordered set E; the set of upper bounds of a given subset A of E*); implicitly two relations between objects (*to be greater than; to be smaller than);*

explicitly two properties of structure (*to be bounded above; to have a minimum*), linked with two binary relations between an object and a structure (*to be an upper bound of; to be the smallest element of*).

As usual, the definition is an open sentence that could be satisfied or not by a given element of E. In the case where it is possible to find an element satisfying the definition, it is unique. It is noticeable that in this definition in natural language, there are no explicit quantifiers, while formalizing it in Predicate calculus shows the presence of three quantifiers, including an AE[4] alternative:

Given E an ordered set and A a subset of E,

- An element α of E is the supremum of A *if and only if* "α is an upper bound for A" and "for all element y of E less than α, y isn't an upper bound for A (i.e. no number less than α is an upper bound for A)".
- An element α of E is the supremum of A *if and only if* $(\forall x \in \mathrm{A}\ x \leq \alpha) \wedge (\forall y < \alpha\ \exists x \in A\ y < x)$.

As underscored by Chellougui (2016) the logical formalization in predicate calculus reveals the logical complexity of the definition. She hypothesizes that giving the definition in natural language without *unpacking* (Selden & Selden 1995) with students the logical underlying structure (which was the case in her institution, a Faculty of Sciences in Tunisia) might prevent an adequate appropriation of this concept by students. To test this hypothesis, she conducted six interviews of pairs of first year university students, asking them to provide a definition of the supremum and questioning them to try to identify their understanding of the logical structure and of the links with the meaning of the concept. During the interviews, it appeared that for all pairs, there was a confusion between the definitions of the relation "to be an upper-bound of" between an object and a subset of an ordered set, and the property "to be bounded above" for such a subset. The effort of the interviewer to help students clarify the logical aspect at stake (open *versus* closed sentence, type of objects, importance of the order of quantifiers in a sentence involving an alternation of quantifiers) was not successful. More precisely, the interviews underline the difficulties met by students in acknowledging the logical structure of the definitions of the different components intertwined in the definition of the supremum, when trying to formalize this definition. The analysis of the students' arguments shows that:

- Definitions do not seem to be identified as providing properties or relationships to be satisfied. For example, when asked to give the definition of an upper bound of a given subset A of E, many students provide a closed sentence, instead of the expected open sentence. Chellougui (2016) hypothesizes that this is related to the lack of explicit work with students on open sentences that model predicates, properties, and relationships.
- Nearly all the interviewed students failed to recognize that the order of the two quantifiers in an alternation of the universal and existential quantifiers has

> an impact on the meaning of the sentence. This is consonant with the results of Dubinsky and Yiparaki (2000) and Durand-Guerrier and Arsac (2005) who underlined the difference between an expert and a novice: experts widely use semantic controls to check the correctness of what they are doing, while a student, as a novice in the field, needs to be able to consider both semantic and syntactic controls to check the validity of their inferences in the process of proof and proving.

The results of this experiment show difficulties both in the appropriation of the meaning of the concept and in the understanding of the links between the syntax of a statement and its semantics (i.e. how to interpret it in terms of objects, properties, relationships). It should be clear for the reader that we do not claim that developing syntactic abilities will be sufficient in itself to improve conceptualization. We rather consider the importance of addressing, with undergraduates, the interplay between syntax and semantics as a crucial tool for contributing to conceptualization and to the development of proof and proving competences (Weber & Alcock, 2004; Alcock 2009; Durand-Guerrier et al., 2012).

Logical complexity in proof by contradiction: the case of "there exist no".

In Brown (2018), the author reports on findings from two analyses of students' engagement in meta-theoretical tasks drawn from a model of the reasoning requirements in a proof by contradiction. The studies aimed to explore students' engagement in the tasks, the extent to which they were successful, and the similarities and/or differences between students' and mathematicians' approaches.

Following Antonini and Mariotti (2008), the author identifies two meta-theoretical requirements (i.e. the use of logical theorems) in a proof by contradiction of a universal conditional statement. The first is at the beginning, when one must produce the negation of the statement to prove, asserting the existence of a counterexample. The second, when one must recognize that having proved that there exists no counter-example (conclusion following the identification of a contradiction) provides a proof of the initial statement. She recalls that Antonini and Mariotti (2008) have pointed out that both points might be difficult to accept for some university students: the first one because it is difficult to accept the idea of starting a valid argument from a supposedly false statement; the second because the equivalence between the fact that there is no counterexample to the initial statement and the initial statement itself might not be easily recognized. These authors also point out that difficulties with indirect proofs may occur either in mathematical theory or in logical theory. Brown (2018) focuses on the second point. She first conducted a survey addressed to 46 students in a university in the United States on the first day of a *Basic Set Theory and Logic* course that served as the university's first course on logic and their "Introduction to Proof" course. In the survey, two tasks concern the identification of the logical equivalence between a universal conditional statement and the assertion that there exists no counter-example: given the statement "For every integer *n*, if *n* is a perfect square, then *n* has an even number of factors", students were asked if this statement could be proved by proving 1. "There exists no integer *n* such that *n* has an even number of factors and *n* is not a perfect square" (the correct answer is *no*); 2. "There exists no integer *n* such that *n* is a perfect square and *n* has an odd number of factors" (the correct answer is *yes*). Only 24% of the students answered correctly to both tasks. This low rate of success called forth a further research question on the persistence of these difficulties after instruction. This was explored by the author in a second study addressed to 21 students (among the 46 surveyed) having completed the *Basic Set Theory and Logic* course, and to 6 mathematicians. The author briefly presents the contents of the course as follows:

As the course focused on set theory and logic in the service of proof writing, the instruction on set theory and logic was limited to basic properties, terms, and definitions, as well as symbolizing practices, and then on specific proof techniques and/or strategies. (Brown, 2018, p.217)

The participants took part in video-recorded interviews during which the task was presented on a large piece of paper. The participants were given as much time as required, and were then asked to explain their answer to the following question:

TABLE 7.1 The interview task (Brown, 2018, p. 217)

Question: Can you prove Theorem 5 by proving Statement A?
Theorem 5. For all positive integers *n*, if *n* mod(3) ≡ 2 then *n* is not a perfect square.
Statement A. There exists no positive integer *n* such that *n* mod(3) ≡ 2 and *n* is a perfect square.

The analysis focused on: 1. the correctness of the answer (the correct answer is *yes*) with or without hesitation; 2. the theory (mathematical or logical) the participant worked in and how they engaged in that theory; 3. the use of syntactic (rule-based) reasoning and/or semantic reasoning (relying on meanings). Responses were considered as *mathematical theory approaches* if the participant was observed: 1. explicitly exploring mathematical statements, definitions and/or terms; and/or 2. constructing a proof of either statement. Responses were considered as *logical theory approaches* if the participant was observed: 1. posing explicit questions of equivalence; 2. constructing truth-tables and/or working with symbolic logic; and/or 3. citing logical theorems or practices. In addition, the use of syntactic and/or semantic reasoning was noted.

Specifically, coding noted participants' use of syntactic and/or semantic reasoning, with semantic referring to reasoning that employs meanings and multiple representational systems and syntactic referring to rule-based reasoning within a representational system. (op. cit. pp. 217–218)

Findings indicate students tend towards syntactic, logical theory approaches while mathematicians gravitate towards semantic, mathematical theory approaches. Drawing on interview data, it is shown that students may use symbols to avoid employing fragile content knowledge, yet encounter further difficulties by viewing quantifiers as additional symbols rather than as variable-binding operators that act on open sentences. The author concludes that instruction had little impact on the students' success with the metatheoretical requirements of a proof by contradiction; indeed the data demonstrate that post-instruction, novices continued to struggle with the metatheoretical requirements of proof by contradiction. As for mathematicians, as expected, they express no doubt about their answer, but three of them show hesitation on their reasoning. To conclude, the author claims that:

The study contributes to the literature by highlighting the logical complexities novices may encounter when producing or comprehending proofs by contradiction, given the approaches they gravitate towards (op. cit. p. 222).

This is in line with previous results in the literature already mentioned in this section concerning: indirect reasoning; quantifications issues; and the role of the interplay between syntax and semantics, echoing the paper of Chellougui presented above.

In the concluding remarks, the author wonders why the students did not progress after the course. She claims that "the lack of progress may be due to an insufficient curricular treatment, or to the cognitive demand of the tasks". Turning to the curricular material supports the former, while the hesitation showed by some mathematicians in their reasoning supports the latter. She finally questions the relevance of inferring instructional recommendations from mathematicians' practises, her data showing that they might generate inference automatically, preventing proper rationalization. This issue was already raised in Durand-Guerrier et al. (2012) in the section entitled "Contexts that may require understanding logic at a syntactic level", which concluded:

> From an educational perspective these examples [given in the section] suggest that syntactic control is not only particularly important for novices but also necessary for more advanced students facing concepts about which their semantic control is still shaky. [...] In such context, a lessening of rigour, considered by experts as an innocuous shortcut, can results in serious misunderstandings by students. (op. cit. p.383)

Analysing proof and proving process at university: case studies.

In this section, we consider two case studies that were presented during the two INDRUM conferences. Both consider means to analyse the links between logical skills and students' activity in proof and proving processes.

Designation at the core of the dialectic between experimentation and proving

It is well known in mathematics education that students experience strong difficulties with elaborating mathematical proofs by themselves, even when they are involved in meaningful research and problem-solving activities. There is research-based evidence that even in settings where the milieu for validation seems rich enough to support the proving process, some students fail to enter it appropriately.

An important aspect for entering a proving process is the *designation* of objects that, according to Tarski (1933), is one of the main characteristics of a semantic approach. In their paper, Gardes and Durand-Guerrier (2016) acknowledge this crucial role of designation, i.e. the representation of a mathematical object by means of a natural language expression or of a symbol (e.g. "**Q**" designates "the set of rational numbers") and provide some empirical results supporting the following hypothesis. Although the designation of objects plays an important role in the heuristic phases, it might not be sufficient to enrol students in elaborating proofs in cases where the properties of these objects and their mutual relationships are not made explicit. The empirical results come from a didactical engineering developed by Gardes (2013), whose main goal was to study the conditions and constraints pertaining to the transposition of professional mathematicians' research activity in the mathematics classroom. The didactical engineering was elaborated from an unsolved problem in Number Theory, the Erdös-Straus conjecture:

> "The equation $\frac{4}{n} = \frac{1}{x} + \frac{1}{y} + \frac{1}{z}$ can be solved in positive integers x, y and z for any integer $n > 1$".

To analyse her data, Gardes (2013) adapted the concept of gesture from Cavaillès (1981) and Châtelet (1993):

> A gesture is an action connecting mathematical objects and which is carried out with intentionality. It is an operation that is accomplished through a combination of signs with respect to the usage rules of these signs. Because they open on possibilities, gestures have the power to enhance mathematical creativity.
>
> *(translation by Gardes and Durand-Guerrier (2016) of Gardes (2013), p. 155)*

Among the seven gestures identified by Gardes (2013), the authors focused on the gesture of "*designating an object*", i.e. "representing a mathematical object by means of a natural language expression or a symbol". The data collected during the research supports the role of this gesture in the reformulation of the conjecture entailing its enrichment: introduction of objects (e.g. prime numbers, congruence), production of intermediate conjectures, formalization of methods, etc. In some cases, there is evidence that this gesture is likely to foster a back-and-forth between semantics and syntax (Gardes 2012). Nevertheless, the *a posteriori* analysis of the data collected during the experiment showed that in some cases, students were unable to make explicit the relationships between the objects that they introduced, which prevented them from moving beyond the successful production of decompositions by trial and error to a more general method. The authors provide the example of two groups who introduced a variable t corresponding to the successor of the integral part of $\frac{n}{4}$, which allowed them to engage in the proving process, but the syntheses they provided at the end of their research show strong differences. In one group, the students explicitly wrote the relation $t = E(\frac{n}{4}) + 1$, and provided a general decomposition that allowed them to engage successfully in the proving process, to reach an important intermediate result and to identify the cases that needed further study. In the other group, the students did not make this relation between *t* and *n* explicit; this prevented them from moving to a general method for decomposition. The comparison between the syntheses of both groups supports the claim that, beyond developing the gesture of *designation of objects* as a key competence, it is also necessary to develop the *capability to identify relationships between the objects that are introduced*, this last point requiring interplay between mathematical knowledge and logical skills.

The authors conclude that

> [this point] should [be taken] into account when elaborating the milieu of such didactical engineering, aiming at fostering the development of students' skills for solving mathematical research problems. The way to do this is an open research question.
>
> *(Gardes & Durand-Guerrier, 2016, p.292)*

Analysing regressive reasoning at university level

Barbero and Gómez-Chacón (2018) focus on the epistemic and cognitive characterization of regressive reasoning in resolving strategic games. The authors explore the use of the *Finer Logic of Inquiry Model*, an adaptation of the Logic of Inquiry by Hintikka (1995), as a tool for examining the way in which university students acquire an ability to use regressive reasoning. They investigate in particular how the Finer Logic of Inquiry Model can be a valid tool to analyse the interplay between the cognitive and epistemic aspects of regressive reasoning. They report the results of a study carried out on 32 undergraduate students preparing a mathematics degree at a university in Spain. For the authors, regressive (or backwards) reasoning refers to the analysis part of the "analysis and synthesis reasoning" as elaborated by Polya (1945) and Hintikka and Remes (1974). They proposed three specific modalities of a logical nature for the deductive components: the detached modality, logical control and deductive modality.

Detached modality is the moment in which a conjecture, which has not arisen immediately after an exploration, is formulated. Logical control is the time when an exploration-control is done without using instruments. It is characterized by the use of formal language. Deductive modality characterizes control phases where instruments are involved. Deductive Steps and Logical Chains are added to the Inquiry Component actions. (Barbero & Gómez-Chacón, 2018, pp. 206–207).

The authors notice that, in problem solving activities, inquiry and deductive components are intertwined. They exemplify the use of the *Finer Logic Inquiry Model* in a case study of a university student solving a strategy game. Triangular Solitaire is a single-player game that requires a board with 15 boxes as the figure shows.

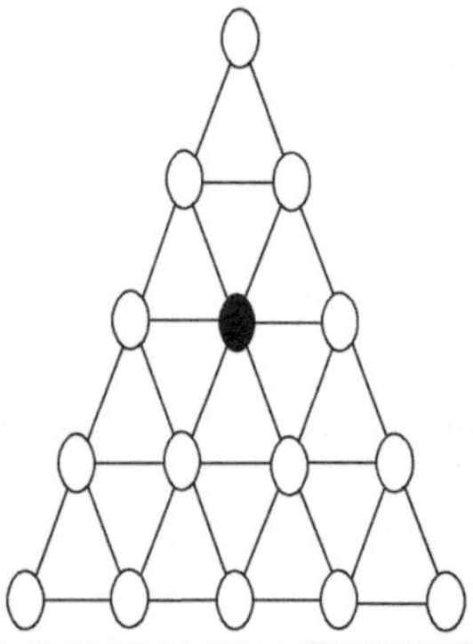

These are the rules:
1. Place the pegs in all boxes, except in the one marked in black.
2. The player can move as many pegs as they like as long as they are able to jump over an adjacent peg and onto an empty space (along the line). At the same time, he "eats" the peg that was jumped over and that peg gets taken out of the game. All pegs move in the same way. Pegs can move around the table in any direction.
Objective: The player wins when there is only one peg on the table.

FIGURE 7.1 Triangular Solitaire (Barbero & Gómez-Chacón, 2018, p. 208)

This strategic game was chosen because it favours the use of regressive reasoning. The experiment was conducted with 32 students, but in the paper a single case is analysed. The authors use the Finer Logic of Inquiry Model to analyse the written work of an individual student. They claim that this student is representative of the group because she was among the 60% of those who used regressive reasoning. In addition, this student introduced a lot of graphical representations to support her resolution process. In the paper, the entire protocol has been translated from Spanish. After a familiarization phase, the student entered an exploration phase and carried out her strategy. The protocol shows consistent use of regressive (backward) reasoning. At the end of the exploratory phase, the student started from a position she conjectured to be successful and progressively filled the triangle in order to get to the initial position. She finally reversed the list of drawings starting from the initial position to the final one.

The authors conclude that analysis using the Finer Logic of Inquiry Model allows to model student's cognitive movement during the problem-solving activity. On the cognitive side, they point out that strategic aspects are more dominant in the ascending and descending modalities, while epistemic ones are prevailing in the neutral modality[5]. They also point out that regressive reasoning involves auxiliary intuition elements that are necessary to achieve the solution, and that abductive reasoning should have been used during the resolution process, but that the protocol does not allow confirming it. (Barbero & Gómez-Chacón, 2018, p. 212).

Conclusion of the second part

The four studies briefly presented in this section support the relevance of considering logical issues in mathematics education. The first and second examples shed light on the logical complexity of concepts' definitions and of proving processes at university level, while the third and fourth ones aim at better understanding the links between logical skills and students' activity in proof and proving. We will elaborate in the last part of the chapter on the logic needs for mathematics and computer science instruction.

Logic needs for mathematics and computer science

In this section, we aim to clarify the role that logic may play in the learning and teaching of mathematics and computer science, taking into consideration different aspects of various proving activities and their dependence on the topics at stake. There are strong parallels between logic, computer science and mathematics. In particular, both computer science and mathematics involve the study and use of formalism, proof and deduction systems, and provide ways of giving them meaning; therefore, the dialectic between syntax and semantics plays an essential role in their teaching and learning (Weber & Alcock, 2004; Durand-Guerrier & Dawkins, 2018; Durand-Guerrier et al., 2019). The content of this section relies mainly on the discussion on logical issues which took place in the working group "Number, Algebra, Logic" at INDRUM 2018.

Logical issues in mathematics instruction

We mentioned earlier in the text the perceived lack of logical skill of young university students. As teachers ourselves, we can attest to the considerable difficulties students face in dealing with the logical issues that arise when they attempt to learn about the process of reasoning and proving, especially when confronted with a higher requirement for accurate formalism.

As claimed by Durand-Guerrier and Dawkins (2018), it is not easy to determine what kind of work with students will help them overcome these difficulties. Looking at professional mathematicians' assumptions on the role of logic in mathematical activity shows a diversity of points of view (Mathieu-Soucy, 2016; Durand-Guerrier et al., 2012). However, considering university students' mathematical activity, things are likely to be different. Indeed, as pointed out by Durand-Guerrier and Arsac (2005), and evidenced in Brown (2018), there are differences between expert mathematicians and novices, the former using semantic controls informed by their knowledge and practices, which is not the case with the latter. In this respect, it seems worthwhile to investigate the knowledge and skills undergraduates need in formal logic in order to solve the mathematical tasks given to them. Mathieu-Soucy (2016) reports on research aiming to address the following question: how does knowledge of formal logic, or a course in formal logic, change the way undergraduate mathematics students produce and validate proofs? She asked eight university students from Quebec, Canada in the second half of a 3-year mathematics program (20–21 years old) with various levels of knowledge in formal logic to produce and validate proofs through a non-routine task. In this paper, the author focuses on an adaptation of the first three of Hilbert's incidence axioms in a non-mathematical context (dogs, robots, friendship relationships). The empirical results suggest that a course in logic changes the way students do mathematical work, in particular by developing alertness. However, the author notices that this alertness is not enough to allow students to proceed towards complete mathematical proof, in particular in cases where the mathematical notions at stake are not well mastered. This is consistent with the claim by Durand-Guerrier and Arsac (2005) that in the proof and proving process, logical and mathematical issues are closely intertwined.

The exploratory and constructive phases of reasoning require frequent practice and a mastery of objects, properties and relationships, and the ability to draw upon relevant conjectures, definitions and theorems. Such practice necessitates an intensive use of inference rules, which may occur subconsciously, but requires controls in order to be developed into an actual proof. Working explicitly on the logical aspects of mathematical reasoning may help students not only reason about logic but also learn to coordinate logical tools with their reasoning about particular mathematical topics (Durand-Guerrier & Dawkins, 2018).

In mathematics, natural deduction provides opportunities for using inference rules from the logic of propositions and from predicate calculus, such as deduction by cut-off, deduction by simplification, *modus ponens, modus tollens*, disjunction of cases, the contradiction rule, the converse rule and so forth. When one reasons with conditional statements that are not equivalences, only two valid inference rules are available (namely *modus ponens* and *modus tollens*). These rules form the very core of the Copi natural deduction. Using Copi's (1954) rules, and based on the work carried out in didactics of mathematics (Chellougui, 2016; Durand-Guerrier & Arsac, 2005), four additional inference rules from predicate logic were identified, which also have crucial importance in logic and mathematic proofs. The presentation below is adapted from Durand-Guerrier et al. (2012, p.374)

1. Given that a result is known to be true for all objects of a certain kind, universal instantiation allows us to conclude that the result is true for any particular object of that kind. This rule is used constantly in the manipulation of algebraic expressions and in providing justifications for most steps in a proof.
2. When an object of a certain type is known to exist, existential instantiation allows one to give it a name. Learning how to use this rule correctly is a significant challenge for most students (Durand-Guerrier & Arsac, 2005).
3. When one knows or hypothesizes that a particular object of a certain type satisfies a property, then one can assert the statement "there exists an object of that type satisfying this property". This rule is named "existential generalization". One uses it, at least implicitly, when asserting the negation of a universal statement "for all x, P(x)" by providing an example of "not P(x)".
4. If one can show that a certain property holds true for a particular but arbitrarily chosen object of a certain type, universal generalization allows us to conclude that the property holds true for all objects of that type. This rule is the basis for almost all mathematical proofs.

We consider that having some level of understanding of the fundamental rules of predicate logic would help students check logical deductions and mathematical statements that are in doubt, to avoid invalid deductions, and to comprehend the basic structures of both proof and disproof by counterexample. Drawing on their research and teaching experience, Epp (2003), Durand-Guerrier and Arsac (2005) and Durand-Guerrier et al. (2012) support the claim that teaching students the basic principles of predicate logic (e.g. Copi, 1954) might provide them with a relevant tool for successful mathematical activity.

Logic and proof in computer science

It is today widely acknowledged that some logic and reasoning skills are required in computer science. There are many examples of situations where logic-inspired objects, tools and notions are used in the various areas of computer science: Boolean expressions in a computer program; logical gates in computer architecture; bit-wise operators and arithmetic modulo 2, etc. Here, elementary propositional logic plays a large role, if only as an inspiration.

Other examples relate to classical first-order (or predicate) logic. The way variable introduction and scope work in many programming languages is similar to the way variables in first-order logic are introduced (or captured) using quantifiers, and their scope delimited using parentheses. Programming languages such as Prolog use logical rules consisting of one or several premises and a conclusion, which may be used to deduce new knowledge.

Finally, one may draw several parallels between programming and proving in a more or less formalized manner. For instance, it is common in mathematics to decompose a large demonstration into lemmas, in order to clarify the text or to better control the level of detail. Similarly, a programmer decomposes a large program into various sub-procedures, functions or modules.

In Durand-Guerrier et al. (2019) the authors list some further examples of situations where computer science draws from mathematics in general, and from logic in particular. We briefly summarize some of these examples, from the most theoretical to the most applied.

Computation models and their links to logic

For almost a century, mathematicians and logicians have been investigating ways to formalize the notion of computability. This can be seen through Hilbert's 1928 *Entscheidungsproblem* asking about the existence of an algorithm to determine the universal validity of a first-order logic statement, and the negative answer independently brought by computer science pioneers Church and Turing in 1936. This type of questions led to the famous Church-Turing thesis, which essentially defines a class of equivalent formalisms corresponding to the intuitive notion of computability (see for instance Hopcroft, Motwani and Ullman (2007), Section 8.2.1 for a brief summary). This result and the work surrounding it are now widely considered as one of the cornerstones of theoretical computer science. Over the years, more restricted classes of computation models were also defined and studied, for instance that of finite automata, taught in most computer science curricula. In the 1960s, a deep connection between finite automata and a fragment of second-order logic was discovered by Büchi, effectively opening a new avenue of research (see for instance Straubing and Weil (2012) for an introduction).

Reasoning about problems, algorithms and programs

As outlined in Part 1, reasoning about programs or algorithms draws upon an assortment of proof techniques and mathematical concepts, including notions attributed to *discrete* mathematics. In this respect, some knowledge of formal logic is useful. The notion of worst-case complexity is a relevant illustration of this claim: to accurately present it to students needs great care, proficiency with several objects and concepts, and a non-trivial use of quantification. It is however an important notion which is a part of almost every computer science syllabus.

Formal rules to describe the semantics and prove the correctness of programs

Beyond short proofs performed by hand about simple programs, it is sometimes necessary to leverage the power of formalism in order to accurately express the meaning of programs, and to prove some of their properties. University textbooks on the formal semantics of programming languages, for instance Winskel (1993), present several mathematical formalisms able to express the meaning of each syntactical construct in a programming language. One such formalism, called *operational semantics*, is presented in the form of rules in which atoms are statements about the effect of a piece of program. Evaluating the overall meaning of a program then comes up to writing down a valid proof using this set of rules and a deduction system similar to natural deduction in classical logic. Another deduction system called *Hoare rules* or *Hoare logic*, provides a tool to prove the partial correctness of imperative programs. Again, it uses deduction to prove assertions about programs, which are essentially written using classical logic and arithmetic predicates.

Through these three examples, which are representative of the discussions held at the last INDRUM meeting, we have tried to make it apparent that the evolution and growth of computer science is tightly related to its pioneers' mathematical culture and proficiency, in this case with a strong emphasis on logic. The ideas these examples convey are general and permeate algorithmic thinking, and are still present in most university computer science curricula today. One may therefore postulate that appropriate training in logic is a requirement for any student in computer science. It is our opinion that further research in this direction is called for.

Conclusion of the third part

Knowledge of the principles of logical reasoning becomes most important when familiarity with the mathematical and computer science subject matter does not suffce by itself to ascertain truth or falsity in a given situation. In addition, following Wilkerson-Jerde and Wilensky (2011) and Modeste (2016), we hypothesize that understanding the role of logic in the activities of expert mathematicians and computer scientists would help in designing relevant activities for improving university students' logical abilities.

Finally, including instruction in logical principles as one part of mathematics and computer science education provides a balance between two extreme positions: that checking the validity of a proof requires that it be completely formalized; and that success in proving requires no explicit knowledge of logical principles (Durand-Guerrier et al., 2012).

Conclusion

In this chapter, we have addressed some of the challenges raised by the teaching and learning of discrete mathematics, computer science, logic, proof and proving, taking in consideration their relationships. Although the research questions evoked in this chapter have already been addressed in the literature, the point of view adopted here, leaning on the papers and discussions from the two INDRUM conferences, open new avenues for research focusing on the relationships between these domains.

In the first part of the chapter, we have highlighted the potential of the teaching and learning of discrete mathematics as a didactical means to make visible the relationships between mathematics and computer science. In the second part, we emphasized the necessity to take into account the logical complexity of the mathematical topics studied at university, and to consider the logical issues underlying mathematics and computer science instruction. There is research evidence of the potentiality of discrete mathematics for developing proof and proving skills, linked to logical issues, and hypotheses regarding a similar role for computer science. Finally, the third part of this text has put forward hypotheses concerning the specific characteristics of formal logic as a teaching topic, in the context of mathematics or computer science curricula, and with a particular emphasis on proof and proving. One may for instance question the expected or possible benefits of training students in formalism and logic on their ability to learn mathematics or computer science. Alternatively, one may try to identify the specific uses of logic in both disciplines as they are taught at undergraduate or graduate levels.

A next step would be to design interdisciplinary didactical engineering aiming to help university students get a deeper understanding of the conceptual relationships between discrete mathematics, computer science, logic and proof. For this purpose, leading further inquiries regarding the dialectic between syntax and semantics seems promising.

Notes

1 Here "proof" refers to both the process of proving and the product.
2 For a presentation in English, see Artigue (1994); for specificities in University mathematics, see González-Martin, Bloch, Durand-Guerrier & Maschietto (2014).
3 The notion of zero-definition was introduced by Lakatos (1961). According to Ouvrier-Buffet (2006, p.266) zero-definitions can assume a denominative function, and are used for

practical purposes at the beginning of the research process before the proper mathematical terms are defined.

4 An AE alternative corresponds to the form: "for all x, there exists y, P(x,y)" (Dubinsky & Yiparaki, 2000)

5 This refers to Soldano (2017) who characterized the cognitive dimension of reasoning through the sequence of actions in three different modalities: ascending, neutral and descending.(Barbgero & Gomez-Chacön, 2018, 206).

References

Abdallah, E. (2018). Contemporary Research Practices in Discrete Mathematics - A way to enrich the understanding of Discrete Mathematics at University Level. *INDRUM 2018 Proceedings*, pp. 194–203.

Alcock, L. (2009). Teaching proof to undergraduates: semantic and syntactic approaches. *Proceedings of the ICMI 19 conference: Proof and Proving in Mathematics Education 1* (pp. 29–34). Taipei: The Department of Mathematics, National Taiwan Normal University.

Antonini, S., & Mariotti, M. A. (2008). Indirect Proof: What is specific to this way of proving? *ZDM: The International Journal on Mathematics Education*, 40, 401–412.

Artigue, M. (1994). *Didactic engineering as a framework for the conception of teaching products*. In R. Biehler, R. W. Scholz, R. Sträßer, & B. Winkelmann (Eds.), *Didactics of mathematics as a scientific discipline* (pp. 27–39). Dordrecht: Kluwer Academic.

Barbero, M., & Gómez-Chacón, I. M. (2018). Analysing regressive reasoning at university level. *INDRUM 2018 Proceedings*, pp. 204–213.

Brown, S. (2018). A Study of Students' Reasoning About "There exists no …". *INDRUM 2018 Proceedings*, pp. 214–223.

Cavaillès, J. (1981). *Méthode axiomatique et Formalisme*. Paris: Hermann.

Châtelet, G. (1993). *Les enjeux du mobile. Mathématique, physique, philosophie*. Paris: Editions du Seuil.

Chellougui, F (2016). Approfondissement du questionnement didactique autour du concept de "borne supérieure". *INDRUM 2016 Proceedings*, pp. 266–275.

Copi, I. (1954). *Symbolic logic* (2nd edition, 1965). New York: Macmillan.

Dubinsky, E. & Yiparaki, O. (2000). On student understanding of AE and EA quantification, *Research in Collegiate Mathematics Education IV, CBMS Issues in Mathematics Education, 8*, pp. 239–289. Providence: American Mathematical Society.

Durand-Guerrier, V., & Arsac, G. (2005). An epistemological and didactic study of a specific calculus reasoning rule, *Educational Studies in Mathematics*, 60(2), 149–172.

Durand-Guerrier, V., Boero, P., Douek, N., Epp, S. S., & Tanguay, D. (2012). Examining the Role of Logic in Teaching Proof. In G. Hanna & M. de Villiers (Eds.), *ICMI Study 19 Book: Proof and Proving in Mathematics Education* (pp. 369–389). New York: Springer.

Durand-Guerrier, V., & Dawkins, P. C. (2018). Logic in University Mathematics Education. In S. Lerman (Ed.) *Encyclopedia of Mathematics Education* (2nd edition). New York: Springer.

Durand-Guerrier, V., Meyer, A., & Modeste, S. (2019). Didactical issues at the interface of mathematics and computer science. In G. Hanna & M. de Villiers (Eds.) *Proof Technology in Mathematics Research and Teaching* (pp. 115–138). New York: Springer.

Durand-Guerrier, V., & Tanguay, D. (2018). Working on proof as contribution to conceptualisation – The case of R-completeness. In A. J. Stylianides & G. Harel (Eds.) *Advances in Mathematics Education Research on Proof and Proving. An international perspective*. ICME-13 Monographs. (pp. 19–34). New York: Springer.

Epp, S. S. (2003). The role of logic in teaching proof. *American Mathematical Monthly* 110(10), 886–899.

Epp, S. S. (2016). Discrete mathematics for computer science. *Proceedings of the 13th International Congress on Mathematical Education*. Hamburg.

Gardes, M.-L. (2012). Intérêts et limites des méthodes algébriques dans un problème de recherche en théorie des nombres. *Actes du 36ème séminaire SFIDA*. Nice2011. https://sites.google.com/site/actessfida/.

Gardes, M.-L. (2013). *Étude de processus de recherche de chercheurs, élèves et étudiants, engagés dans la recherche d'un problème non résolu en théorie des nombres*. Doctoral Dissertation, Université Lyon 1 (France).

Gardes, M.-L., & Durand-Guerrier, V. (2016). Designation at the core of the dialectic between experimentation and proving: a study in number theory. *INDRUM 2016 Proceedings*, pp. 286–295.

González-Martin, A. S., Bloch, I., Durand-Guerrier, V., & Maschietto, M. (2014). Didactic Situations and Didactical Engineering in university mathematics: cases from the study of Calculus and proof. *Research in Mathematics Education*, 16(2), 117–134.

Grenier, D., & Payan, C. (1998). Spécificités de la preuve et de la modélisation en mathématiques discrètes. *Recherches en Didactique des Mathématiques*, 18(1), 59–100.

Hintikka, J. (1995). The Games of Logic and the Games of Inquiry. *Dialectica*, 49, 229–249.

Hintikka, J., & Remes, U. (1974). *The Method of Analysis: Its Geometrical Origin and Its General Significance*. Dordrecht: Reidel.

Hopcroft, J., Motwani, R., Ullman, J. (2007). *Introduction to Automata Theory, Languages, and Computation* (3rd edition). Boston: Addison-Wesley.

Knuth, D. E. (1974). Computer Science and its Relation to Mathematics. *The American Mathematical Monthly*, 81(4), 323–343.

Lakatos, I. (1961). *Essays in the Logic of Mathematical Discovery*, Cambridge: Cambridge University Press.

Lockwood, E. (2013). A model of students' combinatorial thinking. *Journal of Mathematical Behavior*, 32, 251–265.

Lockwood, E., & Reed, Z. (2018). Leveraging Specific Contexts and Outcomes to Generalize in Combinatorial Settings. *INDRUM 2018 Proceedings*, pp. 244–254.

Longo, G. (2005). The Cognitive Foundations of Mathematics: human gestures in proofs and mathematical incompleteness of formalisms. In M. Okada *et al.* (Eds.) *Images and Reasoning*. Tokyo: Keio University Press.

Mason, J., & Pimm, D. (1984). Generic examples: seeing the general in the particular. *Educational Studies in Mathematics*, 15(3), 277–289.

Mathieu-Soucy, S (2016). Should university students know about formal logic: an example of nonroutine problem. *INDRUM 2016 Proceedings*, pp. 316–325.

Modeste, S. (2012). *Enseigner l'algorithme pour quoi? Quelles nouvelles questions pour les mathématiques? Quels apports pour l'apprentissage de la preuve?* Doctoral dissertation, Université de Grenoble (France).

Modeste, S. (2016). Impact of informatics on mathematics and its teaching. On the importance of epistemological analysis to feed didactical research. In F. Gadducci, M. Tavosanis (Eds.) *History and Philosophy of Computing*, IFIP Advances in Information and Communication Technology, New York: Springer.

Ouvrier-Buffet, C. (2006). Exploring mathematical definition construction processes. *Educational Studies in Mathematics*, 63(3), 259–282.

Ouvrier-Buffet, C., Meyer, A., & Modeste S. (2018). Discrete mathematics at university level. Interfacing mathematics, computer science and arithmetic. *INDRUM 2018 Proceedings*, pp. 255–264.

Polya, G. (1945). *How to solve it?*Princeton: Princeton University Press.

Selden, A., & Selden, J. (1995). Unpacking the logic of mathematical statements. *Educational Studies in Mathematics*, 29, 123–151.

Soldano, C. (2017). *Learning with the logic of inquiry. A game-approach within DGE to improve geometric thinking*. PhD Dissertation. Università di Torino.

Straubing, H., Weil, P. (2012). An introduction to finite automata and their connection to logic, in D. D'Souza (Ed.). *Modern applications of automata theory*, IISc Research Monographs (pp. 3–43). Singapore: World Scientific

Tarski, A. (1933). *The concept of truth in the language of deductive sciences*. (English translation 1983)

Weber, K., & Alcock, L. (2004). Semantic and syntactic proof productions. *Educational Studies in Mathematics*, 56(2–3), 209–234.

Wilkerson-Jerde, M. H., & Wilensky, U. J. (2011). How do mathematicians learn math? Resources and acts for constructing and understanding mathematics. *Educational Studies in Mathematics*, 78, 21–43.

Winskel, G. (1993). *The formal semantics of programming languages: an introduction*. Cambridge MA: MIT Press.

8

ABSTRACT AND LINEAR ALGEBRA

Thomas Hausberger, Michelle Zandieh and Yael Fleischmann

Introduction

Abstract Algebra teaching and learning appeared recently as a new entry in the Encyclopedia on Mathematics Education (Hausberger, 2020). This is a sign that University Mathematics Education (UME) research has now reached enough maturity on this topic to account for a substantial body of research-based knowledge but also, as stated in the conclusion of the entry, that "research in this area remains limited and several issues would need to be more systematically investigated".

Linear Algebra has been the focus of more systematic research (Rasmussen & Wawro, 2017), due partly to its importance in the undergraduate curriculum, from freshman courses on matrices, linear spaces and linear transformations, to more advanced topics like eigentheory or applications of Linear Algebra, for instance to quantum mechanics. By contrast, Abstract Algebra "usually designates an upper division undergraduate course typically required for mathematics majors and centred on the structures of groups, rings and fields. In general, it is encountered by students upon completion of courses in Linear Algebra" (Hausberger, 2020).

As a mathematical discipline, Abstract Algebra denotes the new image of algebra that emerged at the turn of the twentieth century through the work of German algebraists (in particular, Noether and her school). The shift of paradigm is well illustrated in the following quote by Corry, who points out an inversion in the conceptual hierarchy of algebra:

> Groups, fields, rings, and other related concepts became the main focus, based on the implicit realisation that all of these concepts were, in fact, instances of a more general, underlying idea: the idea of an algebraic structure. Thus, the main task of algebra became the elucidation of the properties of each of these structures and of the relationships among them […] The systems of real

> numbers, rational numbers and polynomials were studied as particular instances of certain algebraic structures; the properties of these systems depended on what was known about the general structures of which they were instances, rather than the other way round.
>
> *(Corry, 2016, paragraph 1)*

Linear Algebra also inherited this shift towards a more formal, abstract and conceptual presentation, maybe as the product of an even more complex and tortuous historical process. According to Dorier (2000, p. 59):

> the concept of vector space, so elementary in terms of structure, encapsulates, in a very elaborate product, the result of a long and complex process of generalisation and unification. Yet the inevitable risk of reduction that masks the complexity of the modelling process in the treatment of linear problems is the price that has to be paid for the efficiency of the axiomatic structural approach. Because of this, a knowledge of the historical development provides the teacher and the researcher with a field of investigation from which they can better understand the students' difficulties and, more generally, from which they can put some "meat" to the "bare bones" of axiomatic approach.

As is evidenced by this short epistemological investigation of the mathematical topics that are the focus of this chapter, educational research in these fields has to address the challenges related to the mathematical formalism and abstraction processes involved in the conceptualisation of algebraic structures. This gives room to various approaches (cognitive, sociocultural, semiotic, etc.) to the teaching-learning phenomena.

In their review of UME research presented at the CERME conferences, Winsløw et al. (2018) made the distinction between *descriptive research* focusing on UME "as it is" and *research on and for innovation* that includes all forms of design-based research aiming at intervention. With regard to the first category, the authors also noted a shift from "developmental studies" inscribed in the Advanced Mathematical Thinking (AMT; Tall, 1991) trend, to "studies that endorse cognitive as well as sociocultural, discursive and anthropological perspectives" (Winsløw et al., 2018, p. 64). While the former use such theoretical constructs as concept image / concept definition (Tall & Vinner, 1981) or well-known binaries such as procedural/conceptual understanding (Hiebert, 1986), the latter may be anchored in the Anthropological Theory of the Didactic (ATD) initiated by Chevallard and his collaborators (see Bosch & Gascón, 2014) or in the Commognitive Theoretical Framework (CTF) developed by Sfard (2008).

Out of the nine papers that were presented at both INDRUM conferences, six contributions belong primarily to the descriptive research category and three are discussing intervention principles or accounting for a design-based research. Yet, the dichotomy is not as clear-cut as it sounds: although the emphasis may be put in the paper on analysing UME as it is, the study may be part of a wider project that includes intervention as the next step: for instance, the notion of structuralist praxeology introduced by Hausberger (2016) is used to develop a Study and Research Path (Hausberger, 2018), thus a form of intervention in the ATD framework. In fact, the majority of papers that focus explicitly on intervention, namely the papers by Trigueros & Biahchini (2016) and by Papadaki and Kourouniotis (2018), use cognitive approaches in the AMT trend, for instance an APOS-based genetic decomposition, with the exception of Bosch et al. (2016) who present an anthropological approach within ATD. The descriptive studies endorse a plurality of perspectives: two of them (Donevska-Todorova, 2016; Wawro et al., 2018) adopt a cognitive lens and focus on the procedural/conceptual distinction, one paper (Lalaude-Labayle et al., 2018) mixes a systemic approach based on the Theory of Didactical Situations (Artigue et al., 2014) and a semiotic approach, one paper (Ioannou, 2016) uses CTF as theoretical framework (thus a discursive approach), one paper (Hausberger, 2016) ATD, and a last paper (Fleischmann & Biehler, 2018) builds its own local epistemological framework through an a priori mathematical and didactical analysis of tasks.

On the level of mathematical topics, a first group of three papers deal with Abstract Algebra. Among the six papers on Linear Algebra, a second group of four papers focus on elementary aspects of Linear Algebra related to linearity: subspaces, linear spans, linear transformations. The last group comprises two papers that focus on assessing conceptual understanding in the context of either multi-linear forms (Donevska-Todorova, 2016) or eigentheory (Wawro et al., 2018).

Our goal is to account for "burning issues" within UME research on Abstract and Linear Algebra, with its related methodological challenges, and to point out current and new avenues for research. We also aim at cross-analysing results and methodologies, thus answering the question: in which respect do these studies complement each other or contrast with each other? What are the main results, open questions, debates among researchers, elements of convergence/divergence within these studies? We thus decided to organise our synthesis according to the three groups of papers mentioned above (merely according to the topic), and for each group of papers to present a cross-analysis of these papers according to main themes that crystallise those burning issues (in terms of epistemological content, methodological issues, results, in fact the main objects of research that seemed to us appropriate for a vivid, illuminating and contrasted account of our data). We end this chapter by summarising what appeared to us as major advances in research on Abstract and Linear Algebra teaching and learning through the work of the INDRUM network as well as the further avenues for research that have been brought to light.

INDRUM contributions to abstract algebra teaching and learning

All three INDRUM papers related to Abstract Algebra were presented and discussed during the INDRUM 2016 conference in Montpellier. Two of them were later published in an expanded format inside the IJRUME special issue on INDRUM 2016 (Hausberger, 2018; Bosch et al., 2018). The augmented versions will also be considered in this synthesis.

From the point of view of mathematical topics, both Bosch et al. and Ioannou focus on Group Theory, while Hausberger considers Abstract Algebra at large through the analysis of structuralist thinking and praxeologies, yet with examples taken from the context of Ring Theory. From the point of view of theoretical frameworks, both Bosch et al. and Hausberger anchor their research in the framework of the Anthropological Theory of the Didactic (ATD; see Bosch & Gascón, 2014). Ioannou uses the Commognitive Theoretical Framework (CTF) developed by Sfard (2008). Both frameworks can be considered as adopting a sociocultural perspective, while ATD may be described as anthropological or institutional and CTF as discursive. This common sociocultural ground should facilitate the networking (Bikner-Ahsbahs & Prediger, 2014) of both theories as two languages of description of educational practices and two lenses to analyse data and produce results. Discussions in this spirit took place during the working group sessions and will be accounted for in the sequel.

Ioannou focuses on students' difficulties with regard to formalism and abstraction; he provides insights about the conceptualisation of groups viewed through the development of the mathematical discourse. Hausberger starts with an epistemological investigation of algebraic structuralism as a mathematical practice and then raises the ecological question of conditions and constraints for the development of such a praxis (with its corresponding logos, the discourse on the praxis) in an educational context. Finally, Bosch et al. propose an inverse move from didactics to epistemology: the critique of the current teaching paradigm of Abstract Algebra and the didactical quest for a more inquiry-based approach leads to an epistemological inquiry into the rationale of Group Theory. As we may see, this brief account of the content of the three papers already underlines how these contributions complement each other. Indeed, the emphasis lies either on the epistemology of Abstract Algebra, the understanding of students' difficulties, or the development of new instructional approaches.

For each of these three themes, the points of views of the three papers will be analysed and contrasted below.

The epistemological question: what is Abstract Algebra about?

Structuralist thinking and praxeologies (Hausberger)

The starting point of the methodology followed by Hausberger is an epistemological study of the rationale of algebraic structures, based on the work of historians and philosophers: "the goal is to bring to light the anthropological roots of mathematical structuralism in order to be able to build a praxeological model of reference for the actual mathematical practice in Abstract Algebra" (Hausberger, 2018, p. 78).

The methodological dimension of structuralist thinking is then brought to the fore. Bourbaki for instance, a group of French mathematicians who promoted the views of the German algebraists, spoke of the "axiomatic method" in terms of a "standardisation of mathematical techniques". The application of the method relies on a dialectic between the particular and the general, or in other words between objects and structures. The questions and problems are raised to a higher level of generality in order to apply structuralist concepts (e.g., ideal, principal ideal domain, etc.) and tools (e.g., isomorphism theorems, structure theorems, combinatorial of structures, etc.) according to the motto "generalising is simplifying". Abstract Algebra concepts may thus be called FUGS (formalising, unifying, generalising and simplifying) concepts, following Dorier (1995) and Dorier (2000, p.98)[1].

Anchoring his formalisation within ATD, Hausberger (2016, 2018) calls "structuralist praxeologies" these new praxeologies (combined praxis and logos) that originate from the rewriting of classical algebra in terms of structures. The theoretical construct is illustrated in the context of Ring Theory by an analysis of a thread on an online discussion forum. In order to prove that the ring **D** of decimal numbers is a principal ideal domain, the group of learners searches for a proof of the general statement that any subring of **Q** is principal, then investigates whether principality is transferred from a ring to its subrings. These reasonings transform a type of task T into a generalisation T^g that can be treated using structuralist techniques. The objects-structures dialectic is thus visible in the study process of the group and leads to the development of structuralist praxeologies.

The rationale of Group Theory (Bosch et al.)

By contrast, Bosch et al. focus on a given topic within Abstract Algebra: Group Theory (GT). The epistemological investigation is motivated by the aim to "focus, not only on the official raison d'être of GT within university teaching, but also on different possible alternative ones that could motivate or impel the use of GT to solve problematic questions" (Bosch et al. 2018, p. 32). The authors look for external tasks (that is, external to GT) that could lead to the reproduction of a substantial part of GT as a means to ascribe some rationale to it.

This quest is close to that of a fundamental situation, in the sense of TDS: the epistemic game relies on a question suitable to generate an inquiry process that bears the necessity of GT concepts. Bosch et al. refute the argument that the so-called FUGS concepts are antagonistic to the elaboration of fundamental situations (see Dorier 1995, p. 180) since FUGS concepts are not specific to higher education (the notion of number is already FUGS). This debate suggests that further epistemological insight is needed in order to clarify the meaning of FUGS concepts and in particular identify characteristics that hinder the elaboration of fundamental situations. Finally, Bosch and al. argue that a counting problem (such as that of symmetries of a square) may be a suitable candidate for a reconstruction of elementary GT. This choice is justified a priori by the links between GT and the notions of symmetry and invariant in the historical development of GT and by the role played by Lagrange's theorem as a tool to solve the problem.

Nevertheless, Bosch et al. question the feasibility and potential of such an approach: for instance, "is it substantial enough to motivate the study of the isomorphism theorems"? Such a question allows us to draw connections with the structuralist approach conducted by Hausberger: indeed, isomorphism theorems are structuralist tools elaborated by Noether in an approach transversal to Group and Ring theories that elucidated the relationships between both theories and used the meta-concept of structure as an organising principle of the mathematical discourse. This epistemological argument unveils a specificity of Abstract Algebra FUGS concepts: if groups encode symmetries, a substantial part of the rationale of GT relies on its relationship with the structuralist methodology. In other words, the question "What is GT good for?" cannot be separated from another more meta-level question: "Why is GT formalised this way?"

The group structure: a d-object endowed with meta-level aspects (Ioannou)

The latter question on the form of the mathematical discourse certainly meets the point of view of CTF, which sees mathematical objects as discursive constructs. Ioannou identifies the group concept as a *compound discursive object* (or d-object), since it does not relate primarily to a perceptually accessible entity. The process of construction of the d-object is not elucidated further in terms of CTF theoretical constructs (saming, encapsulating, reifying). According to Sfard, algebra may be described as "metaarithmetic, that is the unification of arithmetic with its own metadiscourse" (Sfard, 2008, p. 120), a discourse on expressions. The new question raised by Hausberger's account of algebraic structuralism, to be elucidated, would be: how to analyse, in CTF terms, Abstract Algebra as a meta-level discourse on classical algebra? In particular, how to account for the group concept as a meta-level object with respect to instances of groups such that groups of symmetries, modular arithmetic, etc.? Answering these questions would refine the perspective raised by Ioannou that "GT can be considered as a meta-level development of the theory of permutations and symmetries".

The group axioms are seen by Ioannou as *object-level rules*: they are part of the discourse about groups, as d-objects, and define the regularities in the behaviour of those, for instance in the task "Suppose (G,$\circ$) is a group with the property that $g^2 = e$ for all $g \in$ G. Prove that for all g_1, $g_2 \in$ G, we have $g_1 \circ g_2 = g_2 \circ g_1$ (that is, G is abelian)" considered by Ioannou. The visual mediators are those of symbolic algebra and the expected proof routine is based on algebraic manipulations, the symbolic computations being interpreted as relations satisfied by elements in a group (the key is to note that $(g_1 \circ g_2)^{-1} = g_2^{-1} \circ g_1^{-1}$, every element of order 2 being its own inverse). The task does not show a dialectic between objects and structures (in the sense of Hausberger), thus meta-level aspects of the group structure as narratives (e.g., definitions, propositions) that subsume discourses on lower-level objects (instances of groups). Nevertheless, the group axioms (object-level rules) may be seen as meta-level rules with respect to symbolic algebra.

Understanding students' difficulties: from an analysis of local cognitive problems to the questioning of global institutional choices

Dealing with object-level rules and meta-rules (Ioannou)

Ioannou focuses on "Year 2 mathematics students' conceptual difficulties and the emerging learning and communication aspects in their first encounter with GT". According to commognitive views, such an encounter should be reflected in the learning process by an expansion of the discourse: both an *exogeneous* type of expansion due to the endorsement of group-theoretic narratives and routines as changes in the meta-rules required by the application of the axiomatic method (*meta-level learning*), and an *endogeneous* expansion as a growth in the complexity of the endorsed narratives (*object-level learning*).

The above epistemological account of the task put to the fore in Ioannou's paper has shown that the proof routine takes merely the form of an endogeneous discursive expansion inside symbolic algebra. Nevertheless, a complete proof requires that the syntactical transformations be rigorously justified by means of group axioms and extra properties on G. These narratives may be considered an exogeneous expansion of the discourse, especially in the case of students who have just encountered the axiomatic definition of a group and may thus experience difficulties in endorsing them as object-level rules due to insufficient objectification of the group concept. The data presented by Ioannou support this analysis; in particular, the associativity of the group law remained totally invisible to several students. Their narratives conflict with those of the instructors whose annotations on the students' sheet emphasise the new meta-rules (deduction according to the axiomatic method).

This phenomenon is also related to what Hausberger (2018, p. 81) calls the structuralist dimension of a praxeology: this dimension measures, for instance, the distance between the narrative "$g^2 = e \Rightarrow g = g^{-1}\ \forall g$" produced by a student and a more conceptual narrative such as "any element of order less or equal to 2 in a group is its own inverse". In CTF terms, the latter narrative is the result of a subsequent endogeneous discursive expansion inside GT, which cannot be observed in the students' production at this stage of the learning process.

The access to structuralist thinking: a transition problem (Hausberger)

Hausberger focuses on students' difficulties to access structuralist thinking and analyses this issue in terms of a *transition problem* internal to the university curriculum. The phenomenon should be distinguished from the case of the teaching and learning of Linear Algebra: although both topics offer challenges regarding mathematical formalism and abstraction, Abstract Algebra raises the *level of unification* (Hausberger 2018, p. 75) one step further, which appears as conceptually more demanding. Indeed, algebraic structures are presented in a unified treatment (the same types of questions are raised about different structures and solved using similar tools, highlighting bridges between structures), which goes beyond the unification of different mathematical contexts under a common structure.

Inspired by Winsløw's praxeological formalisation (Winsløw, 2008) of the concrete to abstract transition in analysis (from calculus tasks to more theoretical tasks involving continuity and differentiability of functions as well as the topology of real numbers), Hausberger proposes a model for the epistemological transition to Abstract Algebra, in two phases: the first phase is concerned with the transition from T to T^g described above and leads to the construction of a structuralist praxeology as a fertile strategy to prove properties of concrete objects; the second phase builds on structuralist praxeologies previously developed in order to introduce more abstract and theoretical types of tasks that only consider classes of objects with their structural properties (e.g.: show that a Noetherian integral domain such that every maximal ideal is principal is a principal ideal domain).

This decomposition brings insights into transition issues potentially created by institutional choices: regarding phase 1, a structuralist praxeology may be taught with an artificial praxis block disconnected from the origin T, thus the rationale that motivated the praxeology in history. About phase 2, a transition problem may occur whenever the praxeologies that serve as anchor points to further praxeological development are not available in the praxeological equipment of learners, or the links between praxeologies are too weak. As a methodology, Hausberger suggests conducting praxeological analyses of textbooks and teaching material using this model in order to pinpoint continuities and discontinuities in the mathematical organisations.

A critical view on the standard monumentalist paradigm (Bosch et al.)

Hausberger's institutional perspective on transition issues in Abstract Algebra meets the critique addressed by Bosch et al. towards what is identified as the current dominant teaching paradigm at university and emblematically called the *paradigm of visiting works* in ATD accounts. This paradigm is characterised by *applicationism* (theory precedes applications) as the dominant epistemology and *monumentalism* (contents are rarely questioned and problematised) as the dominant pedagogy. In GT, typical tasks such as the determination of isomorphism classes of groups of given orders are merely internal to the theory and follow the theoretical exposition rather than motivate its development. In ATD terms, "mathematical techniques and technologies do not evolve urged by mathematical types of tasks". As a consequence, GT "exists for its own sake and is its own raison d'être", so that "students risk missing the answers to important questions concerning GT, especially about the motivation of the theory and its use in different domains or disciplines" (Bosch et al., 2018, p. 27). These remarks motivate a further investigation of the rationale of GT as presented above.

Which instructional approach for algebraic structures?

Two propositions from ATD at different stages of its development (Bosch et al.)

Theoretical ideas on intervention are the core of the paper by Bosch et al. Two instructional devices are presented, corresponding to two different stages in the development of ATD. The first one, called "practical workshops", is meant to complement traditional lectures and tutorials by lab sessions centred on the *praxis* (problems and techniques) instead of the *logos* (notions, properties and theorems) which is the primary focus of lectures. The design is based on a praxeological analysis of the mathematical content: the goal is a partial reconstruction of the subject matter into paradigmatic types of tasks and associated main techniques, the study of which is eager to "generate" the subject to be taught. This means that this knowledge will be introduced (by the students or the instructor) in order to fulfil new theoretical needs related to the scope of the techniques, the limits of the types of tasks, etc. For instance, students were given a list of 31 groups, sorted by their order, from 2 to 8, and were asked to determine which ones are isomorphic and which ones are not. This task led to theoretical developments such as the investigation of properties preserved by isomorphism, a classification of groups up to isomorphism for given orders, structural decompositions of groups into simpler ones, etc. This contrasts with applicationism (see above) and contributes to helping students make sense of the taught knowledge: the rationale of notions and theorems is made explicit in relation to these key problems. Yet, Bosch et al. argue that the reasons for studying the initial problem remain absent: why do mathematicians consider isomorphism classes of groups? Instructional proposals that tackle this question may be found in the literature: for instance, Larsen (2013) has designed a "local instructional theory" in order to engage students in the "guided reinvention" (RME) of the concepts of groups and group isomorphism.

The quest of Bosch et al. is more global: following Chevallard's (2015) idea of an inquiry-based counter-paradigm to visiting works, called the *paradigm of questioning the world*, the authors aim for a reorganisation of the curriculum, "not around works to be studied but around questions to be approached". This is why the question of rationale of GT needs to be investigated (see above). To implement the study, ATD proposes an instructional device called Study and Research Paths (SRP; Chevallard, 2015). Nevertheless, contrary to the first device, the proposal of using generating questions such as counting problems on symmetries within an SRP has not been carried out by the authors and therefore remains hypothetical.

Questioning the world and the objects-structures dialectic (Hausberger)

Hausberger (2018) experimented an SRP on arithmetical properties of the ring **D** of decimal numbers. Students worked on a transcript of the thread, extracted from the online forum, which was presented above to illustrate the development of structuralist praxeologies. The theoretical ideas developed by Hausberger in this pilot study meet the project of implementing the paradigm of questioning the world, in the sense that questioning the world "amounts to questioning mathematical objects themselves in such a way that may be developed a fruitful dialectic between objects and structures" (Hausberger, 2018, p. 91). The quest for an external rationale, presented by Bosch et al. as a didactical request, is thus seen by Hausberger as epistemologically grounded in the origin of Abstract Algebra itself as a refoundation of pre-structuralist algebra (thus external to the theory). Many questions about the ecology of such an approach (how to sustain the SRP) are raised by the author. For instance, elementary techniques may hinder the development of more structuralist ones. Therefore, Hausberger relies on the transcript to enrich the didactical milieu, in other word to foster learning by acculturation through the study and sense-making of the discourse of more advanced learners. This certainly meets commognitive views.

Commognitive conflicts and meta-level learning (Ioannou)

CTF assumes that "any substantial change in individual discourse, one that involves a modification in meta-rules or introduction of whole new mathematical objects, must be mediated by experienced interlocutors" (Sfard 2008, p. 254). A proper instructional approach thus relies on the determination of the conditions for effective mediation, which resonates with the ecological approach of ATD (conditions and constraints that hinder or foster the development of praxeologies). We have seen that the transition to Abstract Algebra requires an exogenous change in discourse. CTF further assumes that such a meta-level change is unlikely to be initiated by the learners themselves, but by the experienced interlocutor (in most cases, the instructor), in the context of a *commognitive conflict*. This is a situation in which the different discursants are acting according to different meta-rules, and such a conflict is seen as a potential trigger of meta-level learning. The stage of the research presented by Ioannou (2016) is restricted to the identification of commognitive conflicts and their root causes. A further study of the students' rationalisation of the discursive ways of the expert interlocutor in the expected phase of resolution of the conflict would be needed in order to evaluate the effectiveness of the task and mediation. Nevertheless, the presence of commognitive conflicts is already quite informative: it is a marker for the transition issues identified by Hausberger on epistemological and didactical grounds. It also reminds the researcher in the quest for means to create continuities or meaning through task-design that incommensurable discourses are both unavoidable to some extent and potentially fertile as a lever.

INDRUM contributions to linear algebra teaching and learning

The theme of linear spans and linear transformations

Three of the papers in this section were presented at the INDRUM 2018 conference with one (Trigueros & Biachini, 2016) at the INDRUM 2016 conference. To varying degrees, each of the four papers in this section includes the following components: (1) the creation of innovative tasks, (2) an a priori analysis of the mathematics involved in solving those tasks, (3) an implementation of the tasks in an interactive environment, and (4) an analysis of students' thinking when solving these tasks.

Mathematically, the papers address issues of linearity, with two papers focused on linear transformations and the other two discussing the linearity of a subspace or a linear span. Beyond this overlap, the linear transformation papers touch on a wide range of issues including whether a transformation is a mapping that is an endomorphism, a function, surjective, and/or injective, and with connections to composition of functions, invertibility and matrix representations. The subspace and span papers connect to other content such as systems of equations, geometric representations and linear independence.

In terms of student thinking, each paper has a focus on students' exploration of the ideas on their own as well as the importance of interactions with a tutor or a researcher to perturb their ideas for further learning. The papers cite the importance of both informal reasoning, such as work generating examples and hypotheses, and also reasoning that leads to abstraction, generalisation and making connections. Each paper also discusses different registers or representations of the mathematical constructs used by students. In what follows we provide an overview of results and framings from each paper within the above topic areas and with an emphasis on synthesising across the papers.

The role of an a priori analysis

Each of the four papers in this section analyses the mathematical or didactic aspects of tasks prior to using them with students. Lalaude-Labayle et al. (2018) draw on an *a priori analysis* as one component of their use of the Theory of Didactical Situations (TDS; Artigue et al., 2014). This analysis includes a mathematical analysis of the several ways that the linear transformation problem can be solved as well as a didactic analysis of the role of the student in navigating through the different strata of the didactical *milieu* during the solving process and the difficulties a student may have in solving the given problems.

Fleischmann and Biehler (2018) follow Biehler, Kortemeyer, and Schaper (2015) in calling the a priori analysis a *student expert solution* (SES). This includes an idealised student solution based on the knowledge a student would have at this point in the course. In addition, it provides alternative solution methods and other information such as learning objectives. The methodology is applied to discuss the mathematics and potential student thinking involved in a set of problems about vector spaces in $\mathbf{R}^2$.

Trigueros and Biahchini (2016) employ a related, but distinct type of a priori analysis that is part of their framing of their work using the Action-Process-Object-Schema (APOS) theory (Arnon et al., 2014). In this theory, researchers develop an a priori *genetic decomposition* (GD) that predicts the mental constructions needed to understand the mathematical concept of the study. In this case, Trigueros and Biahchini draw on a slightly modified version of the GD created by Roa Fuentes and Oktaç (2010) for the concept of linear transformation. Their GD focuses on the necessary aspects of checking or understanding that the transformation satisfies the linearity properties as well as the relationship between a linear transformation and its associated matrix transformation.

Although the fourth paper in this group, Papadaki and Kourouniotis (2018) does not provide a TDS a priori analysis nor a GD, it does clearly lay out the rationale for the tasks in ways that overlap information included in the a priori analyses of the other papers. The intention of the tasks is to cause the students to encounter a potential conflict factor that can then be resolved through discussion. To create such a task the authors analyse three aspects of the concept of linear span that their tasks include as well as two conflict factors that may occur in student reasoning. In this way, all four papers include some form of a priori analyses of the mathematics involved in solving the task and potential pedagogical issues.

The role of multiple representations

In analysing students' reasoning, several different symbol systems were prevalent across the four papers. Lalaude-Labayle et al. (2018) frame their discussion of student productions in terms of Pierce's semiotics as well as Duval's (2017) registers of semiotic representations. For the linear transformation questions reported in this study these registers are each symbolic. For example, they ask students to analyse the linear transformation $\phi(P) = P(X + 1) - P(X)$ where P is a polynomial of degree at most n, i.e., the domain is $\mathbf{R}_n[X]$. The problem can be solved in terms of the polynomials directly or by converting the linear transformation into a matrix in the canonical basis of $\mathbf{R}_n[X]$. In each case the student would be working in a different register of semiotic representations, both symbolic.

The other three papers in this group had a stronger focus on connecting graphical representations to the important symbolic representations for the mathematics in their tasks. Trigueros and Biahchini (2016) discussed linear transformations in the context of a cartoonist creating different drawings of a man on a bicycle and needing calculations to convert between them. Students converted between this graphical representation and two different symbolic representations: function notation such as $T(a + b) = \mathrm{T}(a) + \mathrm{T}(b)$ and matrix notation.

In the context of subspaces, Fleischmann and Biehler (2018) created questions that had what Gravesen, Grønbæk and Winsløw (2017) call *linkage potential*. In this case, they stated sets in $\mathbf{R}^2$ in the form of implicitly defined equations of lines and other graphs that they hoped students would connect with their school knowledge about geometric objects and equations. Papadaki and Kourouniotis (2018) expected students to be familiar with geometric representations of 1-, 2-, and 3-dimensional subspaces of $\mathbf{R}^3$ from earlier in the course. If students had trouble remembering them, the tasks gave the opportunity for discussion. In both cases, students were expected to connect a geometric representation in $\mathbf{R}^2$ or $\mathbf{R}^3$ with a symbolic way of representing these sets. In the former paper, sets were given using a relational description, e.g., $\{(x_1,x_2) \in \mathbf{R}: x_1 + 2x_2 = 0\}$, whereas in the latter, the symbolic representation was primarily intended to be linear combinations of vectors.

What is clear from just these few examples is the vast number of relevant representation registers involved in linear algebra problem solving and that different papers on the same mathematical content may focus on different registers. The examples above include systems of linear equations, vector equations, matrix equations, and linear transformation notation, each of which has its own unique geometric representation (Larson & Zandieh, 2013).

The role of examples, modelling and heuristic reasoning

Although each paper indicated the importance of formal deductive reasoning, each paper additionally emphasised the importance of students' use of more informal reasoning to generate problem solving ideas. Fleischmann and Biehler (2018) asked students to determine whether several relationally defined sets were or were not a vector space. Then they asked students to find all subspaces of $\mathbf{R}^2$. In doing so, they expected that students would rely at least in part on the examples and non-examples they had generated. In addition, they described this task as one with *research potential* (Gravesen, Grønbæk, & Winsløw, 2017). In other words, it is a research-like activity based on an open question that allows students to form hypotheses and explore examples of subspaces. These generative activities go beyond a more standard task such as requesting students to make a deductive argument by applying the definition of a subspace.

Other papers also expected students to explore sample spaces and use less formal reasoning. Papadaki and Kourouniotis (2018) expected students to reach for vectors in their *example space* (Mason & Watson, 2008) when trying to determine whether a set of three vectors could be linearly dependent even if the third vector is not in the span of the first two.

The Trigueros and Biahchini (2016) paper focused on student work with a modelling problem. They found that the activities caused students to use models to make predictions and to explore those predictions. In this way, the modelling situations allowed students to construct new knowledge.

Lalaude-Labayle et al. (2018) conclude that it would be better for students' learning if the students engaged in a more heuristic approach. By this, Lalaude-Labayle et al. refer to a heuristic milieu that is grounded in the mathematical task and allows students to test and validate or invalidate their conjectures. The student in their study spent more time in the reference milieu, applying learned procedures to the linear transformation tasks, but struggled when requested to explore the heuristic milieu by looking at examples such as calculating ϕP for various values of P. The observed lack of articulation of the reference milieu and heuristic milieu hindered the student in finding a break-through in resolving the more open-ended aspects of the task.

Across the four papers, there is an expectation that students should use reasoning that is not always deductive and that incorporates exploration and conjecture. This includes creating examples and non-examples.

The role of the teacher

Fleischmann and Biehler (2018) collected data in the form of written work from around 100 students and also video data from 3 groups of 2–4 students each. The students completed the tasks as part of a tutorial session. Although not reported in detail in this paper, the authors were interested in the role of the tutor in students' developing understanding.

Similarly, Papadaki and Kourouniotis (2018) developed tasks intentionally to create conflict factors that would then force a discussion between the students solving the task and others. The interview data that they present does not delve into the role of the interviewer, but they discuss switching these tasks to problem sessions at their institution because of the richness of conversations that they can generate.

Lalaude-Labayle et al. use the notion of milieu as a strong component of the TDS framing. Although TDS plans 7 phases, this paper focuses on a heuristic phase which is based on the student's work in problem situation, through a formulation and validation phase, to an institutionalisation involving the teacher. In this way, the student's work on the problem is initially adidactic, but later involves a didactical moment in interaction with the teacher. The INDRUM paper of Lalaude-Labayle et al. did not have room to go into depth about the role of the teacher in working with the student, but the examples given involve encouraging the student to engage the problem in a more heuristic way and with alternative registers such as a matrix representation.

Trigueros and Biahchini describe a study that took place over a longer period compared with the other studies (4–5 sessions compared with 1 session). Because of this, there were more opportunities for a variety of roles of the teacher to emerge. As part of their work on the modelling tasks, students were sometimes engaged in mathematical ideas that they had not yet learned in a formal way. An example of this highlights a role of the teacher that was unique to the Trigueros and Biahchini paper among the four discussed in this section. The students had been engaged in modelling tasks where they were asked to create matrix representations for linear transformations that had been presented to them in a graphical and applied setting. As part of this process, students began to notice particular properties that some of the transformations had. For example, one student said, "This one changes the form of the bicycle (shear transformation), but this one (translation) does not, it only changes its position in space" (p. 331). The teacher responded to these discussions by defining isometries and asking students to verify which transformations were isometries. In this way, she was able to build on students' current thinking by introducing formal mathematical terminology after the students had experienced the concept through their own engagement in the mathematics.

Although pedagogy was not the primary focus of any of the four papers, each one acknowledged the important role of the teacher. Trigueros and Biahchini had the most detailed discussion of the teacher's role; each paper noted the role teachers could play in encouraging the student to explore the mathematical ideas.

Assessing students' conceptual understanding

In order to evaluate students' conceptual understanding in two different central topics in Linear Algebra, Wawro, Zandieh, and Watson (2018) and Donevska-Todorova (2016) present two different approaches for the design and development of a set of tasks as an assessment instrument. In the following, we will first provide insight into the criteria for the classification of students' learning and reasoning. We then proceed with a description of the methods and results of the task design process in both studies. Finally, we will compare the dominant thinking patterns and solution strategies of the students that were identified in the studies.

Modes of understanding and modes of description in Linear Algebra

Both papers focus on the important concepts, processes and modes of description involved in understanding an aspect of linear algebra. Donevska-Todorova discusses this in terms of *procedural* and *conceptual understanding* (Hiebert, 1986), as well as the *modes of description and thinking* used by students when working with (new) mathematical concepts. The latter can be defined, based on Hillel's (2000) modes of descriptions in Linear Algebra (the *abstract, algebraic* and *geometric modes*) and Sierpinska's (2000) modes of thinking (*synthetic-geometric, analytic-arithmetic* or *analytic-structural*). Donevska-Todorova creates new categories using combinations of these designations to classify the possible solutions of the tasks used in her study, e.g. "abstract/analytic-structural" or "algebraic/arithmetic".

These modes of description/thinking are in fact one of three criteria that she defines to characterise students' work. The other two criteria are *properties of concepts in Linear Algebra* and *subject-specific strategies or solving tools in Linear Algebra.* These thus refine the modes of description by pointing out the main ingredients of the technique used to solve the task, and in particular the conceptual aspects. With respect to the content, she chooses exercises on bi-linear and multi-linear forms, given the lack of studies covering these topics. Her objective is to analyse the students' work with respect to their procedural and conceptual understanding of the subject, based on an attribution of different solution approaches to the three criteria described above.

In Wawro, Zandieh, and Watson (2018), the focus is on eigentheory. This is a rich area for study because of the complexity involved in coordinating the different mathematical concepts (e.g. matrices, vectors, scalars), processes (e.g. matrix multiplication, scalar multiplication) and modes of description that are part of eigentheory.

The two studies have in common that they make efforts to identify the depth of conceptual understanding in terms of the cognitive flexibility needed to deal with abstract mathematical concepts, using homogeneous, written tasks (in contrast to interviews or other individualised assessments of knowledge and skills), that were particularly developed for this purpose. The modes of description as defined by Hillel provide a system for the characterisation of thinking processes that are widely applicable within Linear Algebra. Considering central challenge for students' learning of Linear Algebra, both aspects must be taken into account: Firstly, the necessity of understanding of formal, axiomatic definitions of new concepts and their symbol-based transformations, and secondly the translation of these definitions and procedures to a geometric perspective.

Assessment in Linear Algebra: Discussion of exemplary tasks

An explicit objective of the work of Wawro et al. is to develop and test a set of exercises on eigentheory in order to create an assessment instrument on this topic that allows the evaluation of a student's level of knowledge as well as their preferred way of reasoning. This instrument consists of several exercises in the "Multiple Choice Extended" (MCE) format. In this format, each question begins with a multiple-choice element followed by a list of statements, from which all possible justifications for the answer of the first part must be selected. Central for the study's objective is the design of the provided possible justifications. The starting point was a working model for the understanding of eigentheory, in which three main settings for the framing of eigentheory were identified:

1. relationships indicated by the eigen-equation $Ax = \lambda x$;
2. relationships indicated by the homogeneous form of the eigen-equation $(A - \lambda I)x = 0$
3. relationships indicated by a linear combination of eigenvectors, i.e. relationships between vectors in the same eigenspace.

The working model furthermore is organised in four main interpretations that can be applied when working with the three settings, namely graphical, numeric, symbolic and verbal interpretations. We illustrate in Figure 8.1 these reflections on one of the tasks that were developed to "elicit student's thinking [...] within and across the settings and interpretations" of the working model.

In this task, a numeric interpretation is used, followed by several justifications where e. g., (i) is a symbolic interpretation in the $Ax = \lambda x$ setting, (iii) a symbolic interpretation in the $A - \lambda Ix = 0$ setting and (v) a geometric interpretation in the $Ax = \lambda x$ setting. Similarly, other parts of the MCE instrument were designed in order to provide tasks and justifications using different combinations of settings and interpretations from the working model. Parallel to the use of this MCE-instrument, another group

1. The matrix $A = \begin{bmatrix} -2 & 4 \\ 2 & 5 \end{bmatrix}$ has λ = 6 as one of its eigenvalues. Which of the following vectors is an eigenvector of A with coressponding eigenvalue λ = 6?

(a) $x = \begin{bmatrix} 4 \\ -1 \end{bmatrix}$ ← CORRECT

(b) $x = \begin{bmatrix} 1 \\ 2 \end{bmatrix}$

Because ... (select ALL that could justify your choice)

(i) This vector $\boldsymbol{x}$ makes $A\boldsymbol{x} = 6\boldsymbol{x}$ a true statement.
(ii) This vector $\boldsymbol{x}$ is the only vector in $\mathbb{R}^2$ for which $A\boldsymbol{x} = 6\boldsymbol{x}$.
(iii) This vector $\boldsymbol{x}$ makes $(A - 6I)\boldsymbol{x} = 0$ a true statement
(iv) Subtracting 6 from the diagonal of A yields this vector $\boldsymbol{x}$ as a column vector of the resulting matrix.
(v) The vector $A\boldsymbol{x}$ is 6 times the magnitude and in the same direction as this vector $\boldsymbol{x}$.
(vi) The matrix A also has λ = −3 as an eigenvalue.

FIGURE 8.1 Excerpt from the MCE format (Wawro et al., 2018, p. 278)

of students was asked to solve the same problem, but without a list of possible justifications. These students had to answer an open question and to formulate their own justification for their answer.

Unlike Wawro et al. who contrast closed-ended to open-ended tasks with the same mathematical contents, Donevska-Todorova creates two different types of tasks. Firstly, her study includes (written) open homework exercises, that allow multiple possible solutions ("multiple solution tasks", MST), where she classifies the various solutions according to the three criteria described in the previous section. For example, the solution of MST 1 (see Figure 8.2) is described as follows: (1) The mode of description/thinking presented in the given solution is *algebraic/arithmetic*, (2) the property of concepts used here is that *a determinant of an* $n \times n$*-matrix is a sum of determinants of n sub-matrices of dimension* $(n - 1) \times (n - 1)$ and (3) the applied strategy is based on *Laplace (cofactor) expansion.*

Similarly, she identifies five other solutions that are possible (based on the previous contents of the lecture) for the students to solve the task, including the use of Sarrus rule, geometric considerations and more.

The second type of exercises is intended for (oral) discussion. The tasks are initially closed, since students are asked to determine the truth value of given statements, but in the study, those questions were opened during the discussion when students were asked to justify their answer. An example for such a question is whether the statement $\det(A + B) = \det(A) + \det(B)$ holds for arbitrary $n \times n$-matrices with real entries.

Reasoning processes and thinking patterns of the students

As we have seen in the previous examples, both studies choose different approaches to designing open and closed tasks that are supposed to allow conclusions concerning the students' level of conceptual understanding of the given subjects. Regarding the two dimensions of classification, the findings of both studies seem to confirm and complement each other. In both cases, students showed a distinct favour for algebraic reasonings over geometric arguments. Where Wawro et al. argue that this might be caused by the given descriptions in their exercises,

MST1: Find the determinant of the matrix

$$M = \begin{pmatrix} 2 & 0 & 0 \\ 0 & 2 & 0 \\ 0 & 0 & 2 \end{pmatrix}.$$

writes as many solutions as you can.

Solution 1:

$$detM = det \begin{pmatrix} 2 & 0 & 0 \\ 0 & 2 & 0 \\ 0 & 0 & 2 \end{pmatrix} \underline{\underline{def}}\ 2 \cdot 2 \cdot 2 \cdot$$

$$det \begin{pmatrix} 1 & 0 & 0 \\ 0 & 1 & 0 \\ 0 & 0 & 1 \end{pmatrix} \underline{\underline{def}}\ 8 \cdot 1 = 8.$$

FIGURE 8.2 Excerpt from the MST worksheet (Donevska-Todorova, 2016, p. 280)

Donevska-Todorova deduces that the students have developed mainly procedural understanding and do not show deeper conceptual understanding, which is, in her view, necessary for the transition from algebraic to geometric reasoning.

Both papers emphasise that the geometric mode of description and the related cognitive procedures are likely to be not as intuitive to the students as more algebraic approaches. The influence of the mode of description used to state the task seems to be strong, particularly in the case of open questions. In fact, many students engage in geometric reasoning whenever tasks are presented geometrically, but they do not provide similar answers themselves if no specific request is stated. Wawro et al. moreover conjecture that students might consider algebraic justifications for mathematical statements more acceptable. This effect of the didactical contract, as perceived by the students, may therefore reduce the possibility of deducing from students' answers their ability to use and move between different modes of description. Wawro et al.'s use of the MCE format is meant in part to mitigate this effect. Nevertheless, both papers provide and discuss a format of innovative tasks to assess learning. Both works emphasise the process of reasoning and justification over the reproduction of calculations. In addition, both papers take into consideration potential limitations of the tasks in terms of the challenge the new format of tasks may incur for both students and researchers.

Conclusion

How has research in Linear Algebra teaching and learning advanced since the publication of Dorier's (2000) synthesis? In his conclusion, Dorier emphasised three main points: the diagnosis of epistemological issues that, for instance, resulted in the elaboration of Hillel's (2000) and Sierpinska's (2000) frameworks; the challenge for students of navigating flexibly through the different modes of description and registers of representation; finally, the challenge for UME research of developing and evaluating long-term intervention plans that, according to him, were necessary in the case of FUGS concepts.

The current INDRUM research on Linear Algebra inherits these epistemological guidelines through Hillel's and Sierpinska's frameworks. Much emphasis is still made on the investigation of student thinking through descriptive studies. Attempts are made to systematise this research and develop assessment instruments (Wawro et al.; Donevska-Todorova). Another current path is the refinement of the semiotic analysis by means of Pierce's semiotics (Lalaude-Labayle et al., 2018).

Most of the studies are mainly local, focusing on a single major concept like linear spans, linear transformations or multi-linearity, and restricted to a single teaching session. These studies point out several levers to foster learning and conceptualisation: Papadaki highlights the role of (cognitive) conflict factors and illustrates how lecturers may design tasks, inspired by their observations of students' misconceptions; Trigueros & Biahchini point out the use of modelling situations and design via APOS genetic decompositions; Fleischmann & Biehler conclude that more guidance and preparation is required for the tutorials since, for instance, students experience difficulties in applying geometric knowledge from high school; finally Lalaude-Labayle et al., in terms of TDS, diagnose a lack of articulation between the *heuristic milieu* and the *reference milieu* which is also an indicator of how to organise the epistemic game better in problem-solving activities.

Long-term didactical engineering was not presented at INDRUM. One may hypothesise that Dorier's argument rather than a paper's page restriction explains this fact. The high complexity of implementation and evaluation of long-term teaching design due to numerous global and local choices hinders the possibility for this type of research. Therefore, collaboration with university mathematics teachers is more than ever needed in order to make a breakthrough in the scale of studies as well in the dissemination of the approaches that proved to be fruitful in pilot studies.

The number of INDRUM papers on Abstract Algebra (3) was half that of the Linear Algebra papers and all of them were presented at the first INDRUM. Research in this direction remains limited, but there continue to be new avenues for research. The use of ATD as a theoretical framework puts more emphasis on institutional conditions and constraints; it invites us not to take the knowledge to be taught as a given construct but on the contrary to question this knowledge (its rationale) and envisage other potentially very different mathematical organisations. This raises the need for more epistemological input, for example the description of structuralist praxeologies by Hausberger or the quest for alternative non-official and external (to the theory) raisons d'être for Group Theory. The critique of the current dominant pedagogy at the university and the development of ATD tools for inquiry-based approaches set up the ground for a study by Bosch et al. of a means to offer a complete reorganisation of the GT syllabus around questions and problems. Again, close collaboration with mathematicians and further studies will be necessary to implement such a programme.

In a different approach, Ioannou interprets students' thinking and learning in terms of the development of the mathematical discourse. The challenge of structuralist thinking described by Hausberger in epistemological terms translates into potential commognitive conflicts that CTF sees as a potent lever for the learning. In this respect yet with a different theoretical framework, this study meets the spirit of several INDRUM studies on Linear Algebra.

Our synthesis of INDRUM contributions to Linear and Abstract Algebra teaching and learning divided papers according to the topic. This choice reflects the way papers were presented and discussed at the INDRUM conferences. Yet, this separation is certainly arbitrary on the epistemological point of view, as evidenced by Hausberger who considers Abstract Algebra at large through the study of structuralist thinking. This trend based on the axiomatic method encompasses both Abstract and Linear Algebra, although each topic (and subtopic such as GT, etc.) has different historical roots, in other words, a different history of core problems. Therefore, the mainstream organisation of the curriculum as Linear Algebra followed by Abstract Algebra may also be questioned, either in terms of a transition: what kinds of continuities/discontinuities may be observed in the teaching and learning of these topics (modes of representations and reasoning, structuralist thinking and praxeologies)? Or in terms of possible alternative reorganisations of the curriculum, could one conceive of a more spiralling approach to both subjects? We are calling for such studies and hope that forthcoming INDRUM conferences will offer opportunities for more interrelations of studies on Abstract and Linear Algebra.

Note

1 The theoretical construct of a FUG concept was introduced in mathematics education research by Robert and Robinet in 1987 (Dorier, 1995, p. 175). The simplifying dimension of Linear Algebra was later investigated by Dorier (2000, part I and part II, chap. 1) from an epistemological and didactical perspective.

References

Arnon, I., Cottrill, J., Dubinsky, E., Oktaç, A., Roa Fuentes, S., Trigueros, M., & Weller, K. (2014). *APOS Theory: A framework for research and curriculum development in mathematics education*. New York: Springer Verlag.

Artigue, M., Haspekian, M., & Corblin-Lenfant, A. (2014). Introduction to the Theory of Didactical Situations (TDS). In A. Bikner-Ahsbahs & S. Prediger (Eds.), *Networking of Theories as a Research Practice in Mathematics Education* (pp. 47–65). Berlin: Springer.

Biehler, R., Kortemeyer, J., & Schaper, N. (2015). Conceptualizing and studying students' processes of solving typical problems in introductory engineering courses requiring mathematical competences. In K. Krainer & N. Vondrová (Eds.), *Proceedings of the Ninth Congress of the European Society for Research in Mathematics Education (CERME9, 4–8 February 2015)* (pp. 2060–2066). Prague: Charles University in Prague, Faculty of Education and ERME.

Bikner-Ahsbahs, A., & Prediger, S. (2014). *Networking of Theories as a Research Practice in Mathematics Education*. Berlin: Springer.

Bosch, M., & Gascón, J. (2014). Introduction to the Anthropological Theory of the Didactic (ATD). In A. Bikner-Ahsbahs & S. Prediger (Eds.), *Networking of Theories as a Research Practice in Mathematics Education* (pp. 67–83). Berlin: Springer.

Bosch, M., Gascón, J., & Nicolás, P. (2018). Questioning Mathematical Knowledge in Different Didactic Paradigms: The Case of Group Theory. *International Journal of Research in Undergraduate Mathematics Education*, 4(1), 23–37.

Bosch, M., Gascón, J., & Nicolás, P. (2016). From "monumentalism" to "questioning the world": the case of Group Theory. *INDRUM 2016 Proceedings*, pp. 256–265.

Chevallard, Y. (2015). Teaching mathematics in tomorrow's society: A case for an oncoming counter paradigm. In S. J. Cho (Ed.), *The Proceedings of the 12th International Congress on Mathematical Education* (pp. 173–187). Dordrecht: Springer.

Corry, L. (2016). Algebra. In: *Encyclopædia Britannica*. Accessed 11 March 2018. https://global.britannica.com/topic/algebra/Structural-algebra.

Donevska-Todorova, A. (2016). Procedural and Conceptual Understanding in Undergraduate Linear Algebra. *INDRUM 2016 Proceedings*, pp. 276–285.

Duval, R. (2017). *Understanding the Mathematical Way of Thinking: The Registers of Semiotic Representations*. New York: Springer.

Dorier, J-L. (1995). Meta Level in the Teaching of Unifying and Generalizing Concepts in Mathematics. *Educational Studies in Mathematics*, 29, 175–197.

Dorier, J.-L. (2000). *On the Teaching of Linear Algebra.* Heidelberg: Springer Netherlands

Fleischmann, Y., & Biehler, R. (2018). Students' problems in the identification of subspaces in Linear Algebra. *INDRUM 2018 Proceedings*, pp. 224–233.

Gravesen, K. F., Grønbæk, N., & Winsløw, C. (2017). Task Design for Students' Work with Basic Theory in Analysis: The Cases of Multidimensional Differentiability and Curve Integrals. *International Journal of Research in Undergraduate Mathematics Education*, 3(1), 9–33.

Hausberger, T. (2016). A propos des praxéologies structuralistes en Algèbre Abstraite. *INDRUM 2016 Proceedings*, pp. 296–305.

Hausberger, T. (2018). Structuralist Praxeologies as a Research Program on the Teaching and Learning of Abstract Algebra. *International Journal of Research in Undergraduate Mathematics Education*, 4(1), 74–93.

Hausberger, T. (2020). Abstract Algebra Teaching and Learning. In S. Lerman (Ed.), *Encyclopedia of Mathematics Education* (second edition). London: Springer Nature.

Hiebert, J. (Ed.) (1986). *Conceptual and Procedural Knowledge: The Case of Mathematics*. Hillsdale, NJ: Erlbaum.

Hillel, J. (2000). Modes of description and the problem of representation in linear algebra. In *On the Teaching of Linear Algebra* (pp. 191–207). Heidelberg: Springer Netherlands.

Ioannou, M. (2016). A commognitive analysis of mathematics undergraduates' responses to a commutativity verification Group Theory task. *INDRUM 2016 Proceedings*, pp. 306–315.

Lalaude-Labayle, M., Gibel, P., Bloch, I., & Lévi, L. (2018). A TDS analytical framework to study students' mathematical activity – An example: linear transformations at University. *INDRUM 2018 Proceedings*, pp. 234–243.

Larsen S (2013). A Local Instructional Theory for the Guided Reinvention of the Group and Isomorphism Concepts. *Journal of Mathematical Behavior*, 32(4), 712–725.

Larson, C., & Zandieh, M. (2013). Three Interpretations of the Matrix Equation Ax=b. *For the Learning of Mathematics*, 33(2), 11–17.

Mason, J., & Watson, A. (2008). Mathematics as a constructive activity: Exploiting dimensions of possible variation. In M. Carlson & C. Rasmussen (Eds.) *Making the connection: Research and teaching in undergraduate mathematics education* (pp. 191–204). Washington, DC: MAA.

Papadaki, E., & Kourouniotis, C. (2018). Tasks for enriching the understanding of the concept of linear span. *INDRUM 2018 Proceedings*, pp. 265–274.

Rasmussen, C., & Wawro, M. (2017). Post-calculus research in undergraduate mathematics education. In J. Cai (Ed.), *The Compendium for Research in Mathematics Education* (pp. 551–579). Reston: National Council of Teachers of Mathematics.

Roa Fuentes, S., & Oktaç, A. (2010). Construcción de una Descomposición Genética: Análisis Teórico del Concepto de Transformación Lineal. *Revista Latinoamericana de Matemática Educativa*, 13(1), 89–112.

Sfard, A. (2008). *Thinking as Communicating: Human Development, the Growth of Discourses, and Mathematizing*. New York: Cambridge University Press.

Sierpinska, A. (2000). On some aspects of students' thinking in linear algebra. In J.-L. Dorier (Ed.), *On the Teaching of Linear Algebra* (pp. 209–246). Heidelberg: Springer Netherlands.

Tall, D. (1991). *Advanced Mathematical Thinking*. Dordrecht: Kluwer.

Tall, D., & Vinner, S. (1981). Concept Image and Concept Definition in Mathematics with Special Reference to Limits and Continuity. *Educational Studies in Mathematics*, 12, 151–169.

Trigueros, M., & Biahchini, B. (2016). Learning Linear Transformations using models. *INDRUM 2016 Proceedings*, pp. 326–336.

Wawro, M., Zandieh, M., & Watson, K. (2018). Delineating Aspects of Understanding Eigentheory through Assessment Development. *INDRUM 2018 Proceedings*, pp. 275–284.

Winsløw, C., Gueudet, G., Hochmuth, R., & Nardi, N. (2018). Research on University Mathematics Education. In T. Dreyfus, M. Artigue, D. Potari, S. Prediger & K. Ruthven (Eds.), *Developing Research in Mathematics Education: Twenty Years of Communication, Cooperation and Collaboration in Europe* (pp. 60–74). New Perspectives on Research in Mathematics Education series, Vol. 1. London: Routledge.
Winsløw, C. (2008). Transformer la théorie en tâches: la transition du concret à l'abstrait en analyse réelle. In R. Rouchier *et al.* (Eds.) *Actes de la XIIIème Ecole d'Eté de Didactique des Mathématiques* (pp. 1–12). Grenoble: La Pensée Sauvage.

9

MATHEMATICS AND OTHER DISCIPLINES, AND THE ROLE OF MODELLING

Advances and challenges

Alejandro S. González-Martín, Ghislaine Gueudet, Berta Barquero and Avenilde Romo-Vázquez

Introduction

Ten years ago, Artigue, Batanero, and Kent (2007) stated that a major part of university mathematics education (UME) research had been established "having in mind more or less explicitly the mathematical education of 'pure mathematicians' or mathematics teachers and their particular needs" (p. 1030). Although in recent years more papers have examined a wider variety of student profiles, there is still much to do: in the opening session of the first INDRUM conference, Artigue (2016) stated that the practice of the mathematician is still an implicit reference in most research studies on UME. This indicates that many facets of UME continue to be under-researched. However, research examining how non-specialists learn and are taught university mathematics could yield many benefits, considering the number of students enrolled worldwide in mathematics service courses as compared with specialist mathematics courses. The high number of students in mathematics service courses is an important issue, due to the high number of students who drop out of their programmes after failing calculus (Rasmussen & Ellis, 2013). In particular, many engineering students abandon their programme after failing a mathematics course (Faulkner, Earl, & Herman, 2019); tellingly, some engineering programmes blame mathematics courses for as many as one third of their dropouts (Faulkner et al., 2019).

In recent years, UME researchers have shown a growing interest in mathematics for non-specialists (as reported in various research surveys on UME, e.g., Biza, Giraldo, Hochmuth, Khakbaz, & Rasmussen, 2016; Winsløw, Gueudet, Hochmuth, & Nardi, 2018). Many studies report that non-specialist students encounter difficulties with mathematics, identifying a common phenomenon that may be aggravating these difficulties: there is often no explicit link between the content of mathematics courses and the content of professionally-oriented courses specific to other fields of study (such as engineering). In fact, mathematics course content can differ significantly from the content of specialist courses (e.g., Harris, Black, Hernandez-Martinez, Pepin, Williams, & TransMaths Team, 2015) and often seems disconnected from students' future professional practice (Wake, 2014). Certain pedagogical interventions are therefore needed: teaching mathematics in a way that provides links with other fields (e.g., biology, economics or physics) and that draws on problems encountered in these disciplines or in the workplace. This raises the issue of modelling and the need to investigate how modelling activities can play a central role in such interventions.

The above also applies to engineering, which has attracted more research than other fields in recent years, giving rise to a whole area of study: mathematics education for engineers. Research on the professional practices of engineers (e.g., Artigue et al., 2007; Gainsburg, 2006; Kent & Noss, 2003) shows that, in many cases, engineers do not recognise the mathematics they use. The following statement illustrates this phenomenon:

> It is notable in talking to experienced engineers that the mathematics that is useful and relevant has in many cases long since come to be thought of as part of engineering, whereas mathematics that is *not* used is regarded as "mathematics"
>
> *(Artigue et al., 2007, p. 1034)*

This "invisibility" of mathematics can lead students (and decision-makers) to question the prominence of mathematics in engineering education. However, the aforementioned authors point out that this invisibility is due to the fact that most professional tasks requiring the explicit use of mathematics are carried out by specialists or are performed using sophisticated computer programmes that function as a *black box*. The use of tables and metalanguages that facilitate the implicit use of mathematics, often entwined with knowledge of the engineering sciences, is also evidenced. However, it is important to highlight that the professional needs of engineers are continuing to evolve, with the use of mathematics growing in importance. Examples include interpreting the multiplicity of computer-generated results, or using mathematical knowledge and language to communicate modelling tasks and results to other engineers (Quéré, 2019).

The above issues were raised in many of the works discussed in Thematic Working Group 2 (TWG2, "Mathematics for engineers – Modelling – Mathematics and other disciplines") at the INDRUM 2016 and INDRUM 2018 conferences. TWG2 had seven papers and two posters in 2016 and 10 papers and four posters in 2018; it offered an opportunity to present and discuss ongoing doctoral studies (e.g., Feudel, 2016; Tetaj, 2018) as well as projects at their beginning stages (e.g., Fredriksen & Hadjerrouit, 2016; Tetaj, 2018; Viirman, 2016) that later became CERME or journal publications. In what follows, we summarise some of what we see as the main contributions of INDRUM to the above themes. We focus mainly on contributions that directly address issues specific to the training of non-specialists (including the role of modelling in this context). Keeping in mind the content of TWG2 contributions, we have structured this chapter as follows:

those contributions that mainly address teaching and learning issues (including curriculum issues and the learning of specific topics), and those that mainly address interventions (such as the use of modelling or other types of intervention). We conclude the chapter with some reflections on potential ways forward for research on these themes.

Teaching mathematics to non-specialists: curriculum, resources, and learning

As mentioned above, research on engineering studies is attracting a growing number of mathematics education researchers. For this reason, this section is divided into two subsections: the first is devoted to engineers while the second looks at other cases involving non-specialists. In the context of INDRUM, these "other cases" focus on two fields of study: biology and economics.

Mathematics in engineering training

Around the world, engineering training programmes generally reflect the model of the École Polytechnique, founded in France in 1794 (Romo-Vázquez, 2009). The French mathematician Monge was a key player in its foundation, and he conceived of a curriculum organised like a tree, mathematical and physical knowledge forming two main branches (Belhoste, 1994), with geometry (where analysis was studied through geometrical applications) forming the core of the mathematical branch. In this way, "the curriculum takes on an encyclopaedic structure, where pieces of knowledge are linked one to the other, from the principles of science to practical applications[1]" (p. 13). In 1795, geometry was declining while analysis was on the rise, with Monge, Lagrange, and Prony teaching courses in analysis, which took on greater importance. Most of the characteristics of the analysis course were developed by Lagrange, who saw analysis as "a general method that can be applied to geometry and mechanics" (p. 22). This "reduces all mechanics and geometry to analysis, itself reduced to a purely algebraic calculation, and its implementation, both pedagogical and scientific, calls for a deep reflection on its principles." (p. 22) Thanks to Laplace, the course eventually encompassed series, derivatives, and integrals. It is a model that continues to influence the structure of many engineering training programmes today, with basic and specialised training usually offered in different course blocks (see, for instance, Artigue et al., 2007; Faulkner et al., 2019). Mathematics dominates most foundational courses in engineering programmes, which usually require the use of algebra, linear algebra, calculus, discrete mathematics, differential equations, etc. This educational model has created a gap between the abstract content of mathematics and its application in professionally-oriented courses (Christensen, 2008) and in professional practice, as illustrated decades ago in the ICMI 3 study (Howson, Kahane, Lauginie, & Tuckheim, 1988) and in more recent works (e.g., Wake, 2014).

A first step in addressing the gap mentioned above involves analysing the content of courses in engineering programmes to identify the ruptures and connections between courses and professional engineering practices. The studies considered in this section take this approach, investigating how mathematics is taught to future engineers or used in advanced engineering courses. This raises a number of questions, such as: in mathematics courses, is the content linked to other disciplines and/or to problems encountered in the workplace? In courses of other disciplines, what mathematics is introduced and used, and how does this relate to the content taught in "pure" mathematics courses? In the context of INDRUM, these questions have been addressed using an institutional approach: the mathematics courses, the professionally oriented courses, and the workplace are seen as different institutions, shaping the mathematics taught and used through different practices and needs. We discuss two papers: Gueudet & Quéré (2018) and González-Martín & Hernandes-Gomes (2018). These papers analyse mathematical content in curriculum resources: online courses in the first paper, and textbooks in the second.

Gueudet and Quéré (2018) analysed trigonometry-related content in mathematics courses for engineers and the use of trigonometry in electrical engineering courses. They looked at several electrical engineering courses available online and analysed their use of trigonometry. They then analysed how content concerning trigonometry was presented in two online mathematics courses for future engineers: a MOOC designed to teach mathematical foundations, and an online course entitled Mathematical Tools for Physics. The analysis was conducted in a global way through an institutional lens (Chevallard, 2006) and in a more local way by considering the notion of *connectivity*. This construct of *connectivity* has been defined by Gueudet, Pepin, Sabra, Restrepo, and Trouche (2018) in the context of e-textbooks analysis. They distinguish two types of connectivity: 1) connectivity at a micro-level, which concerns a particular mathematical topic and includes connections between different representations, links with other topics or other disciplines, different strategies for solving problems, etc.; and 2) connectivity at a macro-level, which concerns connections outside the textbook, such as connections between users, between users and the textbook authors, between the textbook and the users' resource system, etc. In the context of teaching mathematics to engineers, the issue of connections – or lack of connections – between mathematics and other disciplines, or between mathematics and workplace scenarios, is central. Hence, the use of the connectivity construct is likely relevant for the analysis of mathematics taught to future engineers.

Gueudet and Quéré (2018) provide examples of their results at the micro-level, for instance concerning the alternating sinusoidal current. The authors show connections between the functional representation of the sine function $s(t) = A\sqrt{2}\sin(\omega t + \varphi)$ used to model the current and voltage signals of a circuit, and its vector representation, associated with the notion of Fresnel vector. Indeed, the voltage $u(t) = U\sqrt{2}\sin(\omega t)$ and the current $i(t) = I\sqrt{2}\sin(\omega t - \varphi)$ may be considered as functions, and represented by graphs (Figure 9.1, left). At the same time, they may be represented as vectors (Figure 9.1, right).

In examining connectivity at the micro-level in the case of trigonometry, Gueudet and Quéré (2018) observe that electrical engineering courses naturally connect mathematics with electrical engineering; these courses also draw links between different mathematical concepts (namely functions and vectors) and various representations.

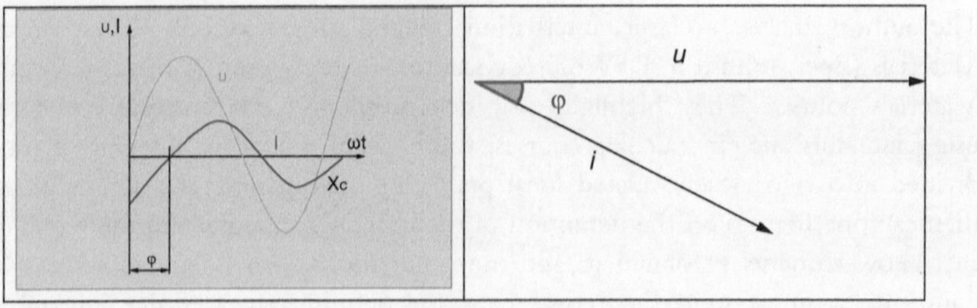

FIGURE 9.1 Signals in the alternating sinusoidal current and phase difference. On the left: a functional frame; on the right, Fresnel vectors in the vectorial frame (Gueudet & Quéré, 2018, p. 139)

In contrast, mathematics courses for future engineers offer fewer connections; the authors point out that: "within mathematics, some exercises are limited to a single register: the study of trigonometrical functions, for example, does not always require thinking in terms of angles. Using trigonometry in engineering courses, on the opposite, always requires such connections." (p. 142)

The way calculus is used in engineering courses has been discussed by González-Martín and Hernandes-Gomes (2018). Specifically, they examined the notion of first moment of an area, which appears in Strength of Materials courses. Like Gueudet and Quéré (2018), these authors also consider elements of the anthropological theory of the didactic (ATD), specifically the notion of praxeology and the circulation of praxeologies between institutions (Castela & Romo Vázquez, 2011; Chevallard, 1999). Through the lens of ATD, mathematics and engineering courses are seen as different institutions (with textbooks seen as a way to have access to what is done in them): mathematical praxeologies are taught in mathematics courses, while in engineering courses these praxeologies are used under different rationales, adapted to the norms and needs of the new institution. This construct serves as a powerful tool to analyse practices in different courses, in particular their different components, both explicit and tacit. It is also worth noting that the authors collaborated with an engineer, who helped pinpoint topics specific to engineering in which calculus is used and the courses where these topics are presented (we return to the issue of collaboration in the next subsection). The notion of first moment of an area can be defined as follows:

> In civil engineering, for example, to solve bending problems one must take into account some specific geometrical characteristics of cross-sections of a bar, which is the general term for structures that include beams (Feodosyev, 1973). In this situation, the notion of the first moment of an area is used to calculate the centroid of an area and the shearing stresses in transverse bending. The centroid of area A is its geometrical barycenter and is the point C of coordinates $\bar{x}$ and $\bar{y}$ such that the following relationships hold true: $\int_A x dA = A\bar{x}$ and $\int_A y dA = A\bar{y}$. (González-Martín & Hernandes-Gomes, 2018, p. 116)

The authors analyse a classic, international engineering textbook – Mechanics of Materials (Beer, Johnston, DeWolf, & Mazurek, 2012) – used in many Strength of Materials courses. They highlight that first moments and centroids are defined using integrals and, in calculus courses, the content related to integrals is usually divided into two closely related local praxeological organisations (i.e. two major themes): one focuses on the definition of the indefinite integral and the techniques that allow students to obtain it, and the other focuses on Riemann sums which constitute a more solid theoretical basis for defining the definite integral and interpreting it as an area, as well as for presenting the Fundamental Theorem of Calculus and the techniques to calculate integrals using Barrow's rule. Regarding the engineering textbook, González-Martín and Hernandes-Gomes (2018) show that although first moments of an area are defined as an integral, their use does not require the calculation of integrals; students are instead prompted to use basic area calculations and ready-to-use formulae to substitute values. The analyses reveal that tasks and techniques are completely different in the Calculus course and the Strength of Materials course (although the notion of integral appears in both), leading to two different praxeologies and two ways of working for the student. Moreover, in solving tasks in the engineering book, students may not realise that the notions they are using are based on the notion of integral. These results are consistent with previous work concerning the use of integrals to define bending moments in beams (González-Martín & Hernandes-Gomes, 2017).

Two significant issues are addressed by Gueudet and Quéré (2018) and by González-Martín and Hernandes-Gomes (2018): 1) there are relationships between mathematics and engineering, at different levels, and they can be identified through the study of mathematics and engineering courses (notions such as *connectivity* and *praxeology* are useful tools for conducting such analyses); 2) The way that mathematical notions (such as integrals, or trigonometry concepts) are presented in upper-level engineering courses (such as Strength of Materials, or Electricity) is such that the rules of the course are imposed on these mathematical notions in order to adapt them to their needs. That is to say, importing a mathematical concept or praxeology into non-mathematics, specialised courses generates, in effect, a transpositive process. Both papers also show how the use of sociocultural and anthropological tools helps address specific issues related to the use of mathematics in service and professionally-oriented courses, bypassing some of the obvious limitations of cognitive approaches to these issues (Artigue et al., 2007).

We end this section by highlighting that recent works (e.g., Faulkner, 2018; Quéré, 2019) have also analysed engineering courses, with similar findings: although mathematical content is present in these courses, their use and the tasks that call for this content can be very distant from what is usually presented in mathematics courses. This calls for more research regarding the professional needs of engineers, to further the debate about what should be taught in mathematics courses, but also to help propose modelling activities that are closer to the students' future professional practice.

Mathematics for other disciplines: the cases of biology and economics

In this section we highlight the use of mathematics in two other fields of study – economics and biology – which, although fairly unexamined in the international literature, have attracted attention in recent years (e.g., Mkhatshwa & Doerr, 2015; Viirman & Nardi, 2017, 2019).

In the case of economics, Feudel (2016, 2018) found that students seemed to consider the economic interpretation of the derivative as less relevant than pure mathematical procedures (such as differentiation rules). Derivatives can have a very specific interpretation in economics, in particular because of their link with the concept of *marginal cost*. Feudel (2016) notes that the derivative $C'(x)$ of a cost function $C(x)$ "is often interpreted as the additional cost while increasing the production from x to $x + 1$ units" (p. 182). Economics textbooks usually introduce derivatives from a purely mathematical point of view, then derive the approximation formula $f(x + h) - f(x) \approx f'(x)\cdot h$ for h close to 0, and finally use the argument that $h = 1$ is small in economics to justify the identification of $C'(x)$ and $C(x + 1) - C(x)$. Based on this, Feudel (2016) investigated which aspects of expected knowledge concerning the derivative are considered important by economics students. He asked students to summarise the important aspects of derivatives, and analysed the responses of 146 economics students, observing in particular:

- Aspects related to differentiation rules and derivatives of elementary functions (algorithms) appeared in 86.3% and 79.5%, respectively, of the summaries.
- Aspects related to definition, slope of the tangent line and rate of change (representations) appeared in 74.7%, 58.2%, and 16.4%, respectively, of the summaries.
- The economic interpretation is the least present category (8.9%), whereas aspects related to the term "marginal," meaning derivative, and the approximation aspect appeared in 48.6% and 15.1%, respectively, of the summaries.

These results led the author to conjecture difficulties in justifying why $C'(x)$ can be identified with $C(x + 1) - C(x)$, which could explain the poor performance of economics students involving the use of derivatives for approximation, as reported by Bingolbali and Monaghan (2008). As a follow-up, Feudel (2018) interviewed eight economics students who had successfully completed their calculus course about the connections they could make between the derivative and its economics interpretation. Using the construct of *concept image* (Tall & Vinner, 1981), the results show that the students had trouble making adequate connections, even though derivatives had been covered in their calculus course. These results reflect those obtained by Mkhatshwa and Doerr (2015), who also showed that many economics majors mistake the *amount of change* and the *rate of change* interpretations, when solving economic problems.

In the case of biology, mathematics is becoming increasingly important, leading to international efforts to improve biology students' mathematical skills (Viirman &

Nardi, 2017). Discussions at INDRUM 2016 and INDRUM 2018 looked at ways to integrate mathematics into the biology curriculum through, for instance, the use of mathematical modelling. Mathematical modelling was chosen in part because research indicates it can foster more positive attitudes towards, and self-perceived competence in, both biology and mathematics (Viirman & Nardi, 2017, p. 2275). Using a commognitive approach, Viirman (2016) presented preliminary results involving 12 first-year biology students who had sessions on modelling tasks in parallel with their one compulsory mathematics course. Some of these tasks included, for instance, estimating the population density of rabbits close to a highway, or estimating the size of an extinct species of bird by comparing data on dimensions of fossilised bones with similar data from contemporary species of birds. The preliminary results reported commognitive conflict concerning the use of assumptions and the fact that the students' engagement with mathematical modelling discourse differed from that of the teacher. It is also important to emphasise the fact that this process led to the development of new routines (for instance, concerning assumption building and graphing) and that the participants were able to shift from ritualised to exploratory routine use (Viirman & Nardi, 2017, 2019). The role of the teacher in facilitating this shift also appeared crucial. The study of biology students' ways of thinking and reasoning while engaging in biology problems that use mathematical models as their representation was the focus of Tetaj's work (2018), who also found that although biology students enrolled in an Evolutionary Biology course engaged with the biological content of the tasks in an exploratory manner, they employed the necessary mathematical routines in a ritualised way (Tetaj & Viirman, 2019). For instance, in a problem involving the *heritability equation*, students had to either remember the correct form of the equation or construct a mathematical model for the problem. In the second case, the students did not remember the biological meaning of any of the variables present in the equation and worked in a purely process-oriented manner. This might be because the course focused more on using mathematical models rather than on constructing them. Therefore, students seem to be able to use mathematical models, but not to argue the validity of a particular model or come up with reasons why a particular model makes sense or not. These results are consistent with those of Hester, Buxner, Elfring, and Nagy (2014), who report that biology students have trouble spontaneously transferring their mathematical knowledge into biology problems. This suggests it would be useful to propose courses in which students can be actively engaged in the construction of the mathematical models.

In the next section we discuss the potential of modelling activities to train non-specialist university students. Regarding the two above examples from biology, it is worth noting that they are the fruit of a collaboration between research institutions in mathematics education (MatRIC) and biology education (bioCEED). Some of the projects discussed (González-Martín & Hernandes-Gomes, 2018; Tetaj, 2018; Viirman, 2016) involve collaboration between specialists from different fields. This exemplifies the fact that research on the use of mathematics by non-specialists could benefit from such collaboration. In the next section we present research that looks at interventions, many of which also involve collaboration.

Designing and assessing interventions in the context of mathematics for non-specialists

In this section we focus on studies that examine the implementation and evaluation of interventions for non-specialists. As mentioned in the introduction, these interventions are most often designed to improve links between the mathematics taught and the speciality of the students by drawing on problems native to a given discipline. This raises several issues, such as the challenge of designing tasks with an appropriate level of difficulty or of developing students' modelling skills. An increasing number of such interventions are based on the use of ATD (Chevallard 1999, 2006), which proposes a theoretical and methodological framework for designing such interventions through the construct of *Study and Research Paths* (SRPs). For this reason, this section starts with a subsection devoted to SRPs. The next two subsections are devoted to other types of interventions: some of these interventions also draw on modelling, while others (keeping a more traditional approach) focus more on an innovative organisation, drawing on technological means. We note that most of the papers we mention examine engineering education, but other degree programmes are also analysed, such as business administration.

Mathematical modelling within ATD: study and research paths

There is international consensus shared by researchers and practitioners on the need for alternative teaching practices that go beyond the mere transmission of mathematical knowledge, and that use mathematics as an essential modelling tool to study real problems. This contrasts with the usual way of organising university mathematics teaching (as evidenced, for instance, by many authors in Holton, 2001). As discussed in the previous section, this traditional way of teaching may create disconnections and gaps between mathematical content and its uses and functionalities in other scientific disciplines. In order to overcome the constraints imposed by this isolation and lack of functionality of mathematics, university institutions usually propose mathematical modelling to help (re)establish links between mathematics and other disciplines, and more importantly, to reposition mathematics as an essential modelling tool for studying extra-mathematical problems.

The integration of modelling into current university systems has been the subject of numerous investigations but remains a major challenge (Burkhardt, 2008). Mathematical models can be found in the mathematics course syllabi of many university-level natural sciences programmes (Barquero, Bosch, & Gascón, 2013); however, the teaching of mathematical models often comes at the end of the process, if at all. Moreover, at university level, the dominant ideology (which the authors characterise as *applicationism*) is such that in most cases, modelling represents a mere "application" of pre-established knowledge, leaving little room for the process of questioning, building, and validating mathematical models. Some important innovations could address this issue.

Considering ATD, the notion of mathematical modelling has been linked to the notion of mathematical activity since the framework was first developed, assuming that doing mathematics essentially consists of producing, transforming, interpreting and arranging mathematical models (Barquero et al., 2013; García, Gascón, Ruiz-Higueras, & Bosch, 2006). Chevallard (2006, 2015) has proposed the construct of SRPs as an epistemological and didactic model to move towards a new pedagogical paradigm, where questions become central to teaching and learning and where mathematics appears as a modelling tool to provide answers to these questions. According to Barquero and Bosch (2015), the starting point of an SRP is a question of real interest for the community that studies it (students and teachers). The study of this *generating question* (Q_0) evolves and opens many other *derived questions* Q_1, Q_2, ..., Q_n. The continuous searching for answers to Q_0 (and to its derived questions) is the main purpose of the study and an end in itself. As a result, the study of Q_0 and its derived questions Q_i leads to successive intermediate answers A_i that can be helpful in elaborating a final answer $A^{\heartsuit}$ to Q_0. This allows the creation of a question-answer map (Jessen, 2014; Winsløw, Matheron, & Mercier, 2013).

In what follows, we discuss different papers concerning SRPs, but with different foci. The first type discusses internal elements that facilitate students' engagement and researchers' work with teachers; the second type discusses the effects of SRPs on content and the way it is usually organised; and the third discusses the possibilities of transferring SRPs to different university settings.

The first type is represented in the research presented by Barquero, Serrano and Ruiz-Munzón (2016) and Monreal, Ruiz-Munzón and Barquero (2018). These authors present the a priori analysis, implementation, and a posteriori analysis of an SRP proposing a comparison between forecast and actual number of active monthly Facebook users in a complete process of modelling, after two successive implementations with first-year business administration students. The implementation took place in a "Math modelling workshop" divided into six sessions, including personal work outside the classroom. The authors sought to identify what instructional strategies can prompt university students to engage in a coherent and self-sustained study and research activity (Barquero et al., 2016; Barquero, Monreal, Ruiz-Monzón, & Serrano, 2018). Two main dialectics play a key role in analysing the *internal dynamics* of the SRP (Chevallard, 2008; Winsløw et al., 2013):

- the questions-answers dialectic helps describe the arborescence of questions and answers proposed by students and teachers, which allow researchers to sketch the modelling trajectories followed by the participants;
- the media-milieux dialectic focuses on what different *media* (such as web resources, content of the course, GeoGebra or Excel simulation tools, etc.) are

available to students and how students and teachers can create the appropriate means to integrate them into their activity.

Barquero et al. (2016) and Monreal et al. (2018) show that an SRP needs to combine open moments of inquiry with more transmissive moments, where new mathematical or extra-mathematical content is presented depending on the questions opened by the SRP. Moreover, Monreal et al. (2018) show that the two dialectics cited above can enrich teaching and learning practices; they also help open discussions with mathematics teachers about the intervention and integrate the teachers in the research.

The second type is represented by Florensa, Bosch, Gascón and Mata (2016), who focus on the a priori design of an SRP to organise a course on general elasticity in a mechanical engineering programme. Their research evidences the "traditional" organisation of this course, mainly described in terms of the elements of content taught (relatively independent from each other) through a predetermined number of lectures and problem sessions. Considering previous research on the role of modelling in engineering professional practices (Gainsburg, 2006; Kent & Noss, 2003), two particular challenges have been identified in engineering training: 1) practitioners have to understand the phenomenon to be modelled (which usually remains inaccessible to engineers); 2) practitioners have to be explicitly aware of the assumptions and hypotheses of the model used. In this complex context, Florensa et al. (2016) propose an SRP that focuses on generating questions about how to choose one material (with unknown mechanical properties) from a set of three, in designing a specific part for a bike (brake lever, crank, gear and bike lock key). This SRP was implemented in eight two-hour sessions over four weeks with two groups of 25 students each (working in teams of three or four). In this paper, the use of question-answer maps plays a prominent role in analysing the expected mobilised knowledge, from the mathematical and engineering points of views, related to the domain of general elasticity. Two elements of the SRP are highlighted in this paper. First, the teachers were explicit about the elements to include in the final reports (such as dimensional plans, estimated loads, choice of material/s, safety factors, computer simulations of the models used, etc.), which raises the issues that students need to consider in order to give appropriate answers, while also providing a simulation that more closely mimics a real-world scenario. Second, the lecturers asked the working groups to deliver weekly reports in which students were responsible for planning a number of elements that the traditional didactic contract does not contemplate: time planning, questions they planned to work with during the week, descriptions of the tasks carried out even when the results obtained were incorrect, and proposed follow-up questions. Note that some of these elements (such as time planning and budgeting) are aspects that usually remain outside scholarly knowledge related to general elasticity, even though they are very much present in the professional practice of engineers.

Finally, the third type concerns the design and successive adaptations of an SRP in different university settings. The main aim of the research presented in Barquero

(2018) is to analyse what the authors define as the "ecological relativity" of modelling practices; that is, the variety of constraints that appear when modelling is implemented in different university settings. The students concerned are enrolled in a Mathematics programme; but this issue is clearly also relevant for non-specialists. This paper presents an SRP based on an urban bike-sharing system launched in Barcelona in 2007, implemented in two different courses in three different university settings. The initial question of this SRP concerns the difficulty of getting a homogeneous distribution of bicycles throughout a city with many sloping streets, requiring models (with matrices) to describe the daily flow of bikes between stations and challenging students to find ways to predict the bikes' redeployment needs. Barquero (2018) discusses how different institutions established different relationships with the teaching of modelling, identifying conditions that may help students. For instance, in the first case study some initial modelling activities were planned with the aim of building, together with students, a common discourse to analyse modelling practices. This was not considered as important in the second case, where students worked on their modelling projects in small groups throughout most of the course, without participating in very many common activities. Consequently, students were less well equipped with appropriate tools to talk about, justify or validate the modelling activity they had to report.

It is important to emphasise that, while it is explicitly stated only in this last paper, the study of conditions (both favourable and unfavourable) for the inclusion of mathematical modelling in teaching and learning practices is an overarching question linking most of the research developed using SRPs. It generally appears as a central axis to which research from different perspectives can contribute, from analysing the ways in which curricula and educational policies can favour the implementation of modelling and applications in school, to analysing teachers' beliefs in relation to the teaching of mathematical modelling.

Modelling-based interventions for non-specialists

Modelling is also central in many other interventions. The idea – present throughout this chapter – of connecting the mathematics taught in mathematics courses with the mathematics taught in other courses or used in the workplace, is still central to these interventions. Earlier we mentioned the work by Viirman (2016) regarding biologists; we now consider other examples.

The first example is strongly driven by APOS theory (Arnon et al., 2014), which has been used over the past ten years in studies proposing modelling activities regarding differential equations and linear algebra. The use of modelling activities intermingled with activities based on a genetic decomposition has proved to be effective in helping students to construct new concepts and in identifying relationships between different concepts. It is also consistent with the APOS theoretical framework since it provides students with opportunities to bring forward their previously constructed schemas – including mathematical knowledge and previously constructed knowledge in other disciplines or life experiences – and use it in constructing new schemas. An example of such a use is provided by Vázquez, Trigueros and Romo-Vázquez (2016), who designed and tested an activity to introduce linear algebra concepts to second-year students (future engineers, economists, and applied mathematicians): matrix of a linear transformation, linear system, and inverse of a matrix. The starting point of the teaching was a signal processing problem proposed to the students, the Blind Source Separation (BSS) problem, which made a connection to a "professional" topic (signals). In this scenario, a meeting taking place in a room has been recorded with several devices. The students had access to the recordings and to a seating map showing the location of the participants and the recording devices in the room. Using this information, the students were asked to match voices with participants. The 24 participating students worked in teams of six. They had access to a specific online tool that produced pure tones at various frequencies, and to GeoGebra software, which was used to display the results from the application of a mixing matrix. In a first phase, the students identified a linear model: an input is a vector of n sources, considered as pure tones, and a mixing matrix transforms it into an output. In a second phase, the students worked to determine mixing matrices in several cases. Finally, in the third phase, they worked on determining the inverse matrix. Student learning was analysed using APOS theory and the proposed conceptualization of a modelling activity. All students found the adequate model and, by working on the activities, they also found (unaided) an algorithm providing the inverse of a given matrix. The authors concluded that these students constructed an action conception of a matrix transformation and its inverse. Some students, in the case of a 2×2 matrix, found the input vector by solving a linear system, deducing the inverse matrix. They demonstrated a process conception of a 2×2 linear system. Confronted with other cases, some of these students were able to generalise the use of a linear system to study the existence of the inverse matrix and to compute it when possible. The authors concluded that the students constructed a process or an object conception of linear systems. All students were motivated and interested by the context of the problem.

This kind of motivation was also observed by Gjesteland, Vos and Wold (2018), who designed a project-based teaching activity (*the tracker project*) concerning the motion of an object. The students (working in teams of two or three) chose a moving object, filmed it with their smartphones, transformed the videos into graphs using tracker software, and proposed a mathematical model for the movement, presenting it on a poster. The authors evidenced that this process increased the students' motivation and that such teaching approaches can be applied to large class sizes (346 students in this study). The latter is significant, considering the classroom conditions of engineering programmes in many countries, which makes the ability to upscale innovations especially critical. In their work, Schmidt and Winsløw (2018) have also discussed an approach used with many students at the Danish Technical University (approximately 1,100 students each year). During the last four weeks of their basic mathematics course, first-year students work in teams on projects, producing a written text (of 20 to 50 pages) and giving an oral presentation. A specific feature of these projects is that they are "authentic problems from Engineering which can be solved using the mathematics to be taught in their course" (ibid. p. 166). The authors studied the conditions and constraints of the transposition process (Chevallard, 1991) that would allow such projects to be integrated into the engineering courses' institution. They defined ten didactic variables concerning, for example, the mathematical content covered (breadth of the content; old/new content), the use of Maple software, and the branches of engineering related to the project. These variables allow for the analysis of existing projects, sorting isolated applications of mathematics from practical uses of mathematics in solving real engineering problems. They are also likely to constitute a useful tool for designing more projects in other institutions, and are linked with the general issue of articulating mathematics and engineering in the curriculum evoked above in the previous section.

Using technology to design innovative teaching approaches for non-specialists

Several of the modelling-based interventions discussed above make significant use of various technologies: GeoGebra, Maple, other specific software for signal processing or motion analysis, smartphones, etc. Yet the use of technology and its effects on teaching and learning processes are not central issues in these studies. Its use seems more or less self-evident: technology is seen as a natural tool in this context. In this section we discuss papers where the use of technology plays a central role: we start with one paper where technology is considered to be a mediator in students' understanding of mathematical content, and continue later with other contributions where the use of technology does not affect the mathematical content.

In the study by Lagrange and Kiet (2016) on the teaching of probabilities for non-specialists at a university in Vietnam, a spreadsheet (Excel) and the statistics software *R* were used by students to design simulations of a probabilistic model. This teaching approach is innovative, compared with traditional approaches to teaching probabilities in Vietnam, and certainly in other countries. The simulations offered by the software help develop students' probabilistic reasoning skills. This reasoning is indeed specific. For example, Batanero, Chernoff, Engel, Lee, and Sanchez (2016) state that "probabilistic reasoning is a mode of reasoning that refers to judgments and decision-making under uncertainty and is relevant to real life" (p. 9). Probability is an interesting example of a mathematical topic taught to many non-specialist students. Making connections with other fields (such as medicine or economics) fosters students' engagement in modelling activity – with the specific features of the probabilistic model. In the teaching approach designed and tested by Lagrange and Kiet, the software plays a central role in the modelling process because it allows for the simulation, and hence the evaluation, of different models.

Other intervention studies, by contrast, consider the use of technology as a means for introducing changes likely to favour the learning of mathematics by non-specialists, but focus on changes concerning the pedagogical organisation. For example, Fredriksen and Hadjerrouit (2016) tested and analysed a Flipped Classroom (FC) organisation in the context of an engineering course at a Norwegian university. Out-of-class sessions with videos and quizzes were proposed to the students before each in-class session, which is quite different from a "traditional" teaching approach. This change leads to different kinds of tensions in the system. Some students had difficulties understanding the mathematical content of the videos (or simply did not watch the videos) and were therefore unable to engage in group work. Although the FC model has the potential to engage students in more active learning, the new proposed work paradigm was not adopted by all the students (Fredriksen, Hadjerrouit, Monaghan, & Rensaa, 2017). Online quizzes on a Moodle platform were also used in the study by Gaspar Martins (2018): the students were assigned weekly online quizzes as homework. The students appreciated this formative assessment, which supported regular personal work. While these innovative didactical organisations seem promising, they are not a priori specifically from the engineering context; studying which specific features could make them efficient for engineering training is an interesting perspective for further studies.

Conclusions and perspectives

In this chapter we present an overview of the INDRUM contributions to the international literature concerning the teaching and learning of university-level mathematics for non-specialists. These contributions include:

- Analysis of resources and curricular material presenting mathematical content in specialist courses and of students' difficulties dealing with mathematical content as presented in different courses. This helps explain some of the gaps students may experience, but can also lead to recommendations for mathematics courses (for instance, which connections should be reinforced) and hint at possible modelling activities (by identifying professional problems where mathematical content is used).
- Designing modelling-based interventions. These contributions provide pathways towards a more meaningful learning of mathematical content and more connections with professional activity. Some of these interventions integrate the use of technology.

We note that these issues are coherent with problems already seen as critical by recent surveys on UME regarding mathematics for other disciplines (e.g., Biza et al., 2016; Winsløw et al., 2018): the need to consider different student profiles, the professional environment where mathematics will be used, and the integration of modelling activities in the teaching of mathematics (which is particularly crucial for non-specialists).

We also note that the papers discussed in this chapter tackle these issues using a variety of approaches: APOS theory, concept image, commognition – with a significant number of papers relying on ATD, using an institutional perspective. The increase in the use of sociocultural approaches in recent years is evident, and is related to the fact that different kinds of courses (in different professional learning pathways) are seen as different groups or institutions, as already pointed out by Artigue et al. (2007). Moreover, sociocultural approaches have provided interesting insights into issues proper to UME (Nardi, Biza, González-Martín, Gueudet, & Winsløw, 2014). We highlight the potential of several constructs to address issues related to non-specialists: praxeologies, connectivity and discursive shift.

Many perspectives for future research on university mathematics education for non-specialists arise from the synthesis presented in this chapter.

Firstly, concerning the theoretical approaches mentioned above, further studies are necessary to deepen our understanding of the contributions of these constructs. Another interesting avenue for future research involves theoretical networking, as illustrated by Gueudet and Quéré (2018), associating praxeologies and connectivity, for instance.

Secondly, the use of mathematics in professionally-oriented courses and in the workplace also needs further study. Only a small number of courses have been analysed in the literature, and studies examining professional needs in the workplace are scarce. We note some recent studies in this direction (e.g., Faulkner, 2018; Quéré, 2019) that call for more systematic research on these issues. Moreover, as emphasised by Winsløw et al. (2018), studies questioning current curricula, the place of mathematics in them, and the content taught are also necessary. The use of technology in the mathematical training of non-specialists also deserves further research: technology is integrated into the daily practice of many professions and its use in teaching can help introduce more realistic activities.

Thirdly, regarding modelling, the second part addresses some issues that merit further research: how to facilitate students' engagement in modelling activities, how to collaborate with teachers to integrate modelling into their courses, how to reorganise "traditional" content with the integration of more realistic tasks (e.g., budget planning), and how to transfer modelling activities to different contexts. The discussed papers provide some insights that can inform future research. In connection with the previous paragraph, we believe that studies analysing the use of mathematics in other fields, as well as epistemological analyses on the origin of this use, are crucial in order to adequately address questions concerning interventions.

Finally, we also mention that the papers synthesised in this chapter show the importance of collective work, bringing together mathematics education researchers, mathematics teachers (who can be mathematicians but also engineers or specialists in other areas), teachers in other disciplines, and professionals in other fields. As discussed, research on the use of mathematics by non-specialists poses challenges to mathematics education researchers, and, in many cases, requires collaboration. We see this as one example of what Artigue et al. (2007) forecast ten years ago:

> We end this chapter with the feeling that research on mathematical learning and thinking at post-secondary level is entering now a new and fascinating phase, with difficult and new challenges to face, challenges that will require to be solved to extend interactions and collaborations beyond the traditional community of research in mathematics education at advanced level. (pp. 1044–1045)

Such collaborations offer an important perspective for future research. It is also crucial to consider the fact that, in the context of mathematics courses for non-specialists, it is common that teachers with different backgrounds teach the same courses (as discussed by Hernandes-Gomes & González-Martín, 2016), and that the consequences of this have not been explored in great depth (Artigue et al., 2007; González-Martín & Hernandes-Gomes, in press).

The synthesis of recent research presented in this chapter suggests many directions for future research, and we have mentioned only some of these. We anticipate that the way mathematics is used for and in other disciplines, and mathematical modelling at university, will be studied in the years to come by an increasing number of researchers at the international level, leading to fruitful implications for practice.

Note

1 In this chapter, all quotations from sources in French are our own translations.

References

Arnon, I., Cottrill, J., Dubinsky, E., Oktaç, A., Roa Fuentes, S., Trigueros, M., & Weller, K. (2014). *APOS Theory. A Framework for Research and Curriculum Development in Mathematics Education*. New York: Springer.

Artigue, M. (2016). Mathematics education research at university level: achievements and challenges. *INDRUM 2016 Proceedings*, pp. 11–27.

Artigue, M., Batanero, C., & Kent, P. (2007). Mathematics thinking and learning at post-secondary level. In F. K. Lester (Ed.), *Second Handbook of Research on Mathematics Teaching and Learning* (pp. 1011–1049). Charlotte, NC: Information Age.

Barquero, B. (2018). The ecological relativity of modelling practices: adaptations of a study and research path to different university settings. *INDRUM 2018 Proceedings*, pp. 85–95.

Barquero, B., Bosch, M., & Gascón, J. (2013). The ecological dimension in the teaching of mathematical modelling at university. *Recherches en Didactique de Mathématiques*, 33(3), 307–338.

Barquero, B., & Bosch, M. (2015). Didactic engineering as a research methodology: From fundamental situations to study and research paths. In A. Watson & M. Ohtani (Eds.), *Task Design in Mathematics Education*. New ICMI Study Series (pp. 249–272). Dordrecht: Springer.

Barquero, B., Monreal, N., Ruiz-Munzón, N., & Serrano, L. (2018). Linking transmission with inquiry at university level through study and research paths: the case of forecasting

Facebook user growth. *International Journal of Research in Undergraduate Mathematics Education*, 4(1), 8–22.
Barquero, B., Serrano, L., & Ruiz-Munzón, N. (2016). A bridge between inquiry and transmission: the study and research paths at university level, *INDRUM 2016 Proceedings*, pp. 340–349.
Batanero, C., Chernoff, E. J., Engel, J., Lee, H., & Sanchez, E. (2016). *Research on Teaching and Learning Probability*. ICME13 topical series. New York: Springer.
Beer, F., Johnston, E. R., DeWolf, T., & Mazurek, D. F. (2012). *Mechanics of Materials* (7th edition). New York: McGraw-Hill Education.
Belhoste, B. (1994). Un modèle à l'épreuve. L'Ecole Polytechnique de 1794 au Second Empire. In B. Belhoste, A. Dahan-Dalmedico & A. Picon (Eds.), *La Formation Polytechnicienne 1774–1994* (pp. 9–30). Paris: Dunod.
Bingolbali, E., & Monaghan, J. (2008). Concept image revisited. *Educational Studies in Mathematics*, 68(1), 19–35.
Biza, I., Giraldo, V., Hochmuth, R., Khakbaz, A., & Rasmussen, C. (2016). *Research on Teaching and Learning Mathematics at the Tertiary Level: State-of-the-art and Looking Ahead*. New York: Springer. https://link.springer.com/book/10.1007%2F978-3-319-41814-8.
Burkhardt, H. (2008). Making mathematical literacy a reality in classrooms. In D. Pitta-Pantazi & G. Pilippou (Eds.), *Proceedings of the Fifth Congress of the European Society for Research in Mathematics Education* (pp. 2090–2100). Larnaca: University of Cyprus.
Castela, C., & Romo Vázquez, A. (2011). Des mathématiques à l'automatique: étude des effets de transposition sur la transformée de Laplace dans la formation des ingénieurs. *Recherches en Didactique des Mathématiques*, 31(1), 79–130.
Chevallard, Y. (1991). *La Transposition Didactique: Du Savoir Savant au Savoir Enseigné* (2nd edition). Grenoble: La Pensée Sauvage.
Chevallard, Y. (1999). L'analyse des pratiques enseignantes en théorie anthropologique du didactique. *Recherches en Didactique des Mathématiques*, 19(2), 221–266.
Chevallard, Y. (2006). Steps towards a new epistemology in mathematics education. In M. Bosch (Ed.), *Proceedings of the Fourth Congress of the European Society for Research in Mathematics Education* (pp. 22–30). Barcelona: FUNDEMI-1QS.
Chevallard, Y. (2008). Un concept en émergence: la dialectique des médias et des milieux. In G. Gueudet & Y. Matheron (Eds.), *Actes du Séminaire National de Didactique des Mathématiques 2007* (pp. 344–366). Paris: IREM de Paris 7 & ARDM.
Chevallard, Y. (2015). Teaching mathematics in tomorrow's society: A case for an oncoming counter paradigm. In S. J. Cho (Ed.), *Proceedings of the 12th International Congress on Mathematical Education* (pp. 173–187). Dordrecht: Springer.
Christensen, O. R. (2008). Closing the gap between formalism and application: PBL and mathematical skills in engineering. *Teaching Mathematics and its Applications*, 27(3), 131–139.
Faulkner, B. E. (2018). *Mathematical Maturity for Engineering Students* (Unpublished doctoral dissertation). University of Illinois at Urbana-Champaign, USA.
Faulkner, B., Earl, K., & Herman, G. (2019). Mathematical maturity for engineering students. *International Journal of Research in Undergraduate Mathematics Education*, 5(1), 97–128.
Feudel (2016). Relevant knowledge concerning the derivative concept for students of economics – A normative point of view and students' perspectives. *INDRUM 2016 Proceedings*, pp. 181–190.
Feudel (2018). Students' connections between the derivative and its economic interpretation in the context of marginal cost. *INDRUM 2018 Proceedings*, pp. 95–104.
Florensa, I., Bosch, M., Gascón, J., & Mata, M. (2016). SRP design in an Elasticity course: the role of mathematic modelling. *INDRUM 2016 Proceedings*, pp. 191–200.

Fredriksen, H., & Hadjerrouit, S. (2016). Using the flipped classroom model of instruction to explore the teaching and learning activities in mathematical education for engineers: an activity theory perspective, *INDRUM 2016 Proceedings*, pp. 251–252.

Fredriksen, H., Hadjerrouit, S., Monaghan, J., & Rensaa, J.R. (2017). Exploring tensions in a mathematical course for engineers utilizing a flipped classroom approach. In T. Dooley & G. Gueudet (Eds.), *Proceedings of the Tenth Congress of the European Mathematical Society for Research in Mathematics Education* (pp. 2057–2064). Dublin: DCU Institute of Education & ERME.

Gainsburg, J. (2006). The mathematical modelling of structural engineers. *Mathematical Thinking and Learning*, 8(1), 3–36.

García, F.J., Gascón, J., Ruiz-Higueras, L., & Bosch, M. (2006). Mathematical modelling as a tool for the connection of school mathematics. *ZDM: The International Journal on Mathematics Education*, 38(3), 226–246.

Gaspar Martins, S. (2018). Weekly homework quizzes as formative assessment for Engineering students are a fair and effective strategy to increase learning?, *INDRUM 2018 Proceedings*, pp. 105–114.

Gjesteland, T., Vos, P., & Wold, M. (2018). Mathematical modelling and activation – a study on a large class, a project-based task and students' flow. *INDRUM 2018 Proceedings*, pp. 175–184.

González-Martín, A. S., & Hernandes-Gomes, G. (2017). How are Calculus notions used in engineering? An example with integrals and bending moments. In T. Dooley & G. Gueudet (Eds.), *Proceedings of the Tenth Congress of the European Society for Research in Mathematics Education* (pp. 2073–2080). Dublin: DCU Institute of Education & ERME.

González-Martín, A. S., & Hernandes-Gomes, G. (2018). The use of integrals in Mechanics of Materials textbooks for engineering students: the case of the first moment of an area. *INDRUM 2018 Proceedings*, pp. 105–114.

González-Martín, A. S., & Hernandes-Gomes, G. (in press). Mathematics in engineering programs: what teachers with different academic and professional backgrounds bring to the table. An institutional analysis. *Research in Mathematics Education*.

Gueudet, G., Pepin, B., Sabra, H., Restrepo, A., & Trouche, L. (2018). E-textbooks and connectivity: proposing an analytical framework. *International Journal of Science and Mathematics Education*, 16(3), 539–558.

Gueudet, G., & Quéré, P.-V. (2018). "Making connections" in the mathematics courses for engineers: the example of online resources for trigonometry. *INDRUM 2018 Proceedings*, pp. 135–144.

Harris, D., Black, L., Hernandez-Martinez, P., Pepin, B., Williams, J., & TransMaths Team. (2015). Mathematics and its value for engineering students: What are the implications for teaching? *International Journal of Mathematical Education in Science and Technology*, 46(3), 321–336.

Hernandes-Gomes, G., & González-Martín, A. (2016). Teaching Calculus in engineering courses. Different backgrounds, different personal relationships?*INDRUM 2016 Proceedings*, pp. 201–210.

Hester, S., Buxner, S., Elfring, L., & Nagy, L. (2014). Integrating quantitative thinking into an introductory biology course improves students' mathematical reasoning in biological contexts. *Life Sciences Education*, 13(1), 54–64.

Holton, D. (Ed.) (2001). *The teaching and learning of mathematics at university level: An ICMI Study*. Dordrecht: Kluwer.

Howson, A. G., Kahane, L., Lauginie, M., & Tuckheim, M. (1988). *Mathematics as a service Subject*. ICMI Studies. Cambridge: Cambridge University Press.

Jessen, B. E. (2014). How can study and research paths contribute to the teaching of mathematics in an interdisciplinary setting? *Annales de Didactique et de Sciences Cognitives*, 19, 199–224.

Kent, P., & Noss, R. (2003). *Mathematics in the University Education of Engineers*. London: The Ove Arup Foundation.

Lagrange, J.-B., & Kiet, B.-A. (2016). Une approche fréquentiste des probabilités et statistiques en première année d'Université au Vietnam dans un cursus non mathématique. *INDRUM 2016 Proceedings*, pp. 221–230.

Mkhatshwa, T. P., & Doerr, H. M. (2015). Students' understanding of marginal change in the context of cost, revenue, and profit. In K. Krainer & N. Vondrová (Eds.), *Proceedings of the 9th Congress of the European Society for Research in Mathematics Education* (pp. 2201–2206). Prague: Charles University in Prague, Faculty of Education & ERME.

Monreal, N., Ruiz-Munzón, N., & Barquero, B. (2018). Central dialectics for mathematical modelling in the experience of a study and research path at university level. *INDRUM 2018 Proceedings*, pp. 155–164.

Nardi, E., Biza, I., González-Martín, A. S., Gueudet, G., & Winsløw, C. (2014). Guest Editorial: Institutional, sociocultural and discursive approaches to Research in university mathematics education. *Research in Mathematics Education*, 16(2), 91–94.

Quéré, P.-V. (2019). *Les mathématiques dans la formation des ingénieurs et sur leur lieu de travail: études et propositions (cas de la France)*. (Unpublished doctoral dissertation). Université de Brest, France. Retrieved November 2019 from https://hal.archives-ouvertes.fr/tel-02281937.

Rasmussen, C., & Ellis, J. (2013). Who is switching out of Calculus and why. In A. M. Lindmeier & A. Heinze (Eds.), *Proceedings of the 37th Conference of the International Group for the Psychology of Mathematical Education* (Vol. 4, pp. 73–80). Kiel: PME.

Romo-Vázquez, A. (2009). *La formation mathématique des futurs ingénieurs*. (Unpublished doctoral dissertation). Université Paris VII, France. Retrieved November 2019 from http://www.theses.fr/2009PA070029.

Schmidt, K., & Winsløw, C. (2018). Task design for Engineering Mathematics: process, principles and products. *INDRUM 2018 Proceedings*, pp. 165–174.

Tall, D., & Vinner, S. (1981). Concept image and concept definition in mathematics with particular reference to limits and continuity. *Educational Studies in Mathematics*, 12(2), 151–169.

Tetaj, F. (2018). Students' understanding of mathematical models in three Norwegian biology courses. *INDRUM 2018 Proceedings*, pp. 191–192.

Tetaj, F., & Viirman, O. (2019). A case study on mathematical routines in undergraduate biology students' group-work. *Proceedings of the Eleventh Congress of the European Society for Research in Mathematics Education* (to appear).

Vázquez, R., Trigueros, M., & Romo-Vázquez, A. (2016). A model from signal analysis to design linear algebra activities. *INDRUM 2016 Proceedings*, pp. 191–200.

Viirman, O. (2016). The development of the mathematical discourse of biology students working with mathematical modelling. *INDRUM 2016 Proceedings*, pp. 253–254.

Viirman, O., & Nardi, E. (2017). From ritual to exploration: the evolution of biology students' mathematical discourse through mathematical modelling activities. In T. Dooley & G. Gueudet (Eds.), *Proceedings of the Tenth Congress of the European Society for Research in Mathematics Education* (pp. 2274–2281). Dublin: DCU Institute of Education & ERME.

Viirman, O., & Nardi, E. (2019). Negotiating different disciplinary discourses: biology students' ritualized and exploratory participation in mathematical modelling activities. *Educational Studies in Mathematics*, 101(2), 233–252.

Wake, G. (2014). Making sense of and with mathematics: the interface between academic mathematics and mathematics in practice. *Educational Studies in Mathematics*, 86(2), 271–290.

Winsløw, C., Gueudet, G., Hochmuth, R., & Nardi, E. (2018). Research on University Mathematics Education. In T. Dreyfus, M. Artigue, D. Potari, S. Prediger & K. Ruthven (Eds.), *Developing Research in Mathematics Education: Twenty Years of Communication, Cooperation and Collaboration in Europe* (pp. 60–74). London: Routledge.

Winsløw, C., Matheron, Y., & Mercier, A. (2013). Study and research courses as an epistemological model for didactics. *Educational Studies in Mathematics*, 83(2), 267–284.

PART III

Teachers' and students' practices at university level

10

TRANSITIONS TO, ACROSS AND BEYOND UNIVERSITY

Reinhard Hochmuth, Laura Broley and Elena Nardi

Introduction

This chapter is about papers from INDRUM 2016 and 2018 considering transition issues. The aim is to give a systematic review of the themes discussed and the major theoretical lenses used. Transition to and across university mathematics has been attracting increasing research effort (e.g. Biza, Giraldo, Hochmuth, Khakbaz, & Rasmussen, C., 2016). Since all learning is somehow a transition from one state to another, something which is also noted in the introduction of the survey about transition by Gueudet, Bosch, diSessa, Kwon & Verschaffel (2016), nearly all papers presented at the INDRUM conferences could be seen as relevant to this chapter. To narrow our focus, we decided to start from the papers assigned to the transition working group at INDRUM 2018 and added further papers from both INDRUM 2016 and 2018. In total, we identified 32 papers that substantially referred to transition issues. Finally, we connected the INDRUM papers to other, pertinent works in the field and structured the chapter in accordance with three types of transition, a classification that follows a timeline across educational levels and the workplace: Type I: From school to university; Type II: Within and across university courses; and Type III: From university to the workplace.

In comparison with the survey by Gueudet et al. (2016), our chapter is limited to the transitions involving university mathematics education, and does not cover, for example, the pre-school to school or the primary-secondary transitions. Our choice regarding transition types resonates with the ICMI Pipeline Project (Hodgson, 2015), which focused on the following four transition points: A: School to undergraduate courses; B: Undergraduate courses to postgraduate courses; C: University into employment; and D: University into teaching. (A) is included in our (I), (B) is part of (II) and (C) and (D) are linked to (III). Our (II) might also include horizontal transitions between topics and transitions between service courses and non-mathematics major courses. Our (III) covers workplaces outside school and university to the limited extent allowed by INDRUM papers. Our stronger focus on research efforts with regard to (II), and the sparsity of works on (III), can be attributed to the backgrounds and priorities of INDRUM participants. Overall, we note that we faced a challenge with the general issue of typification: for example, what counts as school and university mathematics varies significantly across countries. We have been careful to attend to this variability by including notes on context where appropriate.

To illustrate this issue, and briefly introduce each of the transition types on which our chapter is based, we offer some prototypical examples. For instance, an often-investigated issue of transition type I is the mathematical topic of limits of sequences that has turned out to cause major problems for many students. In school, limits are introduced in an informal way and are algebraically calculated for rational terms. If a definition is presented in school, there are typically no tasks questioning or requiring this definition. In university analysis courses, the formal definitions of limits in **R, C**, metric- or Banach-spaces, which are axiomatically introduced, are fundamental for solving tasks and understanding continuity and differentiability. We note that the transition from an informal to a definition-based use of limits does not always occur in the transition from school to university. In countries where calculation-oriented calculus courses are at the beginning of university studies, the limit-issue shifts from transition type I to type II.

With type II in view, a mathematical notion that is treated in both first semester and advanced university mathematics courses is "continuity". In beginner courses, the continuity of functions is essentially based on the definition of a limit. In topology courses, continuity is introduced more generally for functions between topological spaces. Besides such vertical transitions, in which areas of knowledge and courses build on one another in terms of content and follow one another in time, there are also horizontal transitions, in which areas of knowledge can be taught largely independently of each other and therefore courses do not have to follow each other in time. An example is the use, and further development of, the Riemann – or Lebesgue – integral in courses about partial differential equations or stochastics. A mixture of horizontal and vertical mathematical interrelations often appears in the transition from mathematics service courses to non-mathematics major courses. For example, Fourier series are typically considered in mathematics service courses for engineers, but also in electrical engineering signal theory courses.

As an example of the transition from university to the workplace (III), let us briefly consider the issue of teaching mathematics. Student teachers, lecturers of introductory and advanced mathematics courses and lecturers of service courses have the qualifications that suggest they know the mathematics they teach well. But knowing something and knowing how to teach something is not the same. The latter requires one to address students' learning needs, to choose appropriate ways of introducing students to new content, to design and carry out assessment and to be aware of how mathematics is changed when it is transposed into educational institutions. In resonance with Klein's (1908/1939) poignant points on the double discontinuity between mathematics in school, university and the workplace (school and elsewhere), teaching also necessitates an awareness of the various forms of mathematics needed in various workplaces (engineers', economists', psychologists', etc.).

Since INDRUM is a network of university mathematics educators, typically lecturers doing research in mathematics and/or mathematics education, we see it as helpful to begin by identifying and discussing issues that our research aims to address mainly in *mathematical* terms, as we have done above. This said, and as will be shown throughout the rest of our chapter, the INDRUM papers on transition also highlight epistemological, historical, pedagogical, educational, cultural, sociocultural and affective issues. The studies we refer to have a variety of goals: the analysis of transition processes; development, implementation and evaluation of measures that intend to support students coping with major transition steps; and, integrative theoretical efforts. As a whole, they make a broad contribution to research on transition in university mathematics education and point to some areas that require more attention from the INDRUM community. In the following section, we present a short overview of the theoretical frameworks and constructs that underpin these papers. Then, we discuss the papers relating to the three transition types I, II and III. We conclude with a summary, noticing recent trends and major challenges.

Theoretical lenses that capture transition

Artigue (Chapter 1 in this book), Gueudet et al. (2016) and Nardi (2017) all highlight a first period of research in university mathematics education and transitions espousing cognitive approaches, e.g. identifying the necessities of conceptual changes. Under the umbrella term "advanced mathematical thinking" (Tall, 1991), works covered major differences between school and university mathematics in terms of oppositions such as informal vs. formal, concrete vs. abstract, calculation- vs. structure-oriented, etc. There are also Tall's (2013) three worlds (embodied, symbolic and formal), which extend to the transition from primary to secondary school, and Fischbein's (1989) tacit models, which theorize more generally transitions from previous to new knowledge (e.g. Kidron, 2018, referenced in subsection I). All these theoretical constructs and frameworks work well for analysing mathematical differences and related cognitive phenomena with respect to type I transitions. This is not necessarily the case for analysing transitions of type II or III which, as noted by Artigue (2016), Gueudet et al. (2016) and Thomas et al. (2015), may benefit from a general move in mathematics education research that aims to include a more systematic consideration of social, interactional and intersubjective issues (Lerman, 2000). This does not mean that the initial theories are out of use today. As Artigue (2016) noted in her INDRUM plenary, there has been "a reconstruction of their main outcomes, thus making possible some form of incorporation of these outcomes in the new paradigms" (p.19). We now outline theoretical approaches that are frequently used in INDRUM transition papers.

Concept image / concept definition and APOS

The concept image / concept definition distinction appeared in the work of Shlomo Vinner in the 1970s (e.g. 1976). It became a quintessential part of cognitively oriented, developmental analysis of mathematical learning at university level and elsewhere through the much-cited work of Tall & Vinner (1981). In research on transition phenomena, the distinction emphasizes that unique formal definitions and proofs of theorems based on definitions are an important feature of university mathematics and indicates major differences with school mathematics. A more layered cognitive theoretical framework is APOS (action-process-object-schema) (Dubinsky, 1991; Arnon et al., 2013), which is based on Piaget's genetic epistemology and, in particular, his notion of reflective abstraction. APOS theory – given that the advanced phases of "object" and "cognitive schema" extend naturally to university mathematics – is used particularly in investigations of type I transitions (e.g. Syamsuri, 2019, where cognitive processes in proving activities were studied). Regarding transition type II, one INDRUM paper that deploys the concept image / concept definition construct is by Hamza and O'Shea (2016) who used it to analyse interview data concerning open sets in metric spaces (see also subsection II).

Anthropological Theory of the Didactic (ATD)

ATD (Chevallard 1999) considers knowledge as something that lives in institutions (Bosch & Gascón, 2014) and regards the institutionalisation of knowledge as the result of complex transformation processes, referred to as external didactic transpositions (Bosch, Hausberger, Hochmuth & Winsløw, 2019). Moreover, knowledge is regarded as co-determined by an ecological and hierarchical scale of so-called "levels of codetermination" (Chevallard, 2002) – civilisation, society, school, pedagogy, discipline, domain, etc. – which also indicates that teaching and learning are shaped by constraints located at different levels of specificity. At the centre of the ATD lies the 4T-model of a "praxeology" (i.e., task, technique, technology, theory), which enables researchers to build reference models as a basis for further analyses of taught and learned knowledge. In INDRUM papers, transition can be seen as taking place between institutions (e.g. the transition from the institution of calculus to the institution of probability theory considered in subsection I, or from single to multivariable calculus courses considered in subsection II). Considering type I transitions, there is a change in tasks and techniques and, in particular, the formation and relevance of technological and theoretical blocks, which constitute the discourse necessary for interpreting and justifying the practical block (i.e., task and related techniques), including explanations and justifications of the techniques used, especially in terms of theorems and proofs. Type II transitions within mathematics involve further changes from procedural to theoretical practices. One might even identify phases of transitions within university with fundamental changes of the roles of those blocks; for example, theoretical blocks become techniques of new tasks in advanced courses, accompanied by the introduction of new theoretical blocks (see subsection II). Regarding mathematics service courses, ATD allows formalizing the disconnections between the mathematics taught in these courses and its use in the serviced disciplines. Type III transitions also fit very well with the institutional idea (see our examples regarding Felix Klein's double discontinuity in subsection III). In recent years, ATD, in particular its "study and research paths" construct, a variation of inquiry-oriented teaching efforts (Barquero, Serrano & Ruiz-Munzón, 2016; Florensa, Bosch, Cuadros & Gascón, 2018), has also been used in professionalization courses.

Theory of Commognition

In the transition papers utilizing Sfard's (2008) communicational perspective, the focus is on the shifts of mathematical discourses and rules within and between different groups. Commognition considers knowledge as something that comes into the world through communication, which is regarded not as a secondary aspect of teaching and learning, but as intrinsically related to individuals' cognition. Commognitive notions present in INDRUM papers include word use, visual mediators, endorsed narratives and routines, which are further distinguished into rituals and explorations. A key incentive for developing commognition theory lies in the observation that a cognitive take on conceptual development is often binary (e.g. procedural vs. conceptual), whereas a commognitive take resonates with a spectrum-view of learning and teaching. Commognitive constructs enable descriptions and investigations of all three transition types. Regarding type I, for example, a word like "limit" is understood and used in different ways by lecturers and by first semester students arriving from school. Regarding type II, within mathematics, endorsed narratives change, for example they may become more general and abstract from one domain to another. This is also true of transitions of type III: there are changes of routines and endorsed narratives in the transition from pure mathematics to, say, engineering. Furthermore, the theory might be very helpful in sensitizing school teachers and university lecturers to commognitive conflicts inherent in or emerging from teaching-learning situations.

The referenced approaches facilitate not only the identification of mathematical learning obstacles in transitions, but also the development of well-founded teaching interventions. Some specific approaches in mathematics education research for developing, researching and evaluating interventions have also been applied in transition contexts considered in INDRUM papers: for instance, the theory of didactic situations (TDS: Brousseau, 1997), with its a priori and a posteriori analyses, Realistic Mathematics Education (van den Heuvel-Panhuizen & Drijvers, 2014), as well as guided intervention studies and design-based research as a rather general iterative approach (Prediger, Gravemeijer & Confrey, 2015). Variable psychological approaches adapted to mathematical contexts are also used in INDRUM's transition research. Large-scale questionnaires, for example, are applied in type I transition studies evaluating measures that intend to support students regarding anxiety of mathematics, motivation and interest, mathematical self-efficacy, learning strategies, etc. (see, e.g., in subsection I, Biehler et al., 2018, and Kuklinski et al., 2018).

Last but not least, recent success in networking efforts in mathematics education research (Bikner-Ahsbahs & Prediger, 2014) point to the increasing need for more systematic inclusion of societal issues. One INDRUM example is (Hochmuth, 2018), which networks ATD and the subject scientific approach by Holzkamp (1985).

Three types of transition: school-university, within university, from university to the workplace

I. School-university

The transition from secondary to tertiary education is an exciting and often complex experience for young people – and with many facets: mathematical, curricular, academic and social (Nardi et al., 2017). University, as an institution, and university mathematics, can be seen in this sense as a new world governed by new rules (Gueudet et al., 2016), a new discourse that may intrigue as well as alienate newcomers (Sfard, 2014). Within mathematics, despite obvious contextual differences, there are several key features of this transition that are common across educational systems (e.g. Engelbrecht & Harding, 2008): a "gap" or a curricular discontinuity between what is defined in different contexts as "school" and "university" mathematics (Kajander & Lovric, 2005) and academic and social differences in learning environment (Cherif & Wideen, 1992). Recent studies have also been turning to social aspects of the transition such as changes in students' living arrangements, location, friendship groups, as well as relationships with peers and staff (Bampili, Zachariades, & Sakonidis, 2017).

Most emphatically though, the transition from school to university has been seen as requiring a shift in thinking mathematically (Tall, 1992), particularly in relation to increased requirements for rigour and precision when engaging with the formalism and abstraction of university mathematics (Nardi 1996), for moving from instrumental to more relational approaches to doing mathematics (Clark & Lovric, 2009) and for developing more learner autonomy (Solomon, 2006). There is now a non-negligible body of work that discusses students' arrival at university in terms of mathematical under-preparedness (e.g. Gill, O'Donoghue, Faulkner & Hannigan, 2010) with some effect, in some contexts, on student retention and success (e.g. Anthony, 2000). However, a deficit discourse on this issue is not universal (Engelbrecht & Harding, 2008) and a survey of the literature by Thomas et al. (2015) found a mixed picture in this respect. INDRUM papers mirror this diverse landscape. In what follows, we sketch this landscape in three parts; each with a focus on studies that foreground the mathematics, students' approaches to study and institutional support.

Mathematics. Lecorre (2016) explores pragmatic, empirical and theoretical aspects of rationality in relation to the concept of limit, especially at the transition from informal to formal calculus between school and university. With a focus on what means of justification students bring to their mathematical work, he creates a situation that aims to encourage a shift from dynamic to static perspectives on limit, and vice versa, which he sees as a great platform to develop aforementioned types of rationality. The situation includes forms of scientific debate around the conjectures being put forward by students and is trialled in the context of an upper secondary class in France. He observes students' arguments gradually becoming more robust, until a collectively generated proof is finally endorsed by the class. Ghedamsi and Lecorre (2018), combining TDS and APOS, take this work further: eight tasks were trialled in upper secondary classrooms with reported success particularly with regard to creating a need for formalisation, a need that the secondary students had little prior experience of.

Schüler-Meyer (2018) also works in the context of limiting processes. Defining the convergence of sequences is studied from a commognitive perspective as a discursive practice in transition. Upper secondary students reinvent the definition in a guided intervention process that aims to facilitate their transition from experiential to abstract approaches. Prior experiences of convergence from school are activated and secondary school discourses unlock objectified narratives.

Also with a focus on what students may bring with them from school mathematics, Kidron (2018) explores newly arriving students' understanding of irrational numbers through the ways in which they choose to represent non-repeating decimals. Drawing on developmental constructs such as Fischbein's (1987) mental model, she offers a quantitative analysis and classification of secondary students' concept images of decimals and irrationals as evident in questionnaire responses of 91 Year 10 students and 97 Year 11 and 12 students in Israel. She identifies misconceptions about decimals and rational/irrational numbers. There is a strong parallel in this work with Durand-Guerrier and Tanguay (2016) who offer an epistemological analysis of the transition from upper secondary to university mathematics for real numbers and map didactic horizons that may emerge from such an analysis.

Bloch and Gibel (2016) deploy TDS constructs (particularly that of adidactic situations) as well as semiotic approaches to explore complications in students' learning of calculus at the start of university studies. Their focus is on the increased complexity of signs and reasoning processes at this level. They identify challenges evident in responses from 14 university students in a French university to final exams questions about parametric curves and differential equations. Amongst these challenges are: the learning of new techniques; and, selecting, performing and interpreting the results of calculations. Students' relationship with diagrammatic reasoning also emerges as potentially problematic. These challenges are seen as relating to transition, particularly as the students are in the first year of university studies.

Finally, Derouet, Planchon, Hausberger, and Hochmuth (2018) focus on the challenges posed by the structures of school and university mathematics programmes, such as the compartmentalisation of the curriculum both at secondary and tertiary levels. Drawing on ATD constructs, Derouet et al. analyse the praxeological repertoire that students bring from school into university in probability (continuous distributions) and calculus (integrals) – two topics that students are to some extent familiar with from school and that problematize the compartmentalization of secondary school curricula. Analysis of questionnaire data from 82 students with different backgrounds revealed almost no transfer of applicable techniques across the two topics.

Approaches to study. A small but distinct set of works presented at INDRUM revolves around how students' independent work changes as they enter university mathematics studies (Quéré, 2016), in particular with regard to their use of resources (Gueudet & Pepin, 2016; Kock & Pepin, 2018). Quéré (2016) notes the concern, and sometimes hesitation, of engineering students about how to deploy the range of means and resources available to them as well as a deficit of strategy in their approaches to study for doing so. Gueudet and Pepin (2016) explore this potential deficit through comparing how university mathematics teachers expect students to use resources and how students actually use them. In their UK datasets, they survey students, interview students and lecturers, observe lectures and study university documents. In their France dataset (Year 1, Number Theory module), they survey students, interview one lecturer and explore the impact of coursework, also in follow-up interviews. The lecturers seem to expect students to use resources to understand mathematics (epistemic mediation) while students seem to use resources to find examples fully worked by their lecturers that they can emulate in examinations or other forms of assessment (pragmatic mediation). While pragmatic mediation does not preclude the emergence of epistemic mediation (for instance, a student might expand their understanding of mathematics while adapting a worked example), their analysis, which draws on instrumentation approaches (Rabardel & Bourmaud, 2003), stresses the discrepancy between the two.

Analogous findings, but in the context of calculus and linear algebra, are proposed by Kock and Pepin (2018) who juxtapose students' use of resources at school and university. Their group and individual interviews with students, analysis of resources and interviews with lecturers and tutors highlight the greater variety of resources available to students at university level, the diminished role of the teacher as the main resource and the strong association between students' use of resources and course organisation. In some courses, students spend less time interacting individually with a teacher and there is loose, even poor, alignment between how mathematical content is presented in textbooks and in lectures. This implies that students are expected to establish links between the two and thus invest substantial effort in organising their resource systems.

Institutional support. Overall, the work presented at INDRUM sees the transition to the study of mathematics at university as a multifaceted process that requires a shift in the way students think mathematically. We stress that this shift materialises in a new social and academic environment and according to the norms of a new learning community. This shift is also experienced differently by students with different socioeconomic and educational backgrounds and is part of broader and deeper identity formation processes they live concurrently. How universities support this multifaceted process is essential:

student learning support systems, personal tutoring and peer-support systems are amongst the prevalent modes of support.

There are several studies presented at INDRUM that relate to attracting or retaining students and to supporting a successful transition. These focus on: evaluating programmes that aim to raise awareness and appreciation of university mathematics for high-achieving secondary students on the brink of deciding whether they will undertake university studies with a STEM focus and high mathematical component (Bracke et al., 2016); identifying newly-arriving students' interest in different domains and aspects of mathematics such as real-world problems, calculations and proofs, and exploring who may be at risk of dropping out of university mathematics (Geisler & Rolka, 2018; Fuller et al., 2016; Rach et al., 2018; Gradwohl & Eichler, 2018); and, evaluating tailored support for students either on the cusp of or just arriving at university (Biehler et al., 2018; Kuklinski et al., 2018; Landgärds, 2018). For example, Biehler et al. (2018) evaluate pre-university bridging courses that aim to clarify the goals and expectations of study at university level for incoming students: goals of lecturers and courses are juxtaposed and compared to perceived goals by students as well as students' beliefs about how these goals can be achieved. Within the same larger programme (WiGeMath), Kuklinski et al. (2018) evaluate a first-semester course that aims to facilitate the transition to university study. The evaluation includes an analysis of some affective variables such as interest, mathematical self-concept, basic needs and self-efficacy and goal-fulfilment and traces a decline in interest and some shifts in student beliefs e.g. about the "toolbox" nature of mathematics (Grigutsch & Törner, 1998).

Also with a focus on the affective domain, Fuller et al. (2016) explore how newly-arriving students' completion rates may be influenced by anxiety and personality factors. This empirical study explores psychological traits, uses scales such as the Abbreviated Mathematics Anxiety Rating Scale (MARS) as well as elements of the "Big Five" inventory (extraversion, agreeableness, conscientiousness, neuroticism and openness: Poropat, 2009) and combines extensive demographic data and results. 404 students' mathematical behaviour is evidenced in task completion and mapped onto data concerning reported levels of anxiety, personality traits and course completion. Exam anxiety emerges as a significant influence on non-completion, with Big Five's conscientiousness emerging as key in successful completion. In a similar vein, Gradwohl and Eichler (2018) explore predictors of performance in engineering mathematics with a focus on psychological constructs like motivation and learning strategies. Individual characteristics of students such as prior performance, motivation and engagement are studied for correlation and for their impact on study success. The focus is on 182 engineering students and the aim is to identify domain-specific predictors of study success. The questionnaires include items on goal-orientation, interests and self-reporting on learning strategies (and prior performance, e.g. grades). Strong correlations with mathematics self-concept resonate with other studies of affect and, unsurprisingly, the study identifies an impact of school grades and performance in earlier exams at university. With more of a focus on identifying predictors of dropping out as well as success in examinations, Geisler and Rolka (2018) collected questionnaire responses from 209 students reporting their mathematical self-concept, interest in mathematics, beliefs about the nature of mathematics, basic learning needs and general self-efficacy. Data concerning achievement during one semester and exam results were also collected. Students' achievement during the semester emerged as the best predictor for their exam outcomes. Unsurprisingly, the best predictor for lecture and seminar attendance was their interest in a particular topic in mathematics.

Beyond aforementioned valuable systems of support and studies of affective factors, we see two further modes of supporting students' transition from school to university – that do require, we note, deeper systemic shifts – as essential: (1) reconfiguration of teaching practices at university level that accommodate directly the learning needs of incoming students; and, (2) a stronger synergy between secondary and tertiary learning/teaching communities (e.g. secondary mathematics teachers and university mathematics lecturers).

One study presented at INDRUM that begins to address (1) is by Petropoulou et al. (2016). Drawing on Jaworski's (Potari & Jaworski, 2002) teaching triad, as well as a combination of data grounded theory (Charmaz, 2006) and activity theory (Leont'ev, 1978), the study investigates teaching practice, in particular in relation to goals for supporting students in transition. Analyses of lecture observations, field notes, post-lecture interviews and informal discussions with students reveal various forms that sensitivity to students, both cognitive and affective, can take, even in the context of modes of expository teaching that are largely seen as impersonal, such as a lecture to a large cohort.

There is hardly any work presented at INDRUM that accommodates (2): we note this gap and ask for more work with the learning needs of incoming students in mind, especially work that brings together teaching communities from across the secondary/tertiary divide. Where this may become the case (Winsløw, Biehler, Rønning, Jaworski & Wawro, 2020) is within the rapidly changing landscape of university teachers' professional development.

In addition to experiencing an often far from seamless transition from school to university mathematics, students are typically expected to navigate across mathematical domains, and across other disciplines, when at university. Their lecturers also navigate between mathematical research and teaching cohorts of students with diverse needs. We review studies of these navigations, for students and their lecturers, in the sections that follow.

II. Within university

Beyond their first year, students typically do not have to deal with radical changes in their living and learning environments. Furthermore, as they study material more closely linked to their career goals, they may be less likely to experience a lack of interest or anxiety. This is perhaps why the studies summarized in this section focus on cognitive and epistemological aspects of "within university" transitions. Two kinds of transitions were highlighted by INDRUM. First, the transitions between mathematics courses that occur in succession (i.e., one is the prerequisite of the other); in particular, courses within the Analysis Path (i.e., single variable Calculus, multivariable Calculus, Real Analysis, Metric Spaces, …). Second, the transitions from foundational mathematics courses to specialist courses for disciplines that use mathematics; in particular, the discipline of engineering.

From one course to the next in the Analysis Path

Transitions can vary internationally due to the different kinds of educational institutions that exist within different countries. In North America, for example, a student's first encounter with Calculus can occur at the end of secondary school; nevertheless, most university programmes in STEM-related disciplines require their students to take single variable Calculus courses (again) in their first year[1]. The transition from Calculus to Analysis hence occurs "within university".

Two INDRUM papers (Brandes & Hardy, 2018; Broley & Hardy, 2018) relate to this North American context. They both refer to a model by Winsløw (2006), which uses the ATD's notion of praxeology to formalize the observation that, in the Calculus to Analysis transition, students' practices become less procedural and more theoretical. As depicted in Figure 10.1, students' work in introductory Calculus is principally practical (represented by Π_1): that is, students are required to master techniques for solving different types of tasks. In later courses, two transitions may occur. First, students may be expected to gain awareness of the theoretical block (Λ_1) missing from their practical work (e.g. they not only learn how to calculate the derivative of $\sin(x) + x^2$, but also to describe and prove the differentiation rules that justify their calculations). Second, students are invited to engage in new types of tasks (Π_2) involving the more abstract objects from the theoretical block Λ_1 (i.e., they learn how to develop proofs in a more general sense).

Brandes and Hardy (2018) extend the body of literature on single variable Calculus by exploring the nature of students' work in a subsequent course, in multivariable Calculus. For one such course, they studied curricular materials and final examinations to make praxeological models of (a) the knowledge to be taught and

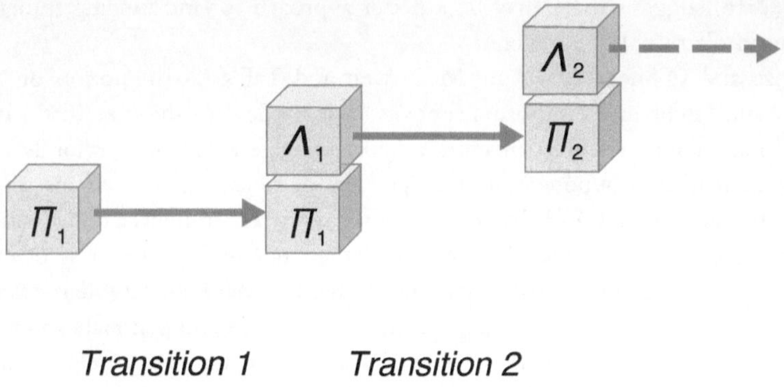

FIGURE 10.1 Transitions in mathematics coursework (adapted from Winsløw, 2006)

(b) the minimal core of the knowledge students must learn to receive a passing grade. They found that, in moving from (a) to (b), the practical blocks were reduced, and the theoretical blocks were transformed into superficial versions. In short, the students were not yet required to go through Transition 1 as depicted in Figure 10.1. Broley and Hardy (2018) concerns what happens in a later course, in Real Analysis, when students are typically invited to start working with and on mathematical theory. The researchers describe a plan to model the knowledge students are expected to learn in the course (according to assignments and past exams), as well as the knowledge some students actually learn (according to task-based interviews), in hopes of gaining insight into how students' practices are evolving under the conditions of the course. They question when (if ever) students' practices reflect the second stage in Figure 10.1 (i.e., (Π_1, Λ_1). One prediction is that students experience the "transition" from Calculus to Analysis more like a "jump" from the first stage (i.e., Π_1) to the third stage (i.e., (Π_2, Λ_2)). Winsløw (2016) investigates this in relation to angles and trigonometry. He notes that the textbooks used in Analysis courses have the potential to assist students in building the theoretical blocks missing from their previously developed practical blocks; however, an interview with a master's student suggests that most students do not (and possibly cannot) engage in this building on their own (i.e., in the absence of supportive didactic activities).

Another ATD-inspired INDRUM paper, by Bergé (2016), adds to the discussion of the importance of activities given to students in their independent work. While studying a sequence of courses at an Argentinian university (Courses I and II, which concern Calculus; and Courses III and IV, which concern Analysis), Bergé (2008) noticed that Courses II, III and IV all included an axiomatic presentation of $\mathbb{R}$ and deduction of related theorems by the professor; but the tasks completed by students varied greatly. For instance, Course II required students to find suprema of concrete sets, while Course IV invited them to use the supremum as a tool in proof tasks. The results of a questionnaire led to expected results: the majority of students in Course II saw the supremum either as not being useful, or as an ordinary upper bound; by Course IV, the majority of students could name situations where the supremum functions as a tool. Based on a historical analysis, Bergé (2016) suggests there may be a better approach for introducing students to the "raison d'être" of completeness.

Hamza and O'Shea (2016) use McGowen and Tall's (2010) notion of "met-before" and Fischbein's (1989) concept of "tacit model" to theorize that a transition into any course involves meeting new knowledge based on previously developed mental images, whose influence (positive or negative) may or may not be known to the learner. Task-based interviews with ten students at the end of a Metric Spaces course in Ireland showed that the students held a variety of images of open sets. As theorized, some students seemed to have tacit models based on previous experience with $\mathbf{R}^n$, which led them to visualize an open ball as a circle or a disc when reasoning about more general spaces. This may have helped students build intuition; but it may also cause difficulties. For instance, the students who routinely based their reasoning on the formal definition of open sets were more successful in task solving. Echoing results about the Calculus to Analysis transition (e.g. Tall, 1992), it seems the transition from Real Analysis to Metric Spaces also requires significant cognitive shifts, including an increased level of abstraction and appreciation of the role of mathematical definitions. Hamza and O'Shea (2016) note that including a range of examples of metric spaces other than $\mathbf{R}^n$, as was done by the lecturer in their study, is not sufficient for assisting students in making the transition.

From introductory mathematics courses to engineering courses

An engineering programme usually starts with foundational courses, which are meant to provide skills needed for later, more specialized courses. Foundational courses typically include mathematics taught by mathematics faculty to large groups of students studying in different programmes (e.g. engineering, computer science, chemistry, biology and physics). Such courses have been found to pose problems for engineering students (Ellis, Kelton & Rasmussen, 2014), leading some to abandon their programme, and others to experience a bumpy transition to the specialized engineering courses. A current interest of mathematics education research, as evidenced by INDRUM, is to identify what kinds of "bumps" may occur and why.

González-Martín and Hernandes-Gomes (2018) and Gueudet and Quéré (2018) adopt an institutional perspective based on the ATD (Chevallard, 1999), which offers one explanation for the bumps. The institution of engineering studies operates according to different ecological conditions when compared to the institution of mathematics studies for mathematics majors; in the least, they correspond to two different Scholarly institutions – Scholarly Mathematics and Scholarly Engineering – where, according to the theory of institutional transposition (see, e.g., Castela, 2016), mathematical knowledge exists in different ways. Mathematics faculty are often educated in the mathematics studies and Scholarly Mathematics institutions, and have little to no knowledge of specialized engineering studies or Scholarly Engineering. When they are responsible for teaching the foundational mathematics courses for engineering students, Schmidt and Winsløw (2018) suggest that a separation of external and internal didactic transposition (in the sense of Chevallard, 1991) may take place: the initial selection of the mathematical contents to be taught (external transposition) is based on the needs of engineering disciplines, but the actual teaching is carried out (internal transposition) according to the standards of teaching mathematics.

González-Martín and Hernandes-Gomes (2018) study (dis)connections that may arise in relation to Calculus. Assuming that post-secondary instructors rely on textbooks to organize their teaching, the researchers performed a comparative analysis of how integrals are presented in the textbooks used in a first-year Calculus course and a second-year course (Strength and Materials) at a university in Brazil. Since many civil engineering notions are defined as integrals, programmes often dictate that engineers learn about them. However, González-Martín and Hernandes-Gomes (2018) found that the knowledge learned in Calculus (e.g. techniques for calculating integrals), is not needed to solve the tasks in Strength and Materials. It is therefore not surprising that engineering students question the need to learn integrals in Calculus. The researchers emphasize the importance of mathematics lecturers becoming more aware of how the content they teach is used in professional courses.

In a similar vein, Gueudet and Quéré (2018) compare the content of existing courses in mathematics and engineering; the difference is that they study online courses (in France), which may offer more flexibility in tailoring contents to the intended audience (i.e., future engineers). These researchers are also interested in connections in multiple senses, not only between mathematics and engineering, but also between mathematical concepts and semiotic registers. They focus on trigonometry since it is rich in such connections and is extensively used in engineering courses. An analysis of course materials showed that electrical engineering courses require students to master connections between concepts and registers (e.g. students must learn to associate an algebraic expression such as $s(t) = A\sqrt{2}\,\sin(\omega t + \phi)$ with a function, a graph and a vector representation). In comparison, the mathematics courses lacked connections, either between concepts and registers or between mathematics and engineering. The researchers conclude that the development of online bridging resources is an important aim for mathematics education research.

III. From university to the workplace

The INDRUM papers related to the third transition type focus mainly on one kind of workplace: where the expectation is to teach mathematics. This is perhaps not surprising given the focus of mathematics education research on the training of future primary and secondary teachers, as provided by universities. This said, the INDRUM community has not only continued to deepen the discussion about "old" issues (i.e., Klein's "double discontinuity"); it has also contributed to the exploration of relatively "new" lines of research, such as the transitions faced by post-secondary teachers (e.g. graduate teaching assistants, college instructors and university professors).

To primary or secondary school teaching

Felix Klein (1908/1939) coined the term "double discontinuity" to represent the disconnections between what education students do at university and what they previously did (as a student) and eventually will do (as a teacher) in schools. Teacher education programmes are often criticized if they do not (a) offer future teachers adequate opportunities to deepen both their content knowledge and pedagogical content knowledge in relation to school mathematics (e.g. when future secondary mathematics teachers follow almost the same curriculum as mathematics majors); and (b) make strong enough connections to actual teaching practices (Gueudet et al., 2016). INDRUM contributions point to several approaches for bridging the gaps.

Biza and Nardi (2016) highlight an approach taken in education courses: inviting teacher students to engage with fictive, albeit realistic classroom situations. The researchers exemplify the potential of asking the students to compose their own situations based on their school placements. Their analysis of the situations and ensuing group discussions is structured around four characteristics – consistency (between beliefs and practice), specificity (to the situation), reification of pedagogical discourse and reification of mathematical discourse – which allow them to study how the content addressed in the education programme interacts with the teacher students' early teaching experiences. Since at this stage in the transition, teacher students have one foot in both university and workplace, there is the possibility not only to analyse the transition process, but also to facilitate it (e.g. by immediately addressing any challenges faced and informing better design of future courses).

Two INDRUM papers focus on bridging gaps that might be experienced in advanced mathematics courses. Hanke and Schäfer (2018) observed that the future teachers at their university often feel they just need to get through these courses and do not naturally gain the "higher standpoint" on school mathematics that is intended by curriculum developers. These researchers are therefore redesigning the courses "in the shape of a Y": At the beginning, all students learn foundational content together, and about two-thirds through the semester, there is a split into two groups based on future professions (i.e., teachers and mathematicians). Mathematics students learn advanced mathematics topics, while teacher students identify notions relevant for secondary school mathematics and create related exploratory learning environments that they test with school pupils. Initial results in a Complex Analysis course suggest that the profession-specific part is more engaging and memorable for the teacher students, though more explicit guidance may be needed to help them connect their pedagogical knowledge (learned in their education courses) and the specialized content knowledge (learned in Complex Analysis).

Stender and Stuhlmann (2018) leverage a theoretical distinction between content and methods (i.e., the ways in which content is created or investigated) to argue that there is a more natural connection between university and school mathematics: if the content seems drastically different, the methods are the same. Moreover, teacher students should eventually be able to make the methods explicit to students when they are teaching. Stender and Stuhlmann (2018) focus on a subset of methods called "heuristic strategies", which are general problem-solving tactics: e.g. understand the problem, change the representation, consider special cases, break the problem into sub-problems and so on (the paper has many more examples). The researchers are designing voluntary seminars (an alternative to the redesign discussed above), which aim to assist teacher students in understanding lecture material (in Linear Algebra and Analysis) and introduce heuristic strategies relevant to that material. Interviews with five students showed some positive results, though it was unclear if the students had learned to explain their strategies to others. While implementing heuristic strategies in lectures is a goal, Stender and Stuhlmann (2018) mention the potential challenge of convincing lecturers to change their habits.

To post-secondary teaching

The separation of teacher students and mathematics majors in the approaches outlined above makes sense: teacher education programmes aim to educate future teachers, while mathematics programmes aim principally to educate future mathematicians. Yet, a large portion of mathematics majors will eventually become teachers at the post-secondary level. One INDRUM paper, by Mathieu-Soucy, Corriveau and Hardy (2018), was driven by the lack of research on this transition and the challenge of transferring results about schoolteachers to the post-secondary level. The researchers frame the transition in terms of changes in a person's relationship with (i.e., visions, opinions, beliefs, attitudes and feelings about) mathematics and its teaching and learning (RWMTL). Interviews with three new post-secondary[2] teachers throughout a semester enabled an identification of pertinent themes. Surprisingly, the teachers did not spontaneously talk about mathematics when asked about their daily life; rather, they spoke of more general issues related to teaching (e.g. classroom methods, assessment and student expectations). Moreover, the teachers' reflections on recent events were not mature enough to have led to significant changes in their RWMTL. The researchers conclude that a different methodological approach is needed if one wishes to document changes in RWMTL. They conjecture, for instance, that inviting new teachers to share significant past experiences in relation to specific themes would lead them to reveal deeper reflections that give more insight into the factors shaping their RWMTL as they transition into their new roles.

Other INDRUM contributions follow the trend of designing and evaluating attempts at easing the transition. Florensa et al. (2018) discuss courses for new (and in-service) university lecturers. At the beginning of four courses, they collected 143 questions that participants indicated they would have (or have had) during their teaching. Like Mathieu-Soucy et al. (2018), they found that the lecturers were concerned mostly with general issues. Florensa et al. (2018) offer possible explanations; notably, the phenomenon of "pedagogical generalism", which separates instructional processes from their content and leads lecturers to accept without question the content they teach. This can be perpetuated by teaching support programmes based solely on pedagogy. Florensa et al. (2018) propose that such programmes introduce certain didactic theories (e.g. from the ATD) to help participants formulate more pertinent questions about their teaching. How such theories are received and used by those who are principally mathematically trained seems to remain an open question.

Huget (2018) reminds us that the transition to post-secondary teaching can start early, when graduate students are asked to engage in teaching-related work. There is increasing interest in the training offered to these students since they not only tend to lack knowledge of teaching, but they may also play a significant role in shaping undergraduate mathematics education, both immediately (for the students in their courses) and in the long run (since they are potential future professors) (Speer, Gutmann & Murphy, 2005). Huget (2018) describes one program for new tutors involving three mandatory modules throughout a term ("introduction to tutoring", "interaction and activation" and "problem solving") and the opportunity to be observed and receive feedback. The program's effects on both tutors and students taking tutorials were under evaluation at the time of INDRUM 2018.

INDRUM has engaged strongly with research about the transition from being a student to teaching mathematics at any level of education. Nonetheless, there are a few papers that touch on the transition from university to other professions. Broley (2016) presents a study of how 14 research mathematicians use computer programming in their research and teaching, and describes the potential gaps between how university mathematics is traditionally taught and the increasingly computational nature of some mathematical research practices. Schmidt and Winsløw (2018) discuss efforts to bridge gaps between engineering students' training (in mathematics) and their future work (as engineers). Their novel approach includes creating authentic engineering problems and integrating them in early mathematics courses. We now conclude with mentioning populations of students (and their transitions to work) that have not yet been significantly represented in INDRUM but are starting to (e.g. economics students) in work presented at a more recent conference on *Calculus in upper secondary and beginning university mathematics* (e.g. Monaghan, Nardi & Dreyfus, 2019).

Summary, conclusion, ways forward

Our overview of the contributions about transition issues demonstrates that this is a vibrant and broadening research area that was well represented at the INDRUM conferences. The internationality of the contributions especially points to the importance of describing the contexts within which transitions take place, since educational institutions vary from country to country. There are still areas that attract little research, even though their importance is recognized in our community.

Research on transition issues within and across university, as well as from university to the workplace is rather new. It is therefore not surprising that there are many unexplored areas. While some contributions go beyond Calculus and Real Analysis, there is still a lack of research on advanced mathematics (beyond Year 1 or 2 of university studies). There are other paths of courses (e.g. in Algebra or Statistics) and other horizontal transitions (i.e., between courses that occur concurrently) that could be considered in future research and at INDRUM conferences. There is also less consideration of the use of mathematics within fields (e.g. psychology, economics, computer science) outside of engineering, all of which could refer important insights back to mathematics service courses. Lastly, there are other populations of students missing from the studies reported at INDRUM: for instance, social sciences and humanities students who are required to take mathematics or statistics courses at university. One might question, for example, how such courses relate to the workforce such students will enter and the mathematical skills (e.g. in numeracy, computational thinking, logic, statistical thinking) that would benefit them most.

For each topic noted, numerous pertinent research questions can be formulated from the point of view of the various theoretical approaches we discussed. For example, from an ATD perspective, the transition model depicted in Figure 10.1 and related issues have mainly been investigated with respect to Calculus and Analysis. But what is the situation for other fields? Or: What does the observed jump from the first to the third stage of the model mean in a longer-term perspective, e.g. taking into account what students (are expected to) learn in more advanced mathematics studies? From a commognitive perspective, transitions are currently studied mostly in terms of changes in meta-discursive rules, changes in word use and emerging commognitive conflicts in the interactions between students and teachers.

There is ample space for data collection and analyses of a different grain size. For example, studies can examine how changes in the curriculum, and in other institutional factors, play out in the way students' discursive activity changes from one educational level to another. Or, they can examine how such changes may play out in different courses as well as in the transition from the world of study to the world of work.

Several papers highlight the need for constructing and implementing interventions that assist students in overcoming difficulties, developing certain kinds of understandings or knowledge, and, ultimately, making smooth transitions. This has already been realised in many countries for the transition from school to university, which has been researched for a long time and could be seen as a classical research topic in university mathematics education. However, there is still not so much research that evaluates implemented interventions or discusses sustainability issues. Regarding all transition types, there is a need to consider more strongly the impact of socially institutionalized contexts (including, e.g., the number of students in courses, exam requirements, or the general role of education in society).

Many of the measures for supporting transitions have appeared in the form of "add-ons" to existing systems (e.g. additional courses or seminars). We note the importance of complementing such efforts with reflections on, and attempts at, deeper systemic changes. This could require more collaboration between communities of teachers and professionals from across transition divides (e.g. school teachers and university lecturers; lecturers of mathematics and specialists in other disciplines; university lecturers and professionals from the workplaces where mathematics is put to use). The INDRUM papers have just begun to hint at both the challenges and possibilities in this regard. University teachers' training and professional development – discussed in chapter 4 of this book (Winsløw et al., 2020) – may be one place where these much-needed synergies can materialise.

Notes

1 For a description of such courses: https://www.maa.org/sites/default/files/pdf/cspcc/InsightsandRecommendations.pdf

2 The teaching context is described thoroughly in the paper. It is within college institutions (so-called CÉGEPs) that serve as intermediary education between secondary schools and universities in the province of Québec, Canada.

References

Anthony, G. (2000). Factors influencing first year students' success in mathematics. *International Journal of Mathematical Education in Science and Technology*, 31(1), 3–14.

Arnon, I., Cottrill, J., Dubinsky, E., Oktaç, A., Roa Fuentes, S., Trigueros, M. & Weller, K. (2013). *APOS Theory: A Framework for Research and Curriculum Development in Mathematics Education*. New York: Springer.

Artigue, M. (2016). Mathematics education research at university level: Achievements and challenges. *INDRUM 2016 Proceedings*, pp. 11–27.

Bampili, A., Zachariades, T. & Sakonidis, C. (2017). The transition from high school to university mathematics: a multidimensional process. In T. Dooley & G. Gueudet (Eds.), *Proceedings of the 10th Conference of European Researchers in Mathematics Education* (pp. 1985–1992). Dublin: Dublin City University.

Barquero, B., Serrano, L. & Ruiz-Munzón, N. (2016). A bridge between inquiry and transmission: the study and research paths at university level. *INDRUM 2016 Proceedings*, pp. 340–349.

Bergé, A. (2008). The completeness property of the set of real numbers in the transition from calculus to analysis. *Educational Studies in Mathematics* 67 (3) 217–235.

Bergé, A. (2016). Le rôle de la borne supérieure (ou supremum) dans l'apprentissage du système des nombres reels. *INDRUM 2016 Proceedings*, pp. 33–42.

Biehler. R., Lankeit, E., Neuhaus, S., Hochmuth, R., Kuklinski, C., Leis, E., Liebendörfer, M., Schaper, N. & Schürmann, M. (2018). Different goals for pre-university mathematical bridging courses – Comparative evaluations, instruments and selected results. *INDRUM 2018 Proceedings*, pp. 467–476.

Bikner-Ahsbahs, A. & Prediger, S. (Eds.) (2014). *Networking of Theories as a Research Practice in Mathematics Education*. Dordrecht: Springer.

Biza, I., Giraldo, V., Hochmuth, R., Khakbaz, A. & Rasmussen, C. (2016).*Research on Teaching and Learning Mathematics at the Tertiary Level: State-of-the-art and Looking Ahead*. ICME-13 Topical Surveys. Cham: Springer.

Biza, I. & Nardi, E. (2016). Consistency, specificity, reification of pedagogical and mathematical discourses in student teacher narratives on the challenges of their school placement experience. *INDRUM 2016 Proceedings*, pp. 425–426.

Bloch, I. & Gibel, P. (2016). A model to analyse the complexity of calculus knowledge at the beginning of University course (Two examples: parametric curves and differential equations). *INDRUM 2016 Proceedings*, pp. 43–52.

Bosch, M. & Gascón, J. (2014). Introduction to the Anthropological Theory of the Didactic (ATD). In A. Bikner-Ahsbahs & S. Prediger (Eds.), *Networking of Theories as a Research Practice in Mathematics Education* (pp. 67–83). Dordrecht: Springer.

Bosch, M., Hausberger, T., Hochmuth, R. & Winsløw, C. (2019): External didactic transposition in undergraduate mathematics. In U. T. Jankvist, M. van den Heuvel-Panhuizen & M. Veldhuis (Eds.), *Proceedings of the Eleventh Congress of the European Society for Research in Mathematics Education* (CERME11, February 6–10, 2019). Utrecht: Freudenthal Group & Freudenthal Institute, Utrecht University & ERME.

Bracke, M., Friedewold, D. & Schnieder, J. (2016). A training concept for supervising self-directed problem-solving in the STEM disciplines. *INDRUM 2016 Proceedings*, pp. 421–422.

Brandes, H. & Hardy, N. (2018). From single to multi-variable calculus: A transition?. *INDRUM 2018 Proceedings*, pp. 477–486.

Broley, L. (2016). The place of computer programming in (undergraduate) mathematical practices. *INDRUM 2016 Proceedings*, pp. 360–369.

Broley, L. & Hardy, N. (2018). A study of transitions in an undergraduate mathematics program. *INDRUM 2018 Proceedings*, pp. 487–496.

Brousseau, G. (1997). *Theory of Didactical Situations in Mathematics* (N. Balacheff, M. Cooper, R. Sutherland & V. Warfield: Eds. and Trans.). Dordrecht: Kluwer.

Castela, C. (2016). When praxeologies move from an institution to another: an epistemological approach to boundary crossing. In R. Göller, R. Biehler, R. Hochmuth & H.-G. Rück (Eds.), *Proceedings of the KHDM Conference: Didactics of Mathematics in Higher Education as a Scientific Discipline* (pp. 153–161). Kassel: Universitätsbibliothek Kassel.

Charmaz, K. (2006). *Constructing Grounded Theory: A Practical Guide Through Qualitative Research*. London: Sage.

Cherif, A. & Wideen, M. (1992). The problems of the transition from high school to university science. *Catalyst*, 36(1), 10–18.

Chevallard, Y. (1999). L'analyse des pratiques enseignantes en théorie anthropologique du didactique. *Recherches en Didactique des Mathématiques*, 19(2), 221–266.

Chevallard, Y. (2002) Organiser l'étude 3. Écologie & régulation. In J.-L. Dorier, M. Artaud, M. Artigue, R. Berthelot, R. Floris (Eds.) *Actes de la 11e Ecole d'Eté de Didactique des Mathématiques* (pp. 41–56). Grenoble: La Pensée Sauvage.

Clark, M. & Lovric, M. (2009). Understanding secondary-tertiary transition in mathematics. *International Journal of Mathematical Education in Science and Technology*, 40(6), 755–776.

Derouet, C., Planchon, G., Hausberger, T. & Hochmuth, R. (2018). Bridging probability and calculus: the case of continuous distributions and integrals at the secondary-tertiary transition. *INDRUM 2016 Proceedings*, pp. 497–506.

Dubinsky, E. (1991). Reflective abstraction in advanced mathematical thinking. In D. Tall (Ed.), *Advanced Mathematical Thinking* (pp. 95–123). Dordrecht: Kluwer.

Durand-Guerrier, V. & Tanguay, D. (2016). Introduire les réels à la transition lycée-université: Questions épistémologiques et horizon didactique. *INDRUM 2016 Proceedings*, pp. 177–179.

Duval, R. (2006). A cognitive analysis of problems of comprehension in a learning of mathematics. *Educational Studies in Mathematics*, 61(1), 103–131.

Ellis, J., Kelton, M. & Rasmussen, C. (2014). Student perceptions of pedagogy and associated persistence in calculus. *ZDM: The International Journal on Mathematics Education*, 46(4), 661–673.

Engelbrecht, J. & Harding, A. (2008). The impact of the transition to outcomes-based teaching on university preparedness in mathematics in South Africa. *Mathematics Education Research Journal*, 20(2), 57–70.

Fischbein, E. (1987). *Intuition in Science and Mathematics: An Educational Approach*. Dordrecht: Reidel.

Fischbein, E. (1989). Tacit models and mathematical reasoning. *For the Learning of Mathematics*, 9(2), 9–14.

Florensa, I., Bosch, M., Cuadros, J. & Gascón, J. (2018). Helping lecturers address and formulate teaching challenges: an exploratory study. *INDRUM 2018 Proceedings*, pp. 373–382.

Fuller, E., Deshler, J., Darrah, M., Trujillo, M. & Wu, X. (2016). Anxiety and personality factors influencing the completion rates of developmental mathematics students. *INDRUM 2016 Proceedings*, pp. 434–443.

Geisler, S. & Rolka, K. (2018). Affective variables in the transition from school to university mathematics. *INDRUM 2018 Proceedings*, pp. 507–516.

Ghedamsi, I. & Lecorre, T. (2018). Students' understanding of ε-statements involving equality and limit. *INDRUM 2018 Proceedings*, pp. 34–43.

Gill, O., O'Donoghue, J., Faulkner, F. & Hannigan, A. (2010). Trends in performance of science and technology students (1997–2008) in Ireland. *International Journal of Mathematical Education in Science and Technology*, 41(3), 323–339.

González-Martín, A. S. & Hernandes-Gomes, G. (2018). The use of integrals in mechanics of materials textbooks for engineering students: the case of the first moment of an area. *INDRUM 2018 Proceedings*, pp. 115–124.

Gradwohl, J. & Eichler. A. (2018). Predictors of performance in engineering mathematics. *INDRUM 2018 Proceedings*, pp. 125–134.

Grigutsch, S. & Törner, G. (1998). *World Views of Mathematics Held by University Teachers of Mathematics Science* (Preprint Series No. 420). Duisburg: University of Duisburg, Department of Mathematics.

Gueudet, G. & Pepin, B. (2016). Students' work in mathematics and resources mediation at entry to university. *INDRUM 2016 Proceedings*, pp. 444–453.

Gueudet, G. & Quéré, P.-V. (2018). "Making connections" in the mathematics courses for engineers: the example of online resources for trigonometry. *INDRUM 2018 Proceedings*, pp. 135–144.

Gueudet, G., Bosch, M., DiSessa, A. A., Kwon, O. N. & Verschaffel, L. (2016). *Transitions in Mathematics Education*. London: Springer International.

Hamza, S. & O'Shea, A. (2016), Concept images of open sets in metric spaces. *INDRUM 2016 Proceedings*, pp. 113–122.

Hanke, E. & Schäfer, I. (2018). Learning complex analysis in different branches – Project Spotlight-Y for future teachers. *INDRUM 2018 Proceedings*, pp. 54–63.

Hochmuth, R. (2018). Discussing mathematical learning and mathematical praxeologies from a subject-scientific perspective. *INDRUM 2018 Proceedings*, pp. 517–526.

Hodgson, B.R. (2015). Whither the mathematics/didactics interconnection? Evolution and challenges of a kaleidoscopic relationship as seen from an ICMI perspective. In S. J. Cho (Ed.), *The Proceedings of the 12th International Congress on Mathematical Education* (pp. 41–61). Cham: Springer.

Holzkamp, K. (1985). *Grundlegung der Psychologie*. Frankfurt: Campus.

Huget, J. (2018). BiMathTutor - The program for further training of new tutors in University Mathematics. *INDRUM 2018 Proceedings*, pp. 458–459.

Kajander, A. & Lovric, M. (2005). Transition from secondary to tertiary mathematics: McMaster University experience. *International Journal of Mathematical Education in Science and Technology*, 36(2–3),149–160.

Kidron, I. (2018). Students' conceptions of irrational numbers. *International Journal of Research in Undergraduate Mathematics Education*, 4(1), 94–118.

Klein, F. (1908/1939). *Elementary Mathematics from an Advanced Standpoint. Part I: Arithmetic, Algebra, Analysis. Part II: Geometry*. (E. R. Hedrick & C. A. Noble, Trans.). New York: Dover.

Kock, Z. J. & Pepin, B. (2018). Student use of resources in Calculus and Linear Algebra. *INDRUM 2018 Proceedings*, pp. 336–345.

Kuklinski, C., Leis, E., Liebendörfer, M., Hochmuth, R., Biehler. R., Lankeit, E., Neuhaus, S., Schaper, N. & Schürmann, M. (2018). Evaluating innovative measures in university mathematics – the case of affective outcomes in a lecture focused on problem-solving. *INDRUM 2018 Proceedings*, pp. 527–536.

Landgärds, I. (2018). Mathematics teaching for economics students, but how?*INDRUM 2018 Proceedings*, pp. 187–188.

Lecorre, T. (2016). Rationality and concept of limit. *INDRUM 2016 Proceedings*, pp. 83–92.

Leont'ev, A. N. (1978). *Activity, Consciousness and Personality*. Englewood Clis, NJ: Prentice Hall.

Lerman, S. (2000). The social turn in mathematics education research. In J. Boaler (Ed.), *Multiple Perspectives on Mathematics Teaching and Learning* (pp. 19–44). Westport, CT: Ablex Publishing.

Mathieu-Soucy, S., Corriveau, C. & Hardy, N. (2018). Exploration of new post-secondary mathematics teachers' experiences: preliminary results of a narrative inquiry. *INDRUM 2018 Proceedings*, pp. 403–411.

McGowen, M. A. & Tall, D. O. (2010). Metaphor or met-before? The effects of previous experience on practice and theory of learning mathematics. *Journal of Mathematical Behavior*, 29(3), 169–179.

Monaghan, J., Nardi, E. & Dreyfus, T. (Eds.) (2019), *Calculus in Upper Secondary and Beginning University Mathematics – Conference Proceedings*. Kristiansand: MatRIC. Retrieved on September 26, 2019 from https://matric-calculus.sciencesconf.org/.

Nardi, E. (1996). *The Novice Mathematician's Encounter with Mathematical Abstraction: Tensions in Concept Image Construction and Formalization*. Doctoral thesis, University of Oxford, UK. https://archive.uea.ac.uk/~m011/thesis/.

Nardi, E. (2017). From advanced mathematical thinking to university mathematics education: A story of emancipation and enrichment. In T. Dooley & G. Gueudet (Eds.), *Proceedings of the 10th Conference of European Research in Mathematics Education (CERME)* (pp. 9–31). Dublin: Dublin City University.

Nardi, E., Bambili, A., Boufi, A., Stouraitis, K. & Triantafyllidis, T. (2017). [translated from Greek] Transitions in mathematics education (primary, secondary, tertiary, schooling): epistemological, psychological, pedagogical and institutional dimensions. In T. Zachariades, D. Potari & G. Psycharis (Eds.), *Proceedings of the 7th Conference of the Greek Association for Research in Mathematics Education (GARME)* (pp. 85–105). Athens.

Nardi, E., Winsløw, C. & Hausberger, T. (Eds.) (2016). *Proceedings of the First Conference of the International Network for Didactic Research in University Mathematics* (INDRUM 2016, 31 March–2 April 2016). Montpellier: University of Montpellier & INDRUM.

Petropoulou, G., Jaworski, B., Potari, D. & Zachariades, T. (2016). Addressing large cohorts of first-year mathematics students in lectures. *INDRUM 2016 Proceedings*, pp. 300–399.

Poropat, A. E. (2009). A meta-analysis of the five-factor model of personality and academic performance. *Psychological Bulletin*, 135(2), 322–338.

Potari, D. & Jaworski, B. (2002). Tackling complexity in mathematics teaching development: using the teaching triad as a tool for reflection and analysis. *Journal of Mathematics Teacher Education* 5(4), 351–380.

Prediger, S., Gravemeijer, K. & Confrey, J. (2015). Design research with a focus on learning processes: an overview on achievements and challenges. *ZDM Mathematics Education*, 47(6), 877–891.

Quéré, P. V. (2016). Une étude de l'autonomie en mathématiques dans la transition secondaire-supérieur. *INDRUM 2016 Proceedings*, pp. 484–493.

Rabardel, P. & Bourmaud, G. (2003). From computer to instrument system: A developmental perspective. *Interacting with Computers*, 15, 665–691.

Rach, S., Ufer, S. & Kosiol, T. (2018). Situational interest in university mathematics courses: Similar for real-world problems, calculations, and proofs?*INDRUM 2018 Proceedings*, pp. 356–365.

Syamsuri, S. (2019). APOS analysis on cognitive process in mathematical proving activities. *International Journal on Teaching and Learning Mathematics*, 1(1), 1–12.

Schmidt, K. & Winsløw, C. (2018). Task design for Engineering Mathematics: process, principles and products. *INDRUM 2018 proceedings*, pp. 165–174.

Schüler-Meyer, A. (2018). Defining as discursive practice in transition – Upper secondary students reinvent the formal definition of convergent sequences. *INDRUM 2018 Proceedings*, pp. 537–546.

Sfard, A. (2008). *Thinking as Communicating. Human Development, the Growth of Discourse, and Mathematizing*. New York, NY: Cambridge University Press.

Sfard, A. (2014). University mathematics as a discourse – why, how, and what for? *Research in Mathematics Education*, 16(2), 199–203.

Solomon. Y. (2006). Deficit or difference? The role of students' epistemologies of mathematics in their interactions with proof. *Educational Studies in Mathematics*, 61(3), 373–393.

Speer, N. M., Gutmann, T. & Murphy, T. J. (2005). Mathematics teaching assistant preparation and development. *College Teaching*, 53(2), 75–80.

Stender, P. & Stuhlmann, A.S. (2018). Fostering heuristic strategies in mathematics teacher education. *INDRUM 2018 Proceedings*, pp. 547–556.

Tall, D. (Ed.) (1991). *Advanced Mathematical Thinking*. Dordrecht: Kluwer.

Tall, D. (1992). The transition to advanced mathematical thinking: Functions, limits, infinity and proof. In D. A. Grouws (Ed.), *Handbook of Research on Mathematics Teaching and Learning* (pp. 495–511). New York: Macmillan.

Tall, D. (2013). *How Humans Learn to Think Mathematically: Exploring the Three Worlds of Mathematics*. Cambridge: Cambridge University Press.

Tall, D. & Vinner, S. (1981). Concept image and concept definition in mathematics with particular reference to limits and continuity. *Educational Studies in Mathematics*, 12(2), 151–169.

Thomas, M., de Freitas Druck, I., Huillet, D., Ju, M. K., Nardi, E., Rasmussen, C. & Xie, J. (2015). *Key mathematical concepts in the transition from secondary school to university*. In S. J. Cho (Ed.), *The Proceedings of 12th International Congress on Mathematical Education*. (pp. 265–284). Cham: Springer.

van den Heuvel-Panhuizen, M. & Drijvers, P. (2014). Realistic Mathematics Education. In S. Lerman (Ed.), *Encyclopedia of Mathematics Education* (pp. 521–525). Dordrecht: Springer.

Vandebrouck, F. & Leidwanger, S. (2016). Students' visualization of functions from secondary to tertiary level. *INDRUM 2016 Proceedings*, pp. 153–162.

Vinner, S. (1976). The naive concept of definition in mathematics. *Educational Studies in Mathematics* 7(4), 413–429.

Winsløw, C. (2006). Transformer la théorie en tâches: la transition du concret à l'abstrait en analyse réelle. In A. Rouchier & I. Bloch (Eds.), *Actes de la XIIIème École d'Été en Didactique des Mathématiques* (cd-rom). Grenoble: La Pensée Sauvage.

Winsløw, C. (2016). Angles, trigonometric functions, and university level analysis. *INDRUM 2016 Proceedings*, pp. 163–172.

Winsløw, C. Biehler, R., Rønning, F., Jaworski, B. & Wawro, M. (2020). Chapter 4. *This volume*.

11

STUDENTS' IN-CLASS AND OUT-OF-CLASS MATHEMATICAL PRACTICES

Chris Rasmussen, Helge Fredriksen, Emma Howard, Birgit Pepin and Johanna Rämö

Introduction

In this chapter, we review the literature on university students' mathematical practices, both in-class and out-of-class. We use the term *practice* to refer to the various ways in which students engage in mathematics. In order to narrow the scope of this broad charge, we focus the literature review on various theoretical perspectives that are relevant for studying students' in-class and out-of-class practices, students' use of resources out-of-class, students' role in assessment practices, students' responses to active learning initiatives, and students' in-class mathematical practices. These themes emerged from our review of the INDRUM 2016 and 2018 proceedings, as well as several recent reviews of the broader terrain of university mathematics education research (e.g., Biza, Giraldo, Hochmuth, Khakbaz, & Rasmussen, 2016; Rasmussen & Wawro, 2017; Winsløw, Gueudet, Hochmuth, & Nardi, 2018; Winsløw & Rasmussen, 2018). Given these multiple sources, the papers highlighted here include contributions to the two INDRUM proceedings as well as publications more broadly. Not all themes were strongly represented in the INDRUM proceedings and hence in these cases we relied more heavily on the broader literature.

While the focus of this chapter is on what students do, we recognise that the ways in which students engage in mathematics does not happen in a vacuum. Instead, it is necessary to situate students' mathematical practices in relation to interactions with other students, the teacher, the mathematics, and resources. These elements constitute the vertices of what Rezat and Sträßer (2012) refer to as the socio-didactic tetrahedron. The first of six sections of this chapter develop in greater detail various theoretical perspectives that further help frame students' mathematical practices, both in-class and out-of-class, as well with respect to the social and institutional milieu. The focus of each of the four sections that follow can be framed in terms of the vertices of the socio-didactic tetrahedron and their respective links, in particular those along the students-mathematics-resources triangular face. Students' use of resources focuses more directly on the interaction between artifacts and the mathematics; students' role in assessment practices highlights the student-teacher and student-mathematics edges; students' responses to active learning initiatives consider more broadly the social and institutional milieu in which various forms of active learning occur; and, finally, students' in-class mathematical practices highlight student-student interaction and the student-mathematics edge of the tetrahedron. For the purpose of this chapter, the socio-didactic tetrahedron provides an orienting frame for seeing one way in which the various sections work together. The final section of this chapter lays our future research directions related to the preceding sections.

Theoretical developments related to student practices

In this section we review the theoretical developments related to research on student practices over the past five years, in particular, those theories that underpin the research reported in the subsequent sections, and the main theoretical frames that were evident in the proceedings of INDRUM 2016 and 2018 and in selected review chapters and articles.

A relatively new, albeit often used theoretical frame, is that of RAME (Resource Approach to Mathematics Education: Trouche, Gueudet, & Pepin, 2019), as evidenced in the following section on students' use of resources out-of-class. RAME is associated with the Documentational Approach to Didactics (DAD: Trouche, Gueudet, & Pepin, 2020). DAD (and RAME) is concerned with students' interactions with resources (Gueudet & Pepin, 2016; Kock & Pepin, 2018). Gueudet and Pepin (2016) define student resources as anything likely to re-source (that is, to source again or differently) students' mathematical practice, leaning on Adler's (2000) definition of mathematics re-sources (in Adler's case used by teachers). Typically, a distinction is made between material resources, human/social resources, and cognitive resources (Pepin & Gueudet, 2020). For material resources, a further distinction has been made between curriculum resources (those resources proposed to students and aligned with the course curriculum), and general resources (which students might find/access randomly online). Curriculum resources are developed, proposed and used by teachers and students for the learning (and teaching) of the course mathematics, inside and outside the classroom. They can include text resources, such as textbooks, readers, websites and computer software, but also feedback on written work. General resources are the non-curricular material resources utilised by students, such as general websites (e.g., Wikipedia, YouTube). In terms of human/social resources, we refer to formal or casual human interactions, such as conversations with friends, peers or tutors. Cognitive resources relate to mathematical concepts that students may use to support their work with mathematics. The aim is to investigate students' practices, activities and learning through the lens of their interactions with resources in/for learning or studying.

Over the past decades, other theories have gained importance in research studies on student practices at university level in higher education mathematics. For example, Anastasakis, Robinson, and Lerman (2017) used Activity Theory to examine engineering students' use of resources. They found that students predominantly used resources provided by the university, and examination goals were the main consideration, with understanding mathematics and acquiring mathematical skills as lesser, albeit important, goals.

Other theories include the Theory of Didactical Situations (TDS) and the Anthropological Theory of the Didactic (ATD). TDS (Brousseau, 1998) investigates relationships that develop in the interactions between students and content (within the teacher-student-content socio-didactic triangle) – the *milieu*. This typically includes student peers, the concepts at stake, and students' previous knowledge and understandings, all of which sets the tone for the milieu. Strømskag (2018) developed a model for instructional design based on TDS, which she used in teacher education. She showed that the model was instrumental in creating student teachers' awareness of the impact of the milieu on the nature of the knowledge developed by the pupils.

Chevallard's (1985) ATD is a model of mathematical knowledge that has been applied to investigate student activity. Simply put, it claims that there are two aspects to student (or human) activity: the practical, and the theoretical – these are the main components of so-called praxeologies. Importantly, ATD draws attention to institutional conditions, the ecology in which the activities take place. Barquero (2018) used this concept to explore the ecology of mathematical modelling practices through the systematic variation of teaching institutions. That is, she explored the variety of constraints in different university classrooms when modelling, where particular proposals of study and research paths were suggested to overcome some of the constraints.

Another widely used theoretical frame for studying student assessment practices is that of Formative Assessment (FA), which underpins the section on students' role in assessment practices (Black & Wiliam, 1998; Hattie & Timperley, 2007; Sadler & Good, 2006). One of the main principles in FA is feedback to support the development of student self-regulation (Butler & Winne, 1995). Another is that it encourages peer and self-assessment. At INDRUM, these theories have been frequently used to investigate meaningful assessment and student involvement in assessment practices (e.g., Copeland, Howard, Meehan, & Parnell, 2018; Gaspar Martins, 2018) as well as self-regulation and self-efficacy (e.g., Lahdenperä, 2018).

Another widely-used framework for the design and analysis/evaluation of mathematical tasks is Realistic Mathematics Education (RME, van den Heuvel-Panhuizen & Drijvers, 2014; Fredriksen, 2018). RME is a domain-specific instruction theory, based on six principles: the activity principle (the importance of doing mathematics); the reality principle (the importance of starting with "real-life" problems that are meaningful for students); the level principle (student levels of understanding); the intertwinement principle (integrated curriculum); the interactivity principle (learning mathematics is a social activity favouring group work and whole-class discussions); the guidance principle (or guided intervention). RME has provided one of the foundations for the development of active learning approaches in mathematics education at both school and university level. For example, Fredriksen (2018) took the perspective of analysing how RME task design could influence the learning experience of a group of first-year students in computer engineering in a flipped classroom setting.

Related to the notion of active learning, the construct of Inquiry-Based Mathematics Education (IBME) (e.g., Artigue & Blomhøj, 2013; Laursen & Rasmussen, 2019) has been prominent. In North America, researchers have successfully used this framing at university level to advance understanding of how students develop knowledge through active learning approaches. This has been explored in the section on students' in-class mathematical practices. At INDRUM, theoretical lenses that are related to inquiry approaches have been prominent (e.g., Gjesteland, Vos, & Wold, 2018; Monreal Galán, Ruiz-Munzón, & Barquero, 2018). For example, Jaworski and colleagues (2018) have used the notion of community of inquiry (consisting of teacher researchers, students, and analytic assistants) to develop tasks in a computer medium to promote students' learning of mathematics – students were seen here as "partners in task design".

Finally, an increasingly used theoretical frame is the theory of commognition (Sfard, 2008). Sfard's commognitive framework conceptualises mathematics as a discourse – a form of communication consisting mainly of its word use, visual mediators, routines, and narratives. It offers researchers the opportunity to employ discourse analysis of the mathematics classroom, in particular how students engage in discourse (with peers, or with the teacher). For example, Schüler-Meyer (2018) explored students' discursive practices of defining convergent sequences in a transition course (between high school and university). Using a commognitive framework, he identified central development stages of the definition from experiential to abstract.

We now turn to the sections on students' use of resources, students' role in assessment practices, students' responses to active learning initiatives, and students' in-class mathematical practices.

Students' use of resources out-of-class

Considering the recent surge of technology being incorporated into mathematics education, we consider the relationship between students and mediating artifacts; where, "mediating artifacts might be mathematics textbooks, digital technologies, as well as tasks and problems, language" (Rezat & Sträßer, 2012, p. 643). Digital resources include online videos (both lecture-capture and video cast), online instructor notes, apps (for example GeoGebra), websites, interactive textbooks, and online quizzes. Much debated is the use of recorded lectures, mainly in terms of students' potential learning benefits and the possible effect of decreased lecture attendance (Morris, Swinnerton, & Coop, 2019).

An area of INDRUM research has been identifying students' choices of artifacts or resources to support their student-mathematics relationship and understanding the reasons for these choices. This research reflects a growing area of interest in general for the mathematics education community.

Recorded lectures

A controversial education topic is the provision of recorded lectures. There is a view by some instructors that the provision of recorded lectures would lead to a decrease in lecture attendance (Witthaus & Robinson, 2015). Danielson, Preast, Bender, and Hassall (2014) examined both instructors' and students' views of recorded lectures in a veterinary science course. They investigated the relationship between learning, students' use of recorded lectures, and course characteristics. Course characteristics are considered as the teaching approach (straight lecture, interactive lecture and mixed lecture with group work), curricular coordination (tight versus loose), and course content (clinical or problem solving, basic science – applied, basic science – research, non-traditional). Danielson et al. (2014) concluded that when lectures are in the traditional format as opposed to an interactive format, students achieve better learning gains from viewing recorded lectures.

Wold (2018) examined students' viewing of recorded lectures in a first-year mathematics course. Her findings include that students watch nearly the entirety of the video rather than selecting parts of the lecture to view, perhaps suggesting that lecture-capture recordings are used as a replacement for lectures. Wold (2018) found that 21% of the class watched the lecture capture recordings regularly with relatively few viewings occurring prior to examinations. However Howard, Meehan, and Parnell (2019) showed in a business mathematics module that the timing of online resources is related to the timing of the assessment, possibly indicating a difference between how students actively engage with lecture-capture recordings and short videos. While Wold (2018) examined students' use of a single resource, recorded lectures, it should be acknowledged that resources are predominantly not used in isolation.

Resource systems

The basis of resource system studies is the identification of which resources students use, and the understanding of why the patterns of usage occur or how students are using the resources. In INDRUM papers, the focus has been on first-year undergraduate students' use of resources, including online resources in a system (Gueudet & Pepin, 2016; Hillesund, 2018; Kanwal, 2018; Kock & Pepin, 2018; Quéré, 2016). These studies consider students as independent self-regulated learners. This interest in first-year undergraduate students may be owing to the increased proportion of mathematics time available for out-of-class practices, or to the effect of transitioning from a teacher-led approach at post-primary level to a self-directed and inquiry approach at university level.

For example, Quéré (2016) analysed open-ended surveys of engineering students for evidence that students had reflected upon the autonomy of the university learner in comparison to the post-primary learner. This self-directed learning was associated with organisation skills, lesson preparation, management of peer-study groups and the responsibility to correct assignments. However, it can be observed that students try to transfer their post-primary level approach of using resources to university education (Gueudet & Pepin, 2016), including a reliance on worked examples.

Kock and Pepin (2018) examined students' use of all available resources in two modules, calculus and linear algebra, with the aim of understanding how students coordinate their learning. They found that a segment of the student population felt lost owing to the variety and volume of resources available to them at university. To explain this, Kock and Pepin (2018, p.1) argue that "the course organisation and the alignment of curriculum materials with the learning goals had an impact on the students' choice and use of resources." They argue that when there is an alignment between these factors, students are able to direct their learning to appropriate resources for understanding (i.e., optimise the student-artifact-mathematics relationship).

In this vein, Kanwal (2018) examined a resource system in an online calculus module which featured a dedicated online interactive system "MyMathLab", tutorial videos and a textbook. MyMathLab contained formative assessments and students' assignments. Kanwal (2018) contends that examination goals drive students' resource choices and techniques, referring to this as students being pragmatic. An example of this pragmatic approach is that students identified videos as being useful for understanding concepts, however, they were also considered as time-consuming in comparison to the support of MyMathLab. Kanwal (2018) also found that websites were being used for explanation of concepts and to check answers. Whether it is the use of university-provided resources through a Learning Management System or external resources such as websites, students are embracing online resources and research shows that students' attitudes towards these are positive even when the academic results are mixed (Trenholm, Alcock, & Robinson, 2012).

INDRUM studies have examined students' use of textbooks in the context of a resource system. For example, students use textbooks to find worked examples which are similar to their assignment questions and see textbooks as less useful than instructors' notes (Gueudet & Pepin, 2016). They may also use textbooks as a repository of questions/examples with answers and as a source of important or relevant formulae (Kanwal, 2018). Quéré (2016) highlighted the role of exercises as central to mathematics, unlike in other disciplines. González-Martín and Hernandes-Gomes (2018) adopted the premise that university instructors draw on textbooks as a main source to guide their teaching. Their work follows from a study by Mesa and Griffiths (2012) which investigated mathematics faculty members use of textbooks. Combining these findings from INDRUM, it could be claimed that students are using online resources provided by instructors for explanation of concepts with textbooks being used primarily for assessment purposes. This aligns with studies which acknowledge a difference in how undergraduate students read mathematics, including textbooks, to the approach taken by mathematicians (Alcock, Hodds, Roy, & Inglis, 2015; Biza et al., 2016).

Qualitative and quantitative approaches to examining student use of resource systems

To analyse resource systems, INDRUM studies have taken a qualitative approach with interviews being a main source of data. For example, Hillesund (2018) examined students' Schematic Representation of Resource System (Pepin, Xu, Trouche, & Wang, 2017). This consists of a student drawing a representation of the resources they used for their module. He surmised that this method allows students to organise their thoughts in addition to imposing minimum interference by the researcher. Analysing the representations, he noted that students organised their resource representation through considering either categories, situations, the purpose of the resources or the features of the resources.

However, this qualitative approach does not capture the data associated with online resources including the date, time, and length of use of a specific resource, and in some settings the data can be per page of a textbook or pdf. An alternative approach to examine students' choice of resources, which can use this data, is cluster analysis. This method can be used to find groups of students who use specific resources or use resources in a similar manner. For example, Inglis et al. (2011) investigated students' use of lecture-capture, live lectures in their university's mathematics learning support centre. They found through clustering that students tended to use predominantly only one of these resources and students who use face-to-face resources achieved higher module marks. Comparatively, Howard, Meehan and Parnell (2018) in their cluster analysis of online videos versus live lectures found four patterns of resource use. These are described as Lecture-Users, Video-Users, Dual-Users and Switchers. Cluster analysis can also be used to investigate the timing of resources. In contrast to this literature, we found far fewer INDRUM papers that relate the types of resources accessed by students to their academic achievement. Rather, INDRUM papers have focused on understanding students' choice of resources in light of the transition from post-primary learners to university learners.

Students' role in assessment practices

Assessment has a key role in shaping learning behaviour. In recent decades, higher education literature has highlighted the relation between assessment and learning by moving from assessment of learning to assessment for learning. At the same time, university mathematics education practices may still emphasise summative assessment and closed-book examinations (e.g. Iannone & Simpson, 2011). Also, mathematics students seem to prefer traditional assessment methods (Iannone & Simpson, 2015).

The influence of assessment on learning is not always positive. Assessment can, for example, lead to grade hunting or poor learning strategies. Negative effects of assessment can be seen in Fuller, Deshler, Darrah, Trujillo, and Wu (2016), who investigated the interaction of students' personal traits and anxiety in a developmental mathematics course. One of their findings was that exam anxiety seemed to be a contributing factor in explaining students not meeting the course requirements.

In this section, we provide selected examples of how assessment can be used to promote learning. The focus is on assessment processes in which students have an active role. We start by presenting studies from the INDRUM proceedings that touch on formative assessment, and then move into themes that embrace the students' active role: engagement with feedback, peer assessment and self-assessment. In terms of the chapter's theoretical framing, the socio-didactic triangle, these themes are located mainly on the edges "student-teacher" and "student–mathematics".

Formative assessment

When emphasising assessment for learning, attention has been paid to formative assessment. Voigt (2016) studied a flipped classroom setting that contained regular quizzes. In his study, students' confidence in their mathematical abilities declined less in a flipped learning format than traditional lecture format. One explanation to this was that the quizzes flipped learning students completed made it possible for the students to make sure that they understand the core topics.

Gaspar Martins (2018) described a model of offering engineering students voluntary weekly quizzes that contributed to their final grade. The students could retry the quizzes until achieving the right answer. The students found the quizzes beneficial to their learning and felt that they made them study more.

Engagement with feedback

A core component of formative assessment is the feedback students receive. However, students might not use or be able to make sense of the teachers' feedback (e.g. Byrne, Hanusch, Moore, & Fukawa-Connelly, 2018). In order to understand students' ability to uptake feedback, Carless and Boud (2018) proposed a framework for feedback literacy. The elements of the framework are students' capability to appreciate the role of feedback, make judgements of the quality of their work, manage affects, and take action in response to feedback information.

Many of the elements of Carless and Boud's framework can be seen in the INDRUM 2018 study of Copeland et al. concerning business mathematics students' engagement with feedback. Students were given the opportunity to resubmit their quizzes, and were asked to identify, explain, and correct their errors. The results indicate that task type may affect students' remediations: conceptual tasks resulted more often in remediations that stated only a solution to the task, and lacked identification and explanation of errors. Students whose grades were lower tended to give solutions as remediations. The hypothesis is that the remediation process is not as beneficial for the lower-achieving students as it is for the higher-achieving students.

Landers and Reinholz (2015) studied students' participation in a feedback reflection activity in a mathematics course at a community college. Students were asked to read the feedback given to them by teachers and keep a written reflection log. According to Landers and Reinholz, students who took part regularly in the reflection activity saw it as an opportunity for self-assessment and grew as effective learners.

Peer assessment

In order to improve feedback processes, Nicol (2010) calls for iterative dialogues, not only between students and teachers but also between students and their peers. Reinholz (2015) has introduced a peer assessment approach called Peer-Assisted Reflection (PAR). In PAR, students attempt a problem, reflect on their work, confer with a peer, and revise and submit a final solution. In Reinholz's study, PAR resulted in improvement of student performance in introductory college calculus.

Alqassab, Strijbos and Ufer (2018) studied the impact of peer-feedback training on pre-service mathematics teachers' ability to provide feedback as well as their beliefs concerning peer-feedback. The effect of peer-feedback training was linked to the level of domain knowledge of the students. They suggest that domain knowledge of students should be taken into account in peer-feedback activities.

Jones and Alcock (2013) used Comparative Judgement (CJ: Pollitt, 2012) as a peer assessment method that requires no assessment criteria. Students were presented pairs of student solutions, and in each case they had to assess which one was better. Based on this, solutions were ranked from lowest to highest. Jones and Alcock argue that CJ can suit assessing tasks that require conceptual understanding.

The benefits of peer assessment for students do not only come from receiving feedback. When producing feedback, students reflect and evaluate their own work (Nicol, Thomson, & Breslin, 2014). To elaborate how peer assessment and self-assessment are connected, Reinholz (2016) presented a theoretical framework, the Assessment Cycle. It builds on Kollar and Fischer's (2010) framework and identifies six possible phases in the peer assessment process: task engagement, peer analysis, feedback provision, feedback reception, peer conferencing, and revision.

Self-assessment

Making qualitative judgements about one's own performance is an essential component in lifelong learning and developing self-assessment skills and can be argued to be core to university studies. Self-assessment has been shown to have a positive influence on learning, for example by improving self-efficacy and self-regulation (Panadero, Jonsson, & Botella, 2017).

There have been several studies on the accuracy of self-assessment. In the context of mathematics, Hosein and Harle (2018) studied students' ability to predict their grades and found that accuracy of predictions was associated with academic performance and mathematics confidence. However, Panadero, Brown and Strijbos (2016) argue that the accuracy in score prediction might not be as essential as the students' ability to describe the quality of their work and detect their strengths and weaknesses.

In INDRUM 2018, Häsä, Nieminen, Rämö, Pesonen, and Pauna described a digital self-assessment model (DISA) in which students regularly evaluated the quality of their learning outcomes, received feedback on their performance, and finally reflected on their learning and decided their own grades. The intended learning outcomes were made transparent through a detailed rubric. According to Häsä et al., students' perceptions of self-assessment indicated increased ownership of learning, but also revealed that self-assessment was a new and strange method to them. A study of the DISA model suggests that students associate self-assessment and the lack of exam with a deep approach to learning (Nieminen, Häsä, Rämö, & Tuohilampi, 2018).

Students' responses to active learning initiatives

Stanberry (2018, p. 960) defined active learning as a "process of education whereby students engage in activities like reading, writing, discussion, or problem solving that encourage analysis, synthesis, reflection and evaluation of class content." The term active learning encompasses many pedagogical strategies, such as collaborative learning, think-pair-share (Prince, 2004), and, lately, flipped classrooms (Bergmann & Sams, 2012). As such, the concept of active learning is quite flexible, as it includes various in-class initiatives engaging students in various forms of reasoning. However, this section will focus mainly on the above-mentioned frameworks of flipped classroom and IBME. There seems to be a certain momentum gaining towards the study of student-active and inquiry teaching approaches between the two conferences of 2016 and 2018. Considering students' practices, INDRUM 2016 included the presentation of two papers relevant to active learning initiatives, while the INDRUM 2018 proceedings presented four. The broader community of ERME also observes this trend. Skott, Mosvold, and Sakonidis (2018, p. 163–4) wrote about "the reform in mathematics education" which "promotes a vision of school mathematics that focuses on students' creative engagement in exploratory and problem-solving activities".

Considering our chapter's instructional tetrahedron theoretical framing, flipped classroom and IBME will be centred mainly along the student-student and student-mathematics axis. However, the teacher-student interaction may be considered to have a vital role in these frameworks, due to the necessity of guidance and facilitation of the teacher.

Flipped classroom

The main advantage of the flipped classroom in connection to active learning is the additional space for in-class learning activities afforded by this pedagogical approach. Since much of the direct instruction can be reserved for out-of-class videos, the teacher may facilitate activities in-class that focus on key concepts with groups of students (Strayer, 2012). Several INDRUM authors have studied the effect of the flipped classroom approach in university mathematics education settings. Voigt (2016) did a statistical comparative analysis of 427 undergraduate pre-calculus students on their experiences, attitudes and mathematical knowledge in a flipped classroom format compared to students in a traditional lecture format. He found that students in the flipped format were more positive about their in-class experience, felt more confident and cooperative in mathematical problem solving, and attained mathematical knowledge slightly better than students in traditional courses.

Fredriksen (2018) examined the learning experience of a group of first-year students in computer engineering in a flipped classroom setting. Following a single session with close attention to two groups of students collaborating, he found that students were able to utilise previous knowledge from the videos, in addition to other mediating means, successfully creating a mathematical model of a simulated double Ferris wheel movement.

Lim, Kim, and Lee (2016) examined the transition from a traditional to a flipped design in calculus and non-linear systems theory. The class size was relatively small, less than 20 learners in each. They found that the intensive in-class activities aided novice learners to gain insight from peer approaches. Furthermore, they claimed to be able to engage more in-depth discussions in class due to the preparatory videos, as they could repeatedly review difficult topics. Jungića, Kaurb, Mulhollanda, and Xinc (2014) described an intervention using flipped classroom in large introductory calculus classes with up to 342 enrolled students. Their in-class strategy was peer-to-peer instruction, utilising clickers to get initial responses on multiple-choice questions highlighted on the auditorium screen. After the initial display of results on-screen, they were told to discuss and attempt to convince their neighbour about their choice, thus spurring discussion. After this discussion phase, students would vote again, and plenary discussions orchestrated by the teacher would eventually converge the answer towards the one aimed at by the teacher. Data collected from a Likert-scale questionnaire showed that students perceived the intervention as beneficiary to their learning experience.

Critical voices were raised among students in these studies towards the increased workload and the lack of interaction possibilities during out-of-class viewing of lecture content. However, the overall impression from students participating in these studies seems to be a positive attitude towards the effect of being better prepared, leading to increased interactions with peers. Furthermore, these results seem to be consistent across a range of class sizes, although ways to implement the flipped format would vary greatly.

Inquiry-based learning approaches

Inquiry-based mathematics education considers ways in which students are actively engaged in working with mathematics in a process similar to how mathematicians and scientists do it. This may involve looking for relationships, making conjectures and generalisations, interpreting solutions and communicating these by various means (Dorier & Maass, 2014; Laursen & Rasmussen, 2019).

Rasmussen and Wawro (2017) performed a comprehensive review of research in undergraduate mathematics education towards such learning approaches. It seems that a large body of research indicates an improved success rate among students exposed to such initiatives (we refer readers to the meta-analysis of Freeman et al., 2014). They also reported on studies indicating that inquiry approaches resulted in significant better achievements among women, a result consistent with findings in other active-learning environments such as Extreme Apprenticeship (XA) (Lahdenperä, 2018). This INDRUM paper presented a quantitative study comparing the XA framework with a more traditional lecture-based course layout on the same cohort of students. Besides the effects of diminishing gender differences, the results indicated enhanced self-efficacy and self-regulation among students in the XA course. Although IBME seems to have an advantageous effect on student practices, there are obstacles particularly related to pedagogical content knowledge among university teachers. Considering how and when to shift between students' discovery and teacher telling in inquiry teaching environments can be a challenge for instructors (Marrongelle & Rasmussen, 2008). Also, inability to facilitate unpacking students' reasoning in whole-class discussions may severely limit the learning potential of these initiatives. As such, there is a need to convey research in inquiry-based teaching during professional development for teaching mathematicians.

Students' in-class mathematical practices

While still not the norm at the university level, there is increasing interest in IBME (Artigue & Blomhøj, 2013; Laursen & Rasmussen, 2019) where students learn mathematics by actively engaging in the authentic practices of mathematics. Our review of the INDRUM proceedings and the recent university mathematics education literature revealed the following two overarching themes related to this emerging trend: (1) Ways in which students engage in doing mathematics in class; and (2) Factors that contribute to student in-class engagement in mathematics. The first theme characterises the nature of students' mathematical activity while the second theme examines factors that enable or constrain students' participation in the authentic practice of mathematics. This second theme is well-represented in INDRUM papers, while the first theme comes primarily from the broader literature. Thus, future INDRUM papers are needed to represent and further develop this first theme. In terms of the chapter's broader theoretical framing of the socio-didactic tetrahedron, these two themes stress primarily the student-mathematics interaction and the student-student interaction.

We use the term mathematical practice to refer to particular types of joint enterprises (Wenger, 1999) in which mathematicians customarily engage. Moschkovich (2002) describes customary activities of mathematicians such as conjecturing, arguing, abstracting, and generalising. Rasmussen, Zandieh, King, and Teppo (2005) describe customary practices of mathematicians such as symbolising, defining, modelling, theoremising, and algorithmatising. Common across these characterisations of practice is the focus on authentic mathematical activity and the reinvention of mathematics. Much of this work is informed by the instructional design theory of Realistic Mathematics Education (RME) (Freudenthal, 1973; Gravemeijer, 1999).

Ways in which students engage in doing mathematics in class

This theme includes studies that investigate the prospects and possibilities for student reinvention of mathematics and their participation in the authentic practices of mathematics. Central to this line of research is the identification of productive, often informal, ways of reasoning that can be leveraged by students and instructional designers to reinvent significant and deep mathematical ideas and the creation of instructional sequences that foster student reinvention. For example, Larsen (2009; 2013) conducted a series of teaching experiments guided by the instructional design theory of RME in which he developed a sequence of tasks in which students generated a minimal set of rules for the set of symmetries of an equilateral triangle to reinvent the group axioms and the concept of isomorphism, as well as an approach to fostering student reinvention of isomorphism by having students determine how many groups could be formed with four elements. Similarly informed by RME, Rasmussen, Dunmyre, Fortune, and Keene (2019) identified key ways of reasoning and an instructional sequence in a first course in differential equations in which students reinvent a bifurcation diagram. Wawro, Rasmussen, Zandieh, Larson, & Sweeney (2012) discovered means by which students can reinvent the ideas core to span and linear (in)dependence based on their informal and intuitive reasoning with vectors. These studies are good examples of the processes and activities that can promote student reinvention of significant mathematical ideas and shows the power of RME as a guiding framework for the design and implementation of instructional sequences that support student reinvention.

The practice of creating and using algorithms (or algorithmatising) is exemplified in the study by Lockwood, Swinyard, and Caughman (2015) in which a pair of undergraduate students successfully reinvented four basic counting formulas. Similarly to the previous studies on reinvention, this study takes seriously Freudenthal's recommendation (1973) to avoid an anti-didactic inversion where symbolic formalism precedes informal reasoning.

The practice of theoremising, which refers to when students are engaged in genuine argumentation where conjectures are made and then justifications are created to support or refute the conjectures, is another area in which researchers are studying and documenting student reinvention (Rasmussen, Wawro, & Zandieh, 2015). Finally, we consider studies that take up students' participation in the practice of defining and the reinvention of definitions. Here we briefly review three of the many studies that have investigated student reinvention of definitions. Zandieh and Rasmussen (2010) for their part detailed the productive ways that students leveraged their concept image in service of creating and using a definition of spherical triangle. Swinyard and Larsen (2012) in their investigation of student reinvention of the limit definition proposed two key developments that were necessary for students to reinvent the formal definition. The first of these is a shift from an "x-first" perspective to a "y-first" perspective. The second key idea is the development of "an arbitrary closeness perspective to operationalise what it means to be infinitely close to a point" (p. 476). Lastly, Chorlay (2019) reported on a teaching experiment carried out within the framework of didactic engineering with year 12 students in France that led to student reinvention of a correct definition of the infinite limit for sequences. This empirical study suggested that some non-standard expressions for the definition may be more consonant with student informal and intuitive reasoning than the standard definition.

Factors that contribute to students' in-class engagement in mathematics

The second theme related to student mathematical practices focuses on factors that either inhibit or positively contribute to students' productive and successful participation in mathematical activity. The range of factors investigated include peer-to-peer discussions, interaction patterns, and affective characteristics.

For example, Hoppenbrock (2016) analysed the role of clicker questions in triggering peer-to-peer discussion and the development of students' conceptual understanding. Informed by Steinberg's (2005) framing of the epistemological triangle in relation to acts of communication, he found that students' understanding of "for all … there exists" (AE) and "there exists… for all" (EA) statements were enriched by a clicker question that invoked high quality discussion of different interpretations of AE and EA expressions. The clicker question was designed to meet four task characteristics that previous research identified as supporting peer-to-peer interaction: meaningful, complex, need for different ability to be solved, and aim of level raising. The notion of level raising fits well with the previously discussed RME perspective on mathematising. Analysis of peer-to-peer discussion of different interpretations of the AE and EA clicker question revealed two main level raising outcomes. The first was students' ability to find connections between their concept image and the symbolic expressions and the positive role of misunderstanding and mistakes in advancing their understanding of definitions.

Discussing and working with peers was also identified by Rämö, Oinonen, and Vihavainen (2016) to be positively associated with higher level of course performance in an introductory mathematics course focused on sets, functions, and proof. Grounded in the theoretical perspective of cognitive apprenticeship (Brown, Collins, & Duguid, 1989), the course was taught using a form of IBME, the Extreme Apprenticeship approach (Rämö, Oinonen, & Vikberg, 2015). In this approach, class time is structured by having students participate in practices that resemble those carried out by mathematicians. These practices align well with the earlier review in this section that focused on the nature of students' engaging in mathematical practices such as conjecturing, defining, modelling, etc. Consistent with the work of Fuller et al. (2016) that examined affective factors, Rämö et al. (2016) found students' motivation and learning strategies to be related to sufficient time and opportunities for peer-to-peer collaboration. As suggested by the previous studies as well as other larger scale studies (e.g., Freeman et al., 2014), peer-to-peer interaction can lead to higher levels of course performance. Rasmussen, Apkarian, Dreyfus, and Voigt (2016) examined ways in which engaging with someone else's reasoning relates to advances in reasoning for the person that decentres from their own way of reasoning. Based on fine-grained, qualitative analysis of individual interviews that approximated classroom peer-to-peer discourse, they found that higher levels of engaging with someone else's reasoning support growth in one's own reasoning. The significance of this work lies in explicating *why* peer-to-peer interaction is repeatedly found to lead to better performance on course outcomes. In a related analysis that used classroom video-recordings from this same course, Dreyfus, Rasmussen, Tabach, and Apkarian (2018) develop a theoretical and methodological approach (networking two different theoretical perspectives, abstraction in context and the emergent perspective) for investigating individual and collective mathematical progress.

Discussing the past and looking forward

We conclude this chapter by looking forward at promising directions for future research and how future research might engage in cross-pollination of the themes of students' use of resources, students' role in assessment practices, students' responses to active learning initiatives, and students' in-class mathematical practices.

The cross-pollination of the four themes that capture students' in-class and out-of-class practices would, in our view, benefit from theoretical networking. While the frameworks discussed in the first section of this chapter have been used in their pure form, combining two or more theories has been shown to be beneficial for developing deeper understandings of particular phenomena (Bikner-Ahsbahs & Prediger, 2014). Moreover, some frameworks, such as Commognition, TDS, and ATD, which are used more broadly in RUME, have not been widely leveraged in studies that examine students' in-class and out-of-class practices, but these theories are well-positioned to examine relationships that develop in the interactions within the teacher-student-content socio-didactic tetrahedron. Hence we see these frameworks as good candidates for networking theories more commonly used in the domain of students' in-class and out-of-class practices. Doing so would add analytic power and provide greater continuity and coherence to the broader RUME field.

For example, the analyses of students' use of online resources outside of the classroom have mostly been based on a single module or a single resource. Research has rarely considered the ways and possibilities that students' out-of-class practices might complement assessment practices and/or in-class inquiry practices. Through examining the student-artifacts-content relationship across a programme, the effect of prior experience of resources and/or module design on this relationship can be established. Studies show that students' out-of-class practices are influenced by assessment goals. Kanwal (2018) suggested that a variety of tasks are needed to encourage students into a knowledge-based approach rather than an assessment-focused approach. Similarly, Gueudet and Pepin (2016) found that a knowledge-based approach was used when a significant alteration to a mathematical method was required. Future research might examine how students' out-of-class practices are impacted by the alteration of module tasks, which intersects well with what we know about students' in-class practices.

Another area for cross-pollination is the use of online resources and interactive textbooks, which nowadays occurs on or off campus and in and out of the classroom. For example, in mobile learning students use their devices wherever they are, and they use them alone and/or with their peers to study the mathematics. These kinds of different study practices, in combination with digital resources, new forms of assessment, and new forms of participating in inquiry learning environments, need to be taken into consideration when investigating student learning of university mathematics. Such investigations seem ripe for cross-pollinating different in-class and out-of-class practices.

In terms of assessment, we regard this a still under-researched area in university mathematics education; this is reflected in the rather small number of INDRUM papers on the topic. More research is needed, as well as implications on how to transfer the research into practice. Attention should be given to preparing students for life-long learning by teaching them to make judgements of their own work and giving them an active role in the assessment process (Boud & Falchikov, 2006). However, we should not assume that assessment is viewed and experienced in the same way in all cultural and educational contexts. Panadero et al. (2016) point out that the effects of self-assessment on students may vary from culture to culture and call for cross-cultural research and awareness of socio-cultural values when studying self-assessment. We suggest the same for assessment in general. Also, from the viewpoint of the socio-didactical tetrahedron, the axis containing the nodes "conventions and norms about being a student and about learning" and "institutions" requires attention. Students and society expect that summative assessment will be used for the purpose of certifying achievement. On the other hand, as Black and McCormick (2010) claim, since summative assessment has a great impact on what and how students study, we should also focus on the learning benefits of summative assessment. This might require radical re-thinking of learning and teaching in higher education that brings together recent advances on IBME, which gives students more power over their own learning. A related assessment question is: how is learning affected if we give students more power over their assessment?

The body of research on flipped classroom and IBME has gained substantial magnitude. Most investigations consider only student satisfaction and performance, leaving little insight into the fundamental aspects that make these initiatives work or not (Fredriksen & Hadjerrouit, 2019). Thus, there is a definite lack of qualitative research considering how the participation in classroom activities is influenced by factors such as task design, out-of-class videos, and institutional settings. The networking of theories may feature a broader understanding of students' learning practices upon being subject to these approaches.

The research to date on students' in-class mathematical practices has shown the power in student reasoning to reinvent mathematics as they participate in mathematical practices. A next step for future research is the theoretical elaboration of learning trajectories that support student reinvention. We conjecture that a research agenda focused on learning trajectories and instruction based on these is likely to result in more generalisable and useful outputs for practitioners and researchers. Moreover, such an agenda has considerable potential to bring together not only the different themes presented here on students' in-class and out-of-class practices, but also the findings and advances on teacher practices (see Chapter 12 in this book).

Finally, we point to an area of research glaringly absent from the INDRUM proceedings – equity and inclusion. A research agenda on equity and inclusion can cut across the four themes of students' use of resources, students' role in assessment practices, students' responses to active learning initiatives, and students' in-class mathematical practices. For example, in what ways do innovative assessment practices and inquiry approaches privilege (or not) particular student demographic groups? Does the use of online platforms appeal more to certain student groups than others and if so, how does that impact learning and persistence in mathematics? While some research suggests that women benefit from IBME (Laursen, Hassi, Kogan, & Westin, 2014), other research (e.g., Johnson et al., in press) points to the possibility that student-centred classrooms may, depending on the instructor, reinforce gendered and racialised implicit biases. Making equity and inclusion a focus of research on students' mathematical practices, and of university mathematics education research at large, is an ethical and moral imperative.

References

Adler, J. (2000). Conceptualising resources as a theme for teacher education. *Journal of Mathematics Teacher Education*, 3(3), 205–224.

Alcock, L., Hodds, M., Roy, S., & Inglis, M. (2015). Investigating and improving undergraduate proof comprehension. *Notices of the American Mathematical Society*, 62(7), 742–752.

Alqassab, M., Strijbos, J. W., & Ufer, S. (2018). Training peer-feedback skills on geometric construction tasks: role of domain knowledge and peer-feedback levels. *European Journal of Psychology of Education*, 33(1), 11–30.

Anastasakis, M., Robinson., C. L., & Lerman, S. (2017). Links between students' goals and their choice of educational resources in undergraduate mathematics. *Teaching Mathematics and its Applications: An International Journal of the IMA*, 36(2), 67–80.

Artigue, M., & Blomhøj, M. (2013). Conceptualizing inquiry-based education in mathematics. *ZDM Mathematics Education*, 45(6), 797–810.

Barquero, B. (2018). The ecological relativity of modelling practices: adaptations of a study and research path to different university settings. *INDRUM 2018 Proceedings*, pp. 85–94.

Bergmann, J., & Sams, A. (2012). *Flip Your Classroom, Reach Every Student in Every Class Every Day*. Washington DC: International Society for Technology in Education.

Bikner-Ahsbahs, A., & Prediger, S. (2014). *Networking of Theories as a Research Practice in Mathematics Education*. Cham: Springer.

Biza, I., Giraldo, V., Hochmuth, R., Khakbaz, A., & Rasmussen, C. (2016). *Research on Teaching and Learning Mathematics at the Tertiary Level: State-of-the-art and Looking Ahead.* ICME-13 Topical Surveys. New York: Springer.

Black, P., & McCormick, R. (2010). Reflections and new directions. *Assessment & Evaluation in Higher Education*, 35(5), 493–499.

Black, P., & Wiliam, D. (1998). Inside the Black Box: Raising Standards through Classroom Assessment, *Phi Delta Kappan International*, 80(2), 139–148.

Boud, D., & Falchikov, N. (2006). Aligning assessment with long-term learning. *Assessment and Evaluation in Higher Education*, 31(4), 399–413.

Brousseau, G. (1998). *La Théorie des Situations Didactiques*, Grenoble: La Pensée Sauvage.

Brown, J. S., Collins, A., & Duguid, P. (1989). Situated cognition and the culture of learning. *Educational Researcher*, 18(1), 32–42.

Butler, D. L., & Winne, P. H. (1995). Feedback and self-regulated learning: A theoretical synthesis. *Review of Educational Research*, 65(3), 245–281.

Byrne, M., Hanusch, S., Moore, R. C., & Fukawa-Connelly, T. (2018). Student interpretations of written comments on graded proofs. *International Journal of Research in Undergraduate Mathematics Education*, 4(2), 228–253.

Carless, D., & Boud, D. (2018). The development of student feedback literacy: Enabling uptake of feedback. *Assessment & Evaluation in Higher Education*, 43(8), 1315–1325.

Chevallard, Y. (1985) *La Transposition Didactique. Du savoir savant au savoir enseigné.* (2nd ed., 1991). Grenoble: La Pensée Sauvage.

Chorlay, R. (2019) A pathway to a student-worded definition of limits at the secondary-tertiary transition. *International Journal of Research in Undergraduate Mathematics Education*, 5 (3), 267–314.

Copeland, C., Howard, E., Meehan, M., & Parnell, A. (2018). Engaging with feedback: How do students remediate errors on their weekly quiz?*INDRUM 2018 Proceedings*, pp. 306–315.

Danielson, J., Preast, V., Bender, H., & Hassal, L. (2014). Is the effectiveness of lecture capture related to teaching approach or content type? *Computers & Education*, 72, 121–131.

Dorier, J.-L., & Maass, K. (2014). Inquiry-based mathematics education. In S. Lerman (Ed.), *Encyclopedia of Mathematics Education* (pp. 300–304). Dordrecht: Springer.

Dreyfus, T., Rasmussen, C., Apkarian, N., & Tabach, M. (2018). The complexity of knowledge construction in a classroom setting. *INDRUM 2018 Proceedings*, pp. 286–295.

Fredriksen, H. (2018). Analysing tasks for the Flipped Classroom from the perspective of Realistic Mathematical Education. *INDRUM 2018 Proceedings*, pp. 296–305.

Fredriksen, H., & Hadjerrouit, S. (2019 – online first). An activity theory perspective on contradictions in flipped mathematics classrooms at the university level. *International Journal of Mathematical Education in Science and Technology*, 1–22.

Freeman, S., Eddy, S. L., McDonough, M., Smith, M. K., Okoroafor, N., Jordt, H., & Wenderoth, M. P. (2014). Active learning increases student performance in science, engineering, and mathematics. *Proceedings of the National Academy of Sciences of the United States of America*, 111(23), 8410–8415.

Freudenthal, H. (1973). *Mathematics as an Educational Task.* Dordrecht: Reidel.

Fuller, E., Deshler, J., Darrah, M., Trujillo, M., & Wu, X. (2016). Anxiety and personality factors influencing the completion rates of developmental mathematics students. *INDRUM 2016 Proceedings*, pp. 274–283.

Gaspar Martins, S. (2018). Weekly homework quizzes as formative assessment for Engineering students are a fair and effective strategy to increase learning?. *INDRUM 2018 Proceedings*, pp. 105–114.

Gjesteland, T., Vos, P., & Wold, M. (2018). Mathematical modelling and activation – a study on a large class, a project-based task and students' flow. *INDRUM 2018 Proceedings*, pp. 175–184.

González, A. S., & Hernandes-Gomes, G. (2018). The use of integrals in Mechanics of Materials textbooks for engineering students: the case of the first moment of an area. *INDRUM 2018 Proceedings*, pp. 115–124.

Gravemeijer, K. (1999). How emergent models may foster the constitution of formal mathematics. *Mathematical Thinking and Learning*, 2, 155–177.

Gueudet, G., & Pepin, B. (2016). Students' work in mathematics and resources mediation at entry to university. *INDRUM 2016 Proceedings*, pp. 444–453.

Hattie, J., & Timperley, H. (2007). The power of feedback, *Review of Educational Research*, 77 (1), 81–112.

Hillesund, E. (2018). Using schematic representation of resource systems to examine how first year engineering students use resources in their studies of mathematics. *INDRUM 2018 Proceedings*, pp. 185–186.

Hoppenbrock, A. (2016). Multiple choice questions and peer instruction as pedagogical tools to learn the mathematical language. *INDRUM 2016 Proceedings*, pp. 254–463.

Hosein, A. & Harle, J. (2018). The relationship between students' prior mathematical attainment, knowledge and confidence on their self-assessment accuracy. *Studies in Educational Evaluation*, 56, 32–41.

Howard, E., Meehan, M., & Parnell, A. (2018). Live lectures or online videos: Students' resource choices in a first-year university mathematics module. *International Journal of Mathematical Education in Science and Technology*, 49(4), 530–553.

Howard, E., Meehan, M., & Parnell, A. (2019). An exploration of the relationship between continuous assessment and resource use in a service mathematics module. *Eleventh Congress of European Research in Mathematics Education*, Utrecht: CERME.

Häsä, J., Nieminen, J. H., Rämö, J., Pesonen, H., & Pauna, M. (2018). Determining your own grade – Student perceptions on self-assessment in a large mathematics course. *INDRUM 2018 Proceedings*, pp. 366–367.

Iannone, P. and Simpson, A. (2015). Students' preferences in undergraduate mathematics assessment. *Studies in Higher Education*, 40(6), 1046–1067.

Iannone, P., & Simpson, A. (2011). The summative assessment diet: how we assess in mathematics degrees. *Teaching Mathematics and its Applications: An International Journal of the IMA*, 30(4), 186–196.

Inglis, M., Palipana, A., Trenholm, S., & Ward, J. (2011). Individual differences in students' use of optional learning resources. *Journal of Computer Assisted Learning*, 27(6), 490–502.

Jaworski, B., Treffert-Thomas, S., Hewitt, D., Feeney, M., Shrish-Thapa, D., Conniffe, D., Dar, A., Vlaseros, N., & Anastasakis, M. (2018). Student partners in task design in a computer medium to promote foundation students' learning of mathematics. *INDRUM 2018 Proceedings*, pp. 316–325.

Johnson, E., Andrews-Larson, C., Keene, K., Keller, R., Fortune, N., & Melhuish, K. (in press). Inquiry and gender inequity in the undergraduate mathematics classroom. *Journal for Research in Mathematics Education.*

Jones, I. & Alcock, L. (2013). Peer assessment without assessment criteria. *Studies in Higher Education*, 39(10), 1774–1787.

Jungića, V., Kaurb, H., Mulhollanda, J., & Xinc, C. (2014). On flipping the classroom in large first year calculus courses. *International Journal of Mathematical Education in Science and Technology*, 46(4), 508–520.

Kanwal, S. (2018). Engineering students' engagement with resources in an online learning environment. *INDRUM 2018 Proceedings*, pp. 145–154.

Kock, Z.-J. & Pepin, B. (2018). Student use of resources in calculus and linear algebra. *INDRUM 2018 Proceedings*, pp. 336–345.

Kollar, I., & Fischer, F. (2010). Peer assessment as collaborative learning: A cognitive perspective. *Learning and Instruction*, 20(4), 344–348.

Lahdenperä, J. (2018). Comparing male and female students' self-efficacy and self-regulation skills in two undergraduate mathematics course contexts. *INDRUM 2018 Proceedings*, pp. 346–355.

Landers, M., & Reinholz, D. (2015). Students' reflections on mathematics homework feedback. *Journal of Developmental Education*, 38(3), 22–36.

Larsen, S. (2009). Reinventing the concepts of groups and isomorphisms: The case of Jessica and Sandra. *Journal of Mathematical Behavior*, 28(2), 119–137.

Larsen, S. (2013). A local instructional theory for the guided reinvention of the group and isomorphism concepts. *Journal of Mathematical Behavior*, 32(4), 712–725.

Laursen, S. L., Hassi, M. L., Kogan, M., & Weston, T. J. (2014). Benefits for women and men of inquiry-based learning in college mathematics: A multi-institution study. *Journal for Research in Mathematics Education*, 45, 406–418.

Laursen, S., & Rasmussen, C. (2019). I on the prize: Inquiry approaches in undergraduate mathematics education. *International Journal of Research in Undergraduate Mathematics Education*, 5(1), 129–146.

Lim, C., Kim, S., & Lee, J. (2016). Designing the flipped classroom in higher education. In J. M. Spector, D. Ifenthaler, D. G. Sampson, & P. Isaias (Eds.), *Competencies in Teaching, Learning and Educational Leadership in the Digital Age: Papers from CELDA 2014* (pp. 245–258). Cham: Springer.

Lockwood, E., Swinyard, C. A., & Caughman, J. S. (2015). Patterns, sets of outcomes, and combinatorial justification: Two students' reinvention of counting formulas. *International Journal of Research in Undergraduate Mathematics Education*, 1(1), 27–62.

Marrongelle, K., & Rasmussen, C. (2008). Meeting new teaching challenges: Teaching strategies that mediate between all lecture and all student discovery. In M. Carlson & C. Rasmussen (Eds.), *Making the Connection: Research and Teaching in Undergraduate Mathematics Education* (pp. 167–178). Washington DC: Mathematical Association of America.

Mesa, V., & Griffiths, B. (2012). Textbook mediation of teaching: An example from tertiary mathematics instructors. *Educational Studies in Mathematics*, 79(1), 85–107.

Monreal Galán, N., Ruiz-Munzón, N., & Barquero, B. (2018). Central dialectics for mathematical modelling in the experience of a study and research path at university level. *INDRUM 2018 Proceedings*, pp. 155–164.

Morris, N. P., Swinnerton, B., & Coop, T. (2019). Lecture recordings to support learning: A contested space between students and teachers. *Computers & Education*, 140.

Moschkovich, J., (2002). An introduction to examining everyday and academic mathematical practices. *Journal for Research in Mathematics Education*. Monograph, Vol. 11, Everyday and academic mathematics in the classroom, p. 1–11.

Nicol, D. (2010). From monologue to dialogue: improving written feedback processes in mass higher education. *Assessment & Evaluation in Higher Education*, 35(5), 501–517.

Nicol, D., Thomson, A., & Breslin, C. (2014). Rethinking feedback practices in higher education: a peer review perspective. *Assessment & Evaluation in Higher Education*, 39(1), 102–122.

Nieminen, J. H., Häsä, J., Rämö, J., & Tuohilampi, L. (2018). Replacing exam with self-assessment: Reflection-centred learning environment as a tool to promote deep learning. *Proceedings of the 20th meeting of the MAA Special Interest Group on Research in Undergraduate Mathematics Education*. San Diego: RUME.

Panadero, E., Brown, G. T., & Strijbos, J. W. (2016). The future of student self-assessment: A review of known unknowns and potential directions. *Educational Psychology Review*, 28 (4), 803–830.

Panadero, E., Jonsson, A., & Botella, J. (2017). Effects of self-assessment on self-regulated learning and self-efficacy: Four meta-analyses. *Educational Research Review*, 22, 74–98.

Pepin, B., & Gueudet, G. (2020). Curricular resources and textbooks. In S. Lerman (Ed.), *Encyclopedia of mathematics education* (2nd edition). Heidelberg: Springer.

Pepin, B., Xu, B., Trouche, L., & Wang, C. (2017). Developing a deeper understanding of mathematics teaching expertise: an examination of three Chinese mathematics teachers' resource systems as windows into their work and expertise. *Educational Studies in Mathematics*, 94(3), 257–274.

Pollitt, A. (2012). The method of Adaptive Comparative Judgement. *Assessment in Education: Principles, Policy & Practice*, 19(3), 281–300.

Prince, M. (2004). Does active learning work? A review of the research. *Journal of Engineering Education*, 93(3), 223–231.

Quéré, P.-V. (2016). Une Etude De L'autonomie en Mathématiques Dans la Transition Secondaire-Supérieur. *INDRUM 2016 Proceedings*, pp. 484–493.

Rasmussen, C., Apkarian, N., Dreyfus, T., & Voigt, M. (2016). Ways in which engaging in someone else's reasoning is productive. *INDRUM 2016 Proceedings*, pp. 504–513.

Rasmussen, C., Dunmyre, J., Fortune, N., & Keene, K. (2019 –online first). Modeling as a means to develop new ideas: The case of reinventing a bifurcation diagram. *PRIMUS*, 29 (6), 509–526.

Rasmussen, C., & Wawro, M. (2017). Post-calculus research in undergraduate mathematics education. In J. Cai (Ed.), *Compendium for Research in Mathematics Education* (pp. 551–581). Reston, VA: National Council of Teachers of Mathematics.

Rasmussen, C., Wawro, M., & Zandieh, M. (2015). Examining individual and collective level mathematical progress. *Educational Studies in Mathematics*, 88(2), 259–281.

Rasmussen, C., Zandieh, M., King, K., & Teppo, A. (2005). Advancing mathematical activity: A view of advanced mathematical thinking. *Mathematical Thinking and Learning*, 7, 51–73.

Rämö, J., Oinonen, L., & Vihavainen, A. (2016). Activity and performance on a student-centred undergraduate mathematics course. *INDRUM 2016 Proceedings*, pp. 494–503.

Rämö, J., Oinonen, L., & Vikberg, T. (2015). Extreme apprenticeship – Emphasising conceptual understanding in undergraduate mathematics. In K. Krainer & N. Vondrová (Eds.), *Proceedings of the Ninth Congress of the European Society for Research in Mathematics Education* (pp. 2242–2248). Prague: Charles University in Prague, Faculty of Education & ERME.

Reinholz, D. (2015). Peer-Assisted Reflection: A design-based intervention for improving success in calculus. *International Journal of Research in Undergraduate Mathematics Education*, 1(2), 234–267.

Reinholz, D. (2016). The assessment cycle: a model for learning through peer assessment. *Assessment & Evaluation in Higher Education*, 41(2), 301–315.

Rezat, S., & Sträßer, R. (2012). From the didactical triangle to the socio-didactical tetrahedron: artifacts as fundamental constituents of the didactical situation. *ZDM Mathematics Education*, 44, 641–651.

Sadler, P. M., & Good, E. (2006). The impact of self-and peer-grading on student learning. *Educational Assessment*, 11(1), 1–31.

Schüler-Meyer, A. (2018). Defining as discursive practice in transition – Upper secondary students reinvent the formal definition of convergent sequences. *INDRUM 2018 Proceedings*, pp. 537–546.

Sfard, A. (2008) *Thinking as Communicating: Human Development, the Growth of Discourses, and Mathematising*. New York, NY: Cambridge University Press.

Skott, J., Mosvold, R., & Sakonidis, C. (2018). Classroom practice and teachers' knowledge, beliefs and identity. In T. Dreyfus, M. Artigue, D. Potari, S. Prediger, & K. Ruthven (Eds.), *Developing Research in Mathematics Education: Twenty Years of Communication, Cooperation and Collaboration in Europe* (pp. 162–180). London: Routledge.

Stanberry, M. L. (2018). Active learning: a case study of student engagement in college Calculus. *International Journal of Mathematical Education in Science and Technology*, 49(6), 959–969.

Strayer, J. F. (2012). How learning in an inverted classroom influences cooperation, innovation and task orientation. *Learning Environments Research*, 15(2), 171–193.

Steinbring, H. (2005). *The Construction of New Mathematical Knowledge in Classroom Interaction* (Mathematics Education Library, Vol. 38). New York: Springer.

Strømskag, H. (2018) A model for instructional design in mathematics implemented in teacher education. *INDRUM 2018 Proceedings*, pp. 422–431.

Swinyard, C., & Larsen, S. (2012). Coming to understand the formal definition of limit: Insights gained from engaging students in reinvention. *Journal for Research in Mathematics Education*, 43(4), 465–493.

Trenholm, S., Alcock, L., & Robinson, C. L. (2012). Mathematics lecturing in the digital age. *International Journal of Mathematical Education in Science and Technology*, 43(6), 703–716.

Trouche, L., Gueudet, B., & Pepin, B. (Eds.) (2019). *The 'Resource' Approach in Mathematics Education (RAME)*. Cham: Springer.

Trouche L., Gueudet G., Pepin B. (2020). Documentational Approach to Didactics. In: S. Lerman (Eds.) *Encyclopedia of Mathematics Education* (2nd edition). Cham: Springer.

van den Heuvel-Panhuizen, M., & Drijvers, P. (2014). Realistic Mathematics Education. In S. Lerman (Ed.). *Encyclopedia of Mathematics Education* (pp. 521–525). Cham: Springer.

Voigt, M. (2016). Comparative analysis of learning gains and students' attitudes in a flipped precalculus classroom. *INDRUM 2016 Proceedings*, pp. 511–519.

Wawro, M., Rasmussen, C., Zandieh, M., Larson, C., & Sweeney, G. (2012). An inquiry-oriented approach to span and linear independence: The case of the magic carpet ride sequence. *PRIMUS*, 22(8), 577–599.

Wenger, E. (1999). *Communities of Practice: Learning, Meaning, and Identity*. Cambridge: Cambridge University Press.

Winsløw, C., Gueudet, G., Hochmuth, R., & Nardi, E. (2018). Research on university mathematics education. In T. Dreyfus, M. Artigue, D. Potari, S. Prediger, & K. Ruthven (Eds.), *Developing Research in Mathematics Education: Twenty Years of Communication, Cooperation and Collaboration in Europe* (pp.82–96). London: Routledge.

Winsløw, C., & Rasmussen, C. (2018). University mathematics education. In S. Lerman (Ed.) *Encyclopaedia of mathematics education*. Cham: Springer.

Witthaus, G. R., & Robinson, C. L. (2015). Lecture capture literature review: A review of the literature from 2012–2015. Centre for Academic Practice, Loughborough University Loughborough. Retrieved from https://repository.lboro.ac.uk/articles/Lecture_capture_literature_review_A_review_of_the_literature_from_2012-2015/9368876.

Wold, M. (2018). Students' usage patterns of video recorded lectures in an undergraduate mathematics course. *INDRUM 2018 Proceedings*, pp. 370–371.

Zandieh, M., & Rasmussen, C. (2010). Defining as a mathematical activity: A framework for characterizing progress from informal to more formal ways of reasoning. *Journal of Mathematical Behavior*, 29, 57–75.

12

MATHEMATICS TEACHING PRACTICES AT UNIVERSITY LEVEL

Nicolas Grenier-Boley, Pedro Nicolás, Heidi Strømskag and Theresia Tabchi

Introduction

In this chapter, the presentations at the first two INDRUM conferences (2016 & 2018) are taken to provide a snapshot of current research into university level mathematics teaching. The papers offer an insight into the variety of theories used to frame research in this specialised field and the issues upon which the research focuses. In recent years, theoretical-methodological issues concerning research in mathematics education at the tertiary level have been considered as a main area of interest (Biza, Giraldo, Hochmuth, Khakbaz, & Rasmussen, 2016). Studies presented at INDRUM focus on understanding and describing teaching and on the development of teaching practices aligned with a chosen theoretical perspective. They mostly take qualitative approaches to study cases in depth. They do not adopt positivist approaches that seek evidence of teaching (as cause) and students' learning outcomes (as effect), which might be generalised from sample to population or from experimental or quasi-experimental designs.

Systematic inquiry into mathematics teaching disseminated through INDRUM reflects the international character of the network. The studies of teaching reported at INDRUM conferences are framed within a variety of theories that have roots in traditions of empiricism and pragmatism, discourse, socio-cultural philosophies and institutional perspectives. Sixteen full research papers have been presented, eight each in 2016 and 2018. These papers reveal a significant bias, about one third of the papers, towards theory rooted in the institutional, didactic transposition developed by Chevallard (1981, 1998), especially the anthropological theory of didactics. The influence of Chevallard's theoretical development is unsurprising given his background in mathematics and focus on didactics. The composition of this chapter reflects the influence of Chevallard and other, mostly francophone, theorists. Another theory developed in the context of mathematics that has gained some ground in higher mathematics education research is Sfard's (2008) adaptation of discourse theory to mathematics education, referred to as "commognition".

Theories with empirical roots in mathematics classrooms, such as the "Teaching Triad" (Jaworski, 1994), the "Knowledge Quartet" (Rowland, Huckstep, & Thwaites, 2005), and "Mathematical Knowledge for Teaching" (Ball, Thames, & Phelps, 2008) also feature. These theories emerge from careful and sustained observation and interpretation of teachers working in school classroom situations. It can be argued that school and university classrooms are different contexts – size of class, maturity of students, expectations of independent learning activity, pace of instruction, motivation to study, and so on. Therefore, it cannot be taken for granted that a theory developed empirically in one context will easily accommodate to another.

Some research is framed within global socio-cultural theories, which have not been developed with direct roots in mathematics education. Examples of these are cultural historical activity theory (Vygotsky, 1986; Leontiev, 1978; Leplat, 1997) and community of practice theory (Wenger, 1998). These well-developed and rather complex theories are grounded in conceptions of "activity" and "practice" respectively.

This chapter includes brief introductory accounts, sufficient to satisfy the needs for discussion of the range of theories included within the INDRUM presentations. Within a single chapter it is impossible to consider such a range of theories with a depth of treatment that could satisfy even a mildly critical reader seeking to develop an understanding the theories from scratch. The discussion is inevitably brief and incomplete. The intention in outlining the theories is primarily to indicate how research into mathematics teaching in higher education is being framed by scholars included within the INDRUM community.

The chapter moves on to consider the theoretical perspectives of the INDRUM papers, accompanied by the substance of these papers. After that comes a section on characteristics of teaching practices which encompasses teacher learning and teacher knowledge, frameworks of teacher knowledge, issues of communicating mathematics, and teacher-researchers' teaching practices. Then follows a section on methodology for instructional/didactic design which encompasses a comparison of didactic engineering and design research, and a discussion of instructional design for inquiry-based learning. The chapter closes with conclusions and thoughts on future research.

Theories framing research into university mathematics teaching

The anthropological theory of the didactic

According to Chevallard (1998), the Anthropological Theory of the Didactic (ATD) allows an understanding of an individual's "practices" through the notion of *praxeology* which can be broken down into two blocks, *praxis* and *logos*. The *praxis* block comprises the "know-how" made of "types of tasks" (things to accomplish) and "techniques" (methods used to carry out tasks of the given type). The *logos* block comprises the discourse that justifies and produces the praxis at two levels: the "technologies" (justifications of techniques) and "theories" (foundations that underlie the technological discourse). Thus, praxeologies are defined by a quadruple: task, technique, technology and theory. The mathematical organisation (MO) describes the mathematical activities of the praxeology, and the didactic organisation (DO) describes the activities supporting the teaching or learning of the MO (Bosch & Gascón, 2006). MOs and DOs are influenced by the institutional organisation of the teaching and the learning processes. Chevallard (2002) takes into account the different aspects of institutional organisation in the didactic analysis of teaching and learning and introduces a scale of levels of didactic co-determination. These levels help distinguish between the conditions and constraints related to the specificities of the discipline taught (discipline, domain, sector, theme, question), and those at a general level of the teaching of any discipline (humanity, civilisation, society, school, pedagogy). *Professional praxeologies* were developed by Chevallard for the analysis of the practices of individuals in a given profession such as the training of teachers or of future engineers.

An *institution* is defined by Chevallard (1998) as an evolving social system imposing on its subjects specific praxeological equipment for a given field of activity. Broley (2016) describes the status of some practices within institutions based on how the integration of computer programming in students' learning activities was related to teachers' research. Six levels were identified of interaction of mathematicians or students with programming that vary from "strictly observe the results of a computer program" to "develop a program, including algorithm development, coding, and verification" (Broley, 2016, p. 363).

The exploratory study by Florensa, Bosch, Cuadros and Gascón (2018) deals with the kinds of questions lecturers point out at the beginning of educational courses. The study reveals that the questions asked remained at a pedagogical level and were poorly connected to the mathematical knowledge to be taught. Based on the experience from the courses presented in the paper, Florensa et al. (2018) postulate that didactics can help university teachers better interpret their practice and question it in a more productive way. They highlight elements from ATD that have shown useful in this respect: praxeology, didactic contract, didactic moments, Herbartian schema, media-milieu dialectics, didactic ecology.

A way to characterise teaching activities is through identifying the learning activities they trigger. Chevallard (2013) distinguishes between two paradigms: the "visit of works" where students are required to study according to a firmly pre-established programme; and "questioning the world" where students would be given "questions needing answers" instead of answers without questions. The new paradigm of "questioning the world" defines new professional praxeologies, where teachers would not be transmitters of knowledge, rather facilitators of learning (Bourgade, 2016). Investigation workshops, introduced by Chevallard (2011), can help investigate the professional praxeologies required when a teacher works in the frame of "questioning the world". Seeking possible answers to an open question in an investigation workshop, without pre-defining the direction that the investigation would take, leads to the development of "study and research paths" (SRPs) (Chevallard, 2011). An SRP with its dialectics is the design heuristic of ATD, starting from a problematic question, where mathematical modelling is at the heart of the teaching and learning processes.

According to Winsløw, Matheron, and Mercier (2013), an SRP emphasises the dialectic between "research" and "study" which characterises any learning activity. "Research" refers to inquiry, problem solving, etc. while "study" designates the consultation of existing knowledge initiated by the teacher. The term "path" designates the possible open trajectories to be followed in an experimentation of the SRP. An SRP starts by raising a "generating question", an open question that provokes the interest of the community of study. Many questions derived from the initial one arise as the study progresses; the answers to these questions trace out the possible paths to be followed in the experimentation of the SRP.

An SRP can also be applied in ways other than as a heuristic device. For instance, Barquero, Serrano, and Ruiz-Munzón (2016) describe the conditions and constraints influencing the integration of SRP in first-year courses of a business and administration degree at university level. The authors analyse the dialectic between inquiry and transmission, emphasising the role of the different didactic devices mobilised in the survival and evolution of the SRPs. The study highlights possible ways of integrating "Study and Research Activities" (SRAs) (Barquero & Bosch, 2015) into SRP, and subsequently links transmissive teaching devices to inquiry-based ones.

Further, Bourgade (2016) aims to answer the question: "To what extent is the elaboration of a professional praxeological equipment dependent on professional difficulties on the one hand, and on the dissemination of this equipment on the other hand?" (p. 351). He initiated and supported the implementation of an investigation workshop at *la prépa des INP*[1] for two years. The paper focuses on the task of "formulating a generating question" from a real-life situation and tries to identify the different didactic moments in the realisation of this task (first encounter, exploration, building of the technological-theoretical block). Analysing the task in terms of didactic moments helps identify professional difficulties and dissemination issues as two catalysers of the *logos* production in the building of a professional praxeology, which usually does not require elaborate *logos*.

A contribution by Romo, Barquero and Bosch (2016) is an adaptation of the methodology of "Study and Research Path" for teacher education (SRP-TE) developed by Sierra (2006) and Ruiz-Olarría (2015). SRP-TE methodology takes as a starting point of the teacher education process a problematic teaching question and approaches it in many phases: searching for available answers in mathematics didactics literature; experiencing one of the findings; analysing the experience from a didactics perspective; and finally, when possible, designing, implementing and developing a class activity adapted to a given institutional context. The study considers a course on mathematical modelling implemented in a Master's Programme for in-service secondary mathematics teachers using an online modality. The design of the course is based on four phases associated with the four stages of didactic analysis: 1) creating a teaching proposal based on the epistemological analysis; 2) elaborating a lesson plan; 3) putting the plan into practice with secondary school students; and 4) redesigning the activity after experimentation taking into account the institutional constraints revealed. Results show that when teachers design modelling activities and implement them in their classrooms, they become aware of the impact institutional constraints have on the implementation of the pedagogical and mathematical forms of these activities.

Instrumental and Documentational approach

The *instrumental approach* focuses on studying the activity of one or more subjects, involved in a goal-directed activity, and mediated by *artefacts*. An artefact is defined by Rabardel (2002) as being a product of human activity, whose design is oriented towards a given goal. An artefact associated with a scheme of use of this artefact in a given situation generates an instrument. The process of development of an instrument is called *instrumental genesis* and has two intricate aspects: *instrumentation* (shaping of the activity to accommodate the tool) and *instrumentalisation* (re-shaping of the tool by the user to adapt to the activity's goals).

A scheme is a relation between knowledge and actions. Vergnaud (1998) defines it as an invariant organisation of conduct for a given set of situations comprising four interacting components (Gueudet, Buteau, Mesa, & Misfeldt, 2014): the *aim* of the activity (intentionality in the organisation of the activity); the *rules of action* (ways of acting to reach a given aim); *operational invariants* (theorems-in-action or concepts-in-action that influence rules of action); and *inferences* (adaptations that can be brought to answer the specificities of a situation). *Instrumental orchestration* (Trouche, 2004) encompasses the intentional and systematic organisation of both the mathematical situation and the artefacts, in order to facilitate the students' instrumental geneses.

The *documentational approach* draws on the instrumental approach, taking into account the broader range of available *resources*. It retains the general notion of resource described by Adler (2000, p. 207): "It is possible to think about resource as the verb re-source, to source again or differently". For a given subject, a resource is thus anything that can possibly intervene in his/her activity. Associating resources with a "scheme of use" of these resources generates a "document". Teachers may select, combine, and design their own resources. They use these resources in class, modify them on the spot or afterwards, share them with colleagues, etc. All this constitutes the teacher's *documentation work*.

The instrumental and the documentational approaches can both be used to describe learning (including teacher learning) and teaching processes analytically, and to design teaching or teacher training devices (Gueudet et al., 2014). A poster contribution by Syrdalen (2018) draws on the instrumental orchestration to explore the role of a lecturer in a flipped classroom. Three elements of orchestration, introduced by Drijvers, Doorman, Boon, Reed and Gravemeijer (2010), are used to explore the lecturers' role in a flipped classroom: the didactic configuration (arrangement of the artifacts and the teaching setting); the exploitation mode (how the didactic configuration is used); and the didactic performance (spontaneous decisions).

From a different perspective, a poster contribution by Hadjerrouit and Gautestad (2016) uses instrumental orchestration in an empirical study on the consequences of integrating an interactive visualisation tool "SimReal+" in mathematics education at university level. The study mobilises instrumental orchestration as defined in the work of Drijvers et al. (2010) and defines new types of orchestrations that emerge in the digital era: 13 different types of orchestration are identified. Results show that SimReal+ can motivate students and enhance their learning of some mathematical topics such as differentiation and integration.

The documentational approach (Gueudet, Pepin, & Trouche, 2012) is often used to study the interactions between a teacher, or a group of teachers, and resources. Tabchi (2018) employs the documentational approach – as developed in (Gueudet, 2017) – to characterise the practices of university mathematics teacher-researchers, particularly the impact of their research activity on their teaching practices, through the lens of their interactions with resources. This exploratory study draws on interviews with two teacher-researchers, whose domain of research is discrete mathematics. They both insisted on the importance of reasoning and proofs, and therefore select contents that contribute to developing students' ability to write proofs. Further, they considered that, despite software's importance in research, it is not essential for teaching purposes. Moreover, Tabchi (2018) highlights the impact of the institutional context (level of education such as Bachelor and Master levels, progressivity in teaching in certain fields, etc.) on teaching practices at university level. Furthermore, in what concerns the relationship between the research activity and the teaching practices of university teacher-researchers, the study shows that the choice to consider the interactions with resources is revealing. But that step should be developed in further studies.

The theory of didactic situations in mathematics

The theory of didactic situations (TDS) offers a systemic framework for studying teaching and learning processes and for carrying out didactic design in mathematics, where the nature of the knowledge at stake is crucial (Brousseau, 1997). The aim of TDS is to investigate the relationship between the design of a teaching situation and the knowledge it generates. This involves studying the conditions for the functioning of the knowledge at stake – that is, how a generic and epistemic subject would be induced to *use* the particular mathematics to make decisions.

A *didactic situation* in mathematics is a specific plan organised so as to cause students to appropriate some piece of mathematical reference knowledge. TDS postulates that every mathematical concept is the solution of at least one specific system of mathematical conditions. This "system" can be interpreted by at least one situation, for example, a game, whose solution (decision, message, argument) is one of the typical manifestations of the concept (Brousseau & Warfield, 2014).

An *adidactic situation* is a situation in which the students take a mathematical task as their own and try to solve it without the teacher's interventions and without trying to figure out the teacher's didactic intention. The *milieu* models the physical and intellectual reality with which the students interact in an adidactic situation. An appropriate milieu acts as a "piece of nature" for the students, that is, something giving objective feedback that tells them whether their responses are adequate or not with respect to the target knowledge. The milieu may be composed of: material or symbolic tools provided (artefacts, informative texts, data, etc.), students' prior knowledge, other students, and arrangement of the classroom and rules for operating in the situation (determinative of who is supposed to interact with whom).

Every didactic process consists of three situations where the teacher's role changes (Brousseau & Warfield, 2014): 1) A *situation of devolution* in which the teacher: a) Engages students in a challenging mathematical situation that can be undertaken without requiring the teacher's help; and b) informs students about the conditions, rules, goal, and the criterion for success. 2) An *adidactic* situation that supports the students in autonomous mathematical activities, both individual and collective, through engaging them in: a) producing "new" statements and discussing their validity; b) making decisions, formulating hypotheses, predicting and judging their consequences, attempting to communicate information, producing and organising models, arguments and proofs, etc.; and c) evaluating and correcting by themselves the consequences of their choices. Students' autonomy depends on the adidactic potential of the milieu. The teacher might need to *regulate* the situation to make the milieu evolve to ensure progression of the knowledge; this can be done by informational jumps or by making references to the didactic contract. 3) A *situation of institutionalisation* in which the teacher informs students about the place, importance and future of the mathematical knowledge reached in the adidactic situation by: a) taking note of the progress of the adidactic situation, of the questions and answers that have been obtained or studied from it and placing them within the perspective of the curriculum; b) distinguishing among the pieces of knowledge that have appeared, and presenting in conventional form those that will serve as reference (the correct ones).

The *didactic contract* is an interpretation of the rules of the interaction between the teacher and the students in a didactic situation. Some of these rules are explicit and related to the mathematics at stake; others are implicit and of more general character related to mathematics teaching and learning. These rules form a set of reciprocal obligations between the teacher and the students. The didactic contract is the source of several didactic phenomena that constrain the meaning of the knowledge taught and learned. Brousseau (1997) has described six such phenomena: the Topaze effect, the Jourdain effect, the improper use of analogy, the metacognitive shift, the metamathematical shift, and the Dienes effect.

In TDS, students' learning is seen as a combination of processes of *adaptation* and *acculturation* (Brousseau, 2000). Independent adaptation to a milieu takes place in an adidactic situation; acculturation into an educational system takes place in a didactic situation with help of a didactic contract. In this model, the devolution ensures the conditions for adaptation, and the institutionalisation ensures the conditions for acculturation.

Didactic engineering is the primary research methodology of TDS. It is based on qualitative analyses and is structured into four phases: preliminary analyses; design and a priori analysis; realisation, observation and data collection; a posteriori analysis and validation (Artigue, 2015). The design heuristic of TDS consists of the *adidactic situation* with its *milieu*: This is designed with a problem (or problems) that has the target knowledge as an (in some sense) optimal solution, where the milieu gives feedback to the students on their responses, with respect to this knowledge.

Although TDS originates experimentally in the context of primary school, numerous aspects of the theory have been used by researchers at secondary and tertiary level. For instance, González-Martín, Bloch, Durand-Guerrier and Maschietto (2014) discuss the use of TDS at university level. They present three research cases where TDS is applied – two on calculus and one on proof. Further, Strømskag Måsøval (2013) used TDS tools in teacher education to identify relationships between the design of an algebraic task and the outcome of student teachers' engagement with the task.

Strømskag (2018) discusses instructional design in teacher education with data from three student teachers' experiments in secondary school. A didactic engineering model developed by Strømskag (2017) was used by student teachers to design, implement and analyse didactic situations. The results show how the student teachers, on the basis of comparison between the a priori and a posteriori analyses, were able to identify relationships between the milieus of the adidactic situations and the knowledge generated by the pupils in those situations. The utility at university level of the applied instructional model involves considering the experimented situations as material milieus for instructional design in teacher education.

Theory of Commognition

The basic principle of Sfard's (2008) theory of *commognition* – an amalgam of *communication* and *cognition* – is that thinking can be regarded as a communicative act. Mathematics can be seen as a discourse and doing mathematics can be seen as engaging in a mathematical discourse. This viewpoint simplifies things, as *thinking* and *thoughts* are traditionally conceived as something less accessible to analysis, than *communicating* and *discourses*.

There is an increasing number of studies in mathematics education at university level that use commognition theory (Nardi, Ryve, Stadler, & Viirman, 2014; Viirman, 2014, 2015; Thoma & Nardi, 2016). Thoma and Nardi's contribution uses commognition to provide insight into closed-book assessment practices. They identify the routines of the assessment discourse of the examination task of a year one compulsory mathematics course in the United Kingdom. We will take a closer look at this contribution in the section devoted to the issues of communicating mathematics.

Cultural-historical Activity Theory

The cultural-historical *Activity Theory* of Vygotsky (1986) and others (e.g. Leont'ev, 1978; Leplat, 1997) is a global theory of cognition that has been used to frame research into university-level mathematics teaching. In this framework, it is hypothesised that students' (possible) learning emerges from their activities as they work on mathematical tasks, where the *activity* is "what a subject engages in during the completion of the task" whereas the task is "the goal to be attained under certain circumstances" (Rogalski, 2013, p. 4). It is interesting to note that, in the context of mathematics education, activity theory has developed with divergent interpretations, as illustrated by Abboud et al. (2018). Two contributions at INDRUM 2016 align with these different interpretations, each of which has specific theoretical-methodological hypotheses related to teachers' practices.

In Grenier-Boley, Bridoux, and Hache (2016), teachers' practices are analysed by using the "Double Approach" framework (Robert & Hache, 2013) which is a didactic and ergonomic approach to practices. Within the Double Approach, mathematics teachers' goals and decisions are described in terms of five components: cognitive, mediative, institutional, personal, social. The authors study discursive activities of the teacher that reflect a *proximity* with students' activities, knowledge or remarks in analogy with the Vygtoskian notion of zone of proximal development (ZPD) (Rogalski, 2013). In Petropoulou, Jaworski, Potari, and Zachariades (2016), teachers' practices during lectures are identified by using the Teaching Triad framework (Jaworski, 1994), an analytic framework oriented by teachers' goals and actions. Its purpose is to describe the complexity of teachers' activities in terms of *management of learning, sensitivity to students* and *mathematical challenge.*

Community of practice theory

Community of practice theory (CPT) has been used as a framework for developmental research that seeks to establish inquiry communities (at university level, e.g., Biza, Jaworski, & Hemmi, 2014). An INDRUM study framed within CPT is (Viirman, 2018), which reports the collaboration between a mathematician and a mathematics education researcher.

Characteristics of teaching practices

Teacher learning and teacher knowledge

Some ongoing studies have focused on the design and implementation of initiatives dedicated to teachers' training at the university. Hamann and Schmidt-Thieme (2018) study the possibility of enriching links between curriculum courses of a preservice training spiral curriculum by attending to basic notions, history and language of mathematics through teaching episodes referred to as "keynotes". The study described in the poster contribution by Viirman, Pettersson, Björklund and

Boustedt (2018) aims to help Swedish student teachers integrate programming into their mathematics classes (as required by a revision of the national curriculum). To do so, the authors propose to implement a training course focusing on programming in a mathematical context with a collaborative teaching approach, and studying the potential transformation of knowledge acquired during teachers' training into teaching practices *in situ*. Huget (2018) reports on the creation of a tutor training program ("BiMathTutor") conceptualised and implemented at the University of Bielefeld, Germany. These contributions make it possible to raise important issues related to the implementation of such initiatives at the university. We may wonder about the extent to which theoretical frameworks can help evaluate the "efficacy" of such an initiative (also pointed out by Hamann & Schmidt-Thieme, 2018).

The question of transition from being a university student to being a teacher in a certain institution has also been addressed. Huget (2018) investigates the effects of "BiMathTutor", both on the possible change in tutors' practice during the training sessions and on the expectations and results of students who attend tutoring classes. Mathieu-Soucy, Corriveau and Hardy (2018) address the main issue of the transition from being a university student to becoming a new teacher in the *cegep*[2] institution in Quebec, Canada. To do so, they plan to investigate *significant experiences* (to be understood from the perspective of Dewey's philosophy (1938)) within these teachers' *relationship with mathematics and its teaching and learning* (RWMTL). The main result of the pilot study is the classification of these teachers' RWMTL into several categories, including personal issues, issues linked to the institution, the relation between teachers and students, and teachers' expectations towards students' mathematical level.

Pre-service and in-service teachers' competences to diagnose mathematics teaching situations have also been studied (Biza & Nardi, 2016, Biza, Nardi, & Zachariades, 2018). Two issues related to teacher learning emerge from these studies. An important matter of interest seems to be the possibility to integrate competences in order to diagnose teaching situations or to address teaching issues in mathematics teachers' pre- and in-service development programmes (Biza & Nardi, 2016; Biza et al., 2018). Another is the question of the role of mathematics education theories in teachers' professional development.

Frameworks of teacher knowledge

Shulman (1987) proposed three components that together constitute the professional knowledge base of teaching: subject matter knowledge (SMK); pedagogical content knowledge (PCK); and curriculum knowledge. Ball et al. (2008) build on Shulman's work and focus simultaneously on defining distinct forms of mathematical knowledge for teaching (MKT) and on developing related measures. They distinguish between forms of SMK consisting of common, specialised, and horizon content knowledge, and forms of PCK comprising knowledge of mathematics and students, mathematics and teaching, and mathematics and curriculum. They describe specialised content knowledge (SCK) as mathematical knowledge that is unique to the work of teaching and distinct from the common content knowledge (CCK) needed and used by teachers as well as others. The works of Shulman (1987) and Ball et al. (2008) form the basis for the Mathematics Teacher's Specialised Knowledge model (MTSK) developed by Carrillo, Climent, Contreras and Muñoz-Catalán (2013).

In Germany, the project Professional Competence of Teachers, Cognitively Activating Instruction, and Development of Students' Mathematical Literacy (COACTIV) developed open response measures of content knowledge (CK, which aligns with SMK) and PCK at the secondary level and used these together with data from a large sample of grade 10 teachers and their students' progress (Baumert et al., 2010). In COACTIV, content knowledge is defined as deep understanding of the mathematics taught in the secondary school – it is associated with the idea of "elementary mathematics from a higher standpoint" in Klein's (2016) textbook, and with Ma's (1999) notion of "profound understanding of mathematics". Results from COACTIV studies showed that while PCK was inconceivable without sufficient CK, the latter could not substitute for PCK.

The above-mentioned frameworks for mathematics teachers' knowledge are developed – and mainly used – in the context of primary or secondary school. Delgado and Zakaryan (2018) build on MTSK (Carrillo et al., 2013) in their attempt to characterise a university teacher's knowledge of the practice of mathematics in the content area of mathematical analysis. They found indicators of the lecturer's knowledge of the practice of mathematics related to ways of *reasoning, validating*, and *proceeding* in mathematics.

Another categorisation of teachers' knowledge is that created by Rowland et al. (2005). On the basis of 24 video-recorded mathematics lessons in primary school, they developed the "Knowledge Quartet" (KQ), comprising foundation, transformation, connection, and contingency knowledge. Foundation knowledge aligns with SMK, and the other three with PCK. The first dimension of the KQ, *foundation*, is composed of teachers' mathematics-related knowledge, beliefs and understanding. The second dimension, *transformation*, is composed of knowledge-in-action as manifested in the planning of teaching and in the act of teaching itself. The third dimension, *connection*, is composed of ways in which the teacher obtains coherence within and between lessons. The fourth dimension, *contingency*, is about the teacher's ability to "think on his/her feet" and deals with unexpected contributions from students.

Breen, Meehan, O'Shea and Rowland (2018) used the Knowledge Quartet framework to analyse the features of university teaching that are highlighted in a set of three university lecturers' (the first three authors of the paper) brief but vivid accounts of memorable incidents that occurred during their teaching over the course of two years. The accounts were written as part of a professional development project using the Discipline of Noticing (Mason, 2002), and focused on troublesome aspects of classes as perceived by the lecturers.

Within the frame of ATD and the paradigm of "questioning the world" (Chevallard, 2015), Bourgade's (2016) study shows how teacher knowledge is developed as a consequence of elaboration and dissemination of a professional praxeological equipment. Bourgade explored how to design a *generating question* (see above) for an investigation workshop in engineering education. He describes the difficulties that a (generic) teacher faces while disseminating a praxeology under construction (initiated by a new type of task) to other subjects of the institution. Furthermore, he explains how these difficulties can contribute towards and are instrumental for the construction of a high-level *logos* of the new type of tasks.

In Syrdalen's (2018) on-going study, the intention is to apply the Systematic Classroom Analysis Notation (Beeby, Burkhardt, & Fraser, 1979) to describe teacher knowledge through evaluation of the depth of demand on, and level of guidance given by, the lecturer during in-class interactions.

Issues of communicating mathematics

The contributions that address the issue of communicating mathematical ideas can be divided in two main directions: those focusing on teaching as a discursive practice and those interested in teachers' discourse in order to address the way students' needs are taken into account.

In the first direction, one possibility is to see teaching as a highly-situated discursive practice as in (Biza & Nardi, 2016) and (Viirman et al., 2018). Another possibility is to see language as a medium and an aim of mathematics lessons (Schmidt-Thieme, 2018). In this work, Schmidt-Thieme proposes to implement a "language curriculum" for pre-service teachers in order to focus on mathematics teachers' language competencies as well as transitions between the registers of mathematical representation (Duval, 2006).

In the second direction, two contributions that adopt an Activity Theory approach to cognition have studied university teachers' discourse in Calculus lectures. First, Grenier-Boley et al. (2016) compare two courses devoted to the introduction of the formal definition of limit of a function at the beginning of university[3]: the course presented in a textbook and the course given by a teacher during a lecture. Several proximities (upward, downward, horizontal)[4] are spotted in the textbook and the teacher's discourses: they correspond to different types of occasions to induce and/or operate a proximity between students' previous knowledge and the new knowledge to be learnt. A finding from the study is that the textbook does not contain occasions of proximities whereas the teacher's discourse contains several proximities that are supposed to address several students' needs. These needs include the use of a progressive formalism, reformulation of part of a definition (words, graph, absolute value, distance, neighbourhood), and recall of logical elements. Second, Petropoulou et al. (2016) study the way in which a lecturer conceptualises students' learning needs and enacts these conceptualisations in his actual teaching. They propose two main categories of this lecturer's goals with respect to students' needs within his practice. The first encompasses introducing students to aspects of advanced mathematical thinking, such as refining a problem, using alternative methods, representing, and justifying. The second is about supporting students' content-related understanding, including providing steps and methods for solving tasks, and highlighting subtle aspects of the mathematics.

The issue of communicating mathematics to non-mathematics students has also been addressed at INDRUM conferences. Martìn-Molina (2016) studies her own adaptation of content and teaching of statistics in several undergraduate programmes. Two outcomes of the study were the importance of disciplinary knowledge that may help contextualise the statistics taught and the need to reduce the level of formalism in instruction. The study by Viirman (2018) focuses on the collaboration between mathematicians and mathematics educators within a project in which modelling activities are proposed to Biology students. The study highlights, in the frame of Community of Practice Theory, the contrasting stances towards students and mathematics by the mathematician and the mathematics education researcher: the former making mathematical content his or her didactic priority, the latter's endeavour being to try to set him/herself in the position of a student and see the mathematics from the student's perspective.

In connection with the above studies, one could ask: how can an instructor's teaching practice be influenced by his/her conceptualisation of students' learning needs? The studies by Grenier-Boley et al. (2016) and Petropoulou et al. (2016) present many similarities[5]: the former formulates hypotheses about students' needs as a consequence of three types of studies (epistemological, curricular and didactical) whereas the starting point of the latter is students' needs as perceived by teachers (cognitive, affective, social).

One of the aims of inquiry-based learning (IBL) is to teach students to construct proofs. Accordingly, one purpose of examinations would be to assess students' ability to generate such a discourse. Thoma and Nardi (2016) show, in their study of a closed-book examination, that students are explicitly guided to justify and construct their responses. This not only fosters a somewhat limited image of mathematics as a predominantly step by step, highly directed activity, as the authors observe, but also reveals a reluctance to ask students to develop a complete and correct reasoning by themselves, the reasoning required in the exams being imitative and not creative.

Teacher-researchers' teaching practices

On an international level, there is an emerging interest in studying university teacher-researchers' practices (Annoot & Fave-Bonnet, 2004). Becher (1994) and Poteaux (2013) affirm the need to take into consideration the nature of the content taught and the possible impact of the research activity of teacher-researchers on their teaching when studying their practices. However, existing research on university teaching practices seldom takes them into account (Henkel, 2004;

Neumann, 2001). In this regard, several contributions to INDRUM show an increasing interest in the practices of university teacher-researchers in mathematics based on didactical approaches. Examples are these studies referred in the previous section: Broley (2016), Florensa et al. (2018), Strømskag (2018), and Tabchi (2018).

Methodologies for instructional/didactic design

Didactic engineering versus design research: four theoretical approaches

According to Artigue (2009), didactic design includes controlled intervention research into the processes of *planning, delivering* and *evaluating* mathematics teaching and learning, and further, it includes the problem of *reproducibility* of results from such interventions. Here, we examine the methodologies for instructional design of four frequently-used theoretical approaches to mathematics teaching and learning: ATD, TDS, Realistic Mathematics Education (RME) and Lesson Study. Whereas the methodology of ATD and TDS is *didactic engineering*, the methodology of RME and Lesson Study is *design research* (even if the research component may not be central in Lesson Study). The design heuristics of ATD and TDS are presented above, in the sections on ATD and TDS respectively.

The design heuristic of RME consists of either *didactic phenomenology* (Freudenthal, 1983) or *emergent modelling* (Gravemeijer, 2004). Didactic phenomenology identifies the phenomena that the target mathematical concept helps organise and understand, whereas emergent modelling starts with identifying the initial problems that refer to a paradigmatic context situation for the target mathematical concept.

The design heuristic of Lesson Study is the *research lesson* (Fernandez & Yoshida, 2004). This is a lesson in which observational data is collected from the test of a teaching-learning experiment. An overview of the four approaches and their methodologies is given in Figure 12.1.

We note that all four approaches involve the design of some educational tool, and they are all informed by, and contribute to the development of, educational theory. Furthermore, they all reject standardised validation processes based on comparison of experiment groups and control groups.

When it comes to the vision of mathematics education and characteristics of the two methodologies, they are different. The vision of didactic engineering is mathematics education as a fundamental science, where the principal aim is understanding of didactic systems and didactic phenomena (an applied component is relevant, but secondary). Didactic engineering is a kind of "phenomenotechnics" that cares for the necessity to *produce* didactic phenomena in order to gain control over them (Bachelard, cited in Chevallard, 1981).

The vision of design research is mathematics education as a design science, where the principal aim is to develop robust teaching sequences through controlled production of educational tools (to understand didactic phenomena is relevant, but secondary). Design research is interventionist and iterative, cyclic in nature.

THEORY	RESEARCH METHODOLOGY	CONSTITUENTS of the methodology
ATD	Didactic engineering SRP (*study and research path*)	- Preliminary analyses - Design and a priori analysis of an SRP - Implementation, observation, data collection, *in vivo* analysis - A posteriori analysis and validation (Barquero & Bosch, 2015)
TDS	Didactic engineering (*adidactic situation*)	- Preliminary analyses - Design and a priori analysis of a didactic situation - Implementation, observation, data collection - A posteriori analysis and validation (Artigue, 2015)
RME	Design research (*emergent modelling*)	Cyclic process of: - Thought experiment - Design of a teaching sequence - Test of the teaching sequence, observation, data collection - A posteriori analysis (Gravemeijer & Cobb, 2013)
Lesson Study	"Open approach method" (*research lesson*)	- Planning of a research lesson - Implementation, observation, data collection - Reflection on the research lesson - Possibly, revision and teaching of the revised research lesson (Murata, 2011)

FIGURE 12.1 Research methodologies of the theories (*design heuristics* in parentheses)

Studies based on didactic design that have been presented at INDRUM and included in the foregoing, are Barquero et al. (2016), Bourgade (2016), Romo et al. (2016), Florensa et al. (2018), and Strømskag (2018).

Instructional design for inquiry-based learning

As noted above, there are several approaches to instructional design that have been theorised. Despite the theorisation, there appears to be a broad common ground when considering the intended learning activities of the students. This common ground is devoted to achieving so-called *inquiry-based learning*, or IBL in short.

From the perspective of studies conducted within ATD, the idea of inquiry is implicit in the notion of *study and research path* (SRP) (Chevallard, 2015). Several contributions to INDRUM 2016 and 2018 are related to this, for instance the papers of Barquero et al. (2016); Romo et al. (2016) and Gascón and Nicolás (2018). SRP was introduced earlier in the chapter, we briefly recall that a SRP consists in a process that starts with a question of genuine interest for the community of study and entails looking for satisfactory answers to this question. Along this quest, as would be the case for real mathematicians, students are allowed to "research" (inquiry, problem solving, problem posing, etc.)

and "study" (consulting existing knowledge, attending lectures where the teacher acts as the main means to provide mathematical knowledge, etc.). In this way, students progressively examine different possible answers until they finally get what the study community (formed by teacher plus students) regards as a suitable answer to the initial question.

A priori, one can consider two types of IBL. On the one hand, we can consider inquiries where there is no predetermined knowledge to be rebuilt by students. The only important point would be to find an "admissible" answer to an initial question (Chevallard, 2015). To adopt this kind of teaching entails a radical change in the conception of education, as the emphasis in this case would be on considering questions and starting with inquiries rather than learning a piece of knowledge explicitly anticipated by the teacher.

On the other hand, inquiries can be considered in which the students are expected to rediscover a certain piece of knowledge previously chosen by the teacher. This kind of IBL is still groundbreaking and has to face many challenges of different nature: political, cultural, epistemological, concerning the teacher's training and the teacher's view of their own profession, etc. Gascón and Nicolás (2018) consider some of these challenges by applying Hintikka's conceptualisation of inquiry (Hintikka, Halonen, & Mutanen, 2002) as a useful tool not only to describe but also to plan inquiries. Among the challenges they address, they consider the problem of finding criteria in order to attach meanings to the objects involved in the expected inquiries. If, in a proposed inquiry, some the meanings are dismissed, then students are most probably deprived of relevant information. Furthermore, one can raise a question about the possible ways in which an inquiry in which things have several meanings can be planned. This is the first challenge of IBL considered by Gascón and Nicolás (2018): the conundrum of implementing semantics in mathematics syntax. The second challenge these researchers consider is concerned with how to plan inquiries. Every inquiry is materialised in the quest of an answer to an initial question. If it is desired that students discover a certain piece of knowledge through an inquiry process, the question to initiate the inquiry must be chosen with great care.

The questions considered in (Gascón & Nicolás, 2018) are still open problems in the study of IBL, but, of course, they are not the only ones. Many issues were not addressed in this contribution, among them the problem of student's assessment in an IBL context. We are used to assessing students' answers to our questions, or students' observance of our requests or command. But how should we assess an inquiry? Should examinations be inquiries? Should examinations disappear?

Conclusion

Teaching mathematics at university is an interesting and compelling area of study for mathematics education researchers that is only just beginning to receive close attention. The special abstract nature of mathematical truths and the big ideas such as function, variability, change, limit and structure create a unique context for the study of mathematical meaning-making, communication and instruction. The studies presented at INDRUM conferences by posters and papers bear witness to the truth of the assertion: the more we learn, the more there is to find out and to investigate further and deeper; the closer we look, the more there is to see.

In higher education, the majority of students who study mathematics are required to do so because it is part of another programme such as engineering or economics. In these programmes, mathematics is often blamed for poor performance and retention, high failure and drop-out; this is a global problem which mathematics education research must address. Presentations at INDRUM reflect both the attraction of research into teaching mathematics and the societal value of the research.

Researching university mathematics education presents a great opportunity for mathematics education researchers and mathematicians to collaborate and thus bridge the divide between the disciplines of mathematics and mathematics education. Research evidence can challenge the, often naïve, epistemological assumptions of teaching by introducing theoretical lenses and frameworks to provide means for analysing, interpreting and making sense of classroom practices. It is essential that mathematics education researchers communicate with mathematics teachers the results of research and the possibilities for improving practice. The presentations at INDRUM are framed in a plethora of theories: ATD, Instrumental and Documentational approaches, TDS, Commognition, CHAT and CPT – it is possible to get a glimpse of how mathematicians react to this situation in the work of Nardi (2008) and in a scathing review by the mathematician Steen (1999) of an ICMI study of mathematics education research. Do the theories drive a wedge between mathematics education researchers and mathematicians/mathematics teachers? For example, in Viirman (2018), there is no mention of "theories" of mathematical instruction entering the discourse of the collaboration between mathematics education researchers and mathematics teachers reported on. The papers that refer to the "teaching triad" (Petropoulou et al., 2016) and the "knowledge quartet" (Breen et al., 2018) reveal the value of less complex models that have empirical roots in mathematics classes, albeit at school level.

The use of models of teaching and learning practices developed in school-based research presents a challenge to INDRUM researchers: are these models transferable to learning in higher education? The differences between the school and higher education contexts are significant. Students study mathematics in higher education because it is a part of a programme of study they have chosen, rather than a compulsory part of the curriculum. In higher education, students are expected to have greater independence and agency, and we assume maturity. The teaching and learning contexts differ radically in terms of class size, intensity and pace of content covered in lectures and sometimes use of minimally "trained" student teaching/learning assistants. The teachers or lecturers in higher education are likely to have a rather different relationship with mathematics than teachers in school, and the workloads and working days of schoolteachers and lecturers are likely to be quite different. Thus, when taking models developed from school-based empirical studies and applying them in the context of higher education, researchers need to consider the utility and applicability of the model, this may not be taken for granted.

It has to be recognised that the presentations at conferences such as INDRUM cover only part of the research field at large. Consequently, the discussion of the theories included in this chapter can only reflect the research presented at INDRUM. What might be concluded from the proportion of papers framed within ATD? Does it reveal something of the utility of the theories, or perhaps the nature of the issues that interest researchers of didactics of university mathematics, or is it about national characteristics and the maturity of research in this field in different national communities (or communities united by sharing a common language)? INDRUM conferences serve a purpose by bringing together these research communities using these theories.

Contrasting theoretical perspectives can offer alternative approaches for exploring and developing mathematics teaching at university. For example: TDS and the Double Approach offer developmental and analytic tools for teachers/researchers. TDS (with its concepts of adidactic situation and milieu) can be used as a tool to design and analyse the implementation of mathematics teaching sequences with respect to (generic) students' appropriation of some particular knowledge. The Double Approach (with its concept of proximities) can be used to analyse the distance between what students do and know and the teacher's goals for the students, and how students' responses influence the actions of the teacher in trying to reduce this distance. Also, there is a difference in the nature of the didactic devices: TDS aims at adidactic functioning of the knowledge, and its evolution, by designing and managing an appropriate milieu; the Double Approach's concept of proximities aims at didactic actions that the teacher can use to bridge the gap between students' existing knowledge and the new knowledge aimed at (Mangiante-Orsola, Perrin-Glorian, & Strømskag, 2018).

More can be developed about teacher-researchers' teaching practices. What is the relationship between teacher practice and researcher practice? To what extent does the researcher's practice influence the teacher and vice versa? In other words, what balance exists between mathematical practice and research in the field? In Bridoux et al. (2019), the purpose of a comparative and interdisciplinary research based on three academic disciplines (Mathematics, Chemistry, Physics) is to question the research discipline's imprint on teaching practices at tertiary level. More generally, the influence of research practices on teacher-researchers learning practices (already stressed by Biza et al., 2016) or the study of teacher-researchers' (possible) training should be a main point of interest. Attention could be focused on the practices of teacher-researchers from the point of view of their interaction with resources and with other disciplines.

Another area calling for research is to relate teaching practices to student performance. However, valid and reliable research that is likely to expose correlations and more importantly cause-effect relationships is at the moment not covered by the qualitative studies mostly reported at INDRUM conferences. Without large-scale research using experimental and quasi-experimental approaches an important area of knowledge will remain unexplored. Is this a challenge the INDRUM community should confront?

INDRUM serves a purpose for mathematics education researchers to negotiate the value of theoretical frameworks in structuring and making sense of mathematics teaching at university level. The network provides an opportunity for critical reflection on the application of theory to teaching practice, and the development of teaching. It is difficult, and perhaps foolish, to try to predict the direction of research in this area. One thing that can be asserted by these contributions though is the enthusiasm of INDRUM researchers for sustaining and continuing to create knowledge of, and for, the teaching of mathematics at university.

Notes

1 "A preparatory class that leads to engineering schools after two years of intense training in sciences and humanities." (Bourgade, 2016)
2 Cegep institutions refer to general or vocational colleges and constitute the first step in post-secondary education.
3 See also Bridoux, Grenier-Boley, Hache, and Robert (2016).
4 Roughly speaking: *upward proximities* are inductive and used during a generalisation or decontextualisation process, *downward proximities* are deductive and are used between the "general" (course) and the "particular" (examples, exercises). *Horizontal proximities* do not change the level of discourse and correspond to analogies or translations.
5 The general framework and the study of Calculus notions related to the learning of advanced mathematics (abstraction, formalism…) accentuated by the context of the secondary-tertiary transition, the type of issue.

References

Abboud, M., Goodchild, S., Jaworski, B., Potari, D., Robert, A., & Rogalski, J. (2018). Use of activity theory to make sense of mathematics teaching: A dialogue between perspectives. *Annales de Didactique et de Sciences Cognitives, Volume Spécial English-French*, 61–92.

Adler, J. (2000). Conceptualising resources as a theme for teacher education. *Journal of Mathematics Teacher Education*, 3, 205–224.

Annoot, E., & Fave-Bonnet, M. F. (2004). *Pratiques Pédagogiques dans L'enseignement Supérieur: Enseigner, Apprendre, Evaluer* [*Pedagogical Practices in Higher Education: Teach, Learn, Evaluate*]. Paris: L'Harmattan.

Artigue, M. (2009). Didactical design in mathematics education. In C. Winsløw (Ed.), *Nordic Research in Mathematics Education* (pp. 7–16). Rotterdam: Sense.

Artigue, M. (2015). Perspectives on design research: The case of didactical engineering. In A. Bikner-Ahsbahs, C. Knipping, & N. Presmeg (Eds.), *Approaches to Qualitative Research in Mathematics Education* (pp. 467–496). Dordrecht: Springer.

Ball, D. L., Thames, M. H., & Phelps, G. (2008). Content knowledge for teaching: What makes it special? *Journal of Teacher Education*, 59, 389–407.

Barquero, B., & Bosch, M. (2015). Didactic engineering as a research methodology: From fundamental situations to study and research paths. In A. Watson & M. Ohtani (Eds.), *Task Design in Mathematics Education* (pp. 249–272). Dordrecht: Springer.

Barquero, B., Serrano, L., & Ruiz-Munzón, N. (2016). A bridge between inquiry and transmission: the study and research paths at university level. *INDRUM 2016 Proceedings*, pp. 340–349.

Baumert, J., Kunter, M., Blum, W., Brunner, M., Voss, T., Jordan, A., … Tsai, Y.-M. (2010). Teachers' mathematical knowledge, cognitive activation in the classroom, and student progress. *American Education Research Journal*, 47(1), 133–180.

Becher, T. (1994). The significance of disciplinary differences. *Studies in Higher Education*, 19(2), 151–161.

Beeby, T., Burkhardt, H., & Fraser, R. (1979). *Systematic Classroom Analysis Notation for Mathematics Lessons*. Nottingham: Shell Centre for Mathematical Education.

Biza, I., Jaworski, B., & Hemmi, K. (2014). Communities in university mathematics. *Research in Mathematics Education*, 16, 161–172.

Biza, I., Giraldo, V., Hochmuth, R., Khakbaz, A., & Rasmussen, C. (2016). *Research on Teaching and Learning Mathematics at the Tertiary Level: State-of-the-art and Looking Ahead*. ICME-13 Topical Surveys. Cham: Springer.

Biza, I., & Nardi, E. (2016). Consistency, reification of pedagogical and mathematical discourses in student teachers' narratives on the challenges of their school placement experience. *INDRUM 2016 Proceedings*, pp. 425–426.

Biza, I., Nardi, E., & Zachariades, T. (2018). Competences of mathematics teachers in diagnosing teaching situations and offering feedback to students: Specificity, consistency and reification of pedagogical and mathematical discourses. In T. Leuders, K. Philipp, & J. Leuders (Eds.), *Diagnostic Competence of Mathematics Teachers* (pp. 55–78). New York: Springer.

Bosch, M., & Gascón, J. (2006). Twenty-five years of didactic transposition. *ICMI Bulletin*, 58, 51–64.

Bourgade, J.-P. (2016). On the diffusion of professional praxeologies at university level: The case of investigation workshops. *INDRUM 2016 Proceedings*, pp. 350–359.

Breen, S., Meehan, M., O'Shea, A., & Rowland, T. (2018). An analysis of university teaching using the Knowledge Quartet. *INDRUM 2018 Proceedings*, pp. 383–392.

Bridoux, S., de Vleeschouwer, M., Grenier-Boley, N., Khanfour-Armalé, R., Lebrun, N., Mesnil, Z., & Nihoul C. (2019). L'identité professionnelle des enseignants-chercheurs en mathématiques, chimie et physique [The professional identity of teacher-researchers in Mathematics, Chemistry, Physics]. In M. Abboud (Ed.), *Actes du colloque EMF 2018* (pp. 540–547). Paris: IREM de Paris.

Bridoux, S., Grenier-Boley, N., Hache, C., & Robert, A. (2016). Les moments d'exposition des connaissances, analyses et exemples [Presenting knowledge and teaching moments, analyses and examples]. *Annales de Didactique et de Sciences Cognitives*, 21, 187–234.

Broley, L. (2016). The place of computer programming in (undergraduate) mathematical practices. *INDRUM 2016 Proceedings*, pp. 360–369.

Brousseau, G. (1997). *The Theory of Didactical Situations in Mathematics: Didactique des Mathématiques, 1970–1990* (N. Balacheff, M. Cooper, R. Sutherland, & V. Warfield, Eds. & Trans.). Dordrecht: Kluwer.

Brousseau, G. (2000). Educacion y didactica de las matematicas. *Educacion Matematica*, 12(1), 5–38. Available in French: https://hal.archives-ouvertes.fr/hal-00466260.

Brousseau, G., & Warfield, V. (2014). Didactic situations in mathematics education. In S. Lerman (Ed.), *Encyclopedia of Mathematics Education* (pp. 163–170). Dordrecht: Springer.

Carrillo, J., Climent, N., Contreras, L. C., & Muñoz-Catalán, M. C. (2013). Determining specialised knowledge for mathematics teaching. In B. Ubuz, Ç. Haser, & M. A. Mariotti (Eds.), *Proceedings of the Eighth Congress of the European Society for Research in Mathematics Education* (pp. 2985–2994). Ankara: Middle East Technical University & ERME.

Chevallard, Y. (1981). The didactics of mathematics: Its problematic and related research. *Recherches en Didactique des Mathématiques*, 2(1), 146–158.

Chevallard, Y. (1998). *Analyse des Pratiques Enseignantes et Didactique des Mathématiques: L'approche Anthropologique* [*Analysis of Teaching Practices and Didactics of Mathematics: The Anthropological Approach*]. Retrieved from yves.chevallard.free.fr/spip/spip/IMG/pdf/Analyse_des_pratiques_enseignantes.pdf.

Chevallard, Y. (2002). Organiser l'étude. Écologie & régulation [Organising study. Ecology & regulation]. In J.-L. Dorier, M. Artaud, M. Artigue, R. Berthelot, & R. Floris (Eds.), *Actes de la XIème École d'Été de Didactique des Mathématiques* (pp. 41–56). Grenoble: La Pensée Sauvage.

Chevallard, Y. (2011). La notion d'ingénierie didactique, un concept à refonder. Questionnement et éléments de réponse à partir de la TAD [The notion of didactic engineering, a concept that needs to be redesigned. Questioning and answers from the TAD]. In C. Margolinas, M. Abboud-Blanchard, L. Bueno-Ravel, N. Douek, A. Flückiger, P. Gibel, … F. Wozniak (Eds.), *En Amont et en Aval des Ingénieries Didactiques, Actes de la XVème École d'Été de Didactique des Mathématiques* (pp. 81–108). Grenoble: La Pensée Sauvage.

Chevallard, Y. (2013). Enseñar Matemáticas en la Sociedad de Mañana: Alegato a Favor de un Contraparadigma Emergente [Teaching Mathematics in tomorrow's society: Plea for an emerging counter paradigm]. *Journal of Research in Mathematics Education*, 2, 161–182.

Chevallard, Y. (2015). Teaching mathematics in tomorrow's society: A case for an oncoming counter paradigm. In S. J. Cho (Ed.), *Proceedings of the 12th International Congress on Mathematical Education* (pp. 173–187). Berlin: Springer.

Delgado, R., & Zakaryan, D. (2018). Knowledge of the practice in mathematics in university teachers. *INDRUM 2018 Proceedings*, pp. 393–402.

Dewey, J. (1938). *Experience and Education*. New York: Collier Books.

Drijvers, P., Doorman, M., Boon, P., Reed, H., & Gravemeijer, K. (2010). The teacher and the tool: Instrumental orchestrations in the technology-rich mathematics classroom. *Educational Studies in Mathematics*, 75, 213–234.

Duval, R. (2006). A cognitive analysis of problems of comprehension in a learning of mathematics. *Educational Studies in Mathematics*, 61, 103–131.

Fernandez, C., & Yoshida, M. (2004). *Lesson Study: A Japanese Approach to Improving Mathematics Teaching and Learning*. New York: Routledge.

Florensa, I., Bosch, M., Cuadros, J., & Gascón, J. (2018). Helping lecturers address and formulate teaching challenges: An exploratory study. *INDRUM 2018 Proceedings*, pp. 373–382.

Freudenthal, H. (1983). *Didactical Phenomenology of Mathematical Structures*. Dordrecht: Reidel.

Gascón, J., & Nicolás, P. (2018). Inquiry-based learning and pre-service teachers' education. *INDRUM 2018 Proceedings*, pp. 412–421.

González-Martín, A. S., Bloch, I., Durand-Guerrier, V., & Maschietto, M. (2014). Didactic Situations and Didactical Engineering in university mathematics: Cases from the study of Calculus and proof. *Research in Mathematics Education*, 16, 117–134.

Gravemeijer, K. (2004). Local instruction theories as means of support for teachers in reform mathematics education. *Mathematical Thinking and Learning*, 6(2), 105–128.

Gravemeijer, K., & Cobb, P. (2013). Design research from the learning design perspective. In T. Plomp & N. Nieveen (Eds.), *Educational Design Research Part A: An Introduction* (pp. 72–113). Enschede: SLO.

Grenier-Boley, N., Bridoux, S., & Hache, C. (2016). Moments d'exposition des connaissances à l'université: Le cas de la notion de limite [Presenting knowledge at the university: The case of the notion of limit]. *INDRUM 2016 Proceedings*, pp. 380–389.

Gueudet, G. (2017). University teachers' resources systems and documents. *International Journal of Research in Undergraduate Mathematics Education* 3(1), 198–224.

Gueudet, G., Buteau, C., Mesa, V., & Misfeldt, M. (2014). Instrumental and documentational approaches: From technology use to documentation systems in university mathematics education. *Research in Mathematics Education*, 16, 139–155.

Gueudet, G., Pepin, B., & Trouche, L. (2012). *From Text to 'Lived resources': Mathematics Curriculum Material and Teacher Development*. New York: Springer.

Hadjerrouit, S., & Gautestad, H. (2016). Using the interactive visualization tool SimReal+ to teach mathematics at the university level: An instrumental approach. *INDRUM 2016 Proceedings*, pp. 429–430.

Hamann, T., & Schmidt-Thieme, B. (2018). Cross-linking maths – using keynotes to structure a curriculum for future teachers. *INDRUM 2018 Proceedings*, pp. 456–457.

Henkel, M. (2004). La relation enseignement-recherche [The relationship between teaching and research]. *Politiques et Gestion de L'enseignement Supérieur*, 16(2), 21–36.

Hintikka J., Halonen, I., & Mutanen, A. (2002). Interrogative logic as a general theory of reasoning. In D. M. Gabbay, R. H. Johnson, H. J. Ohlbach, & J. Woods (Eds.), *Handbook of the Logic of Argument and Inference: The Turn Towards the Practical* (Vol. 1, pp. 295–337). Amsterdam: Elsevier.

Huget, J. (2018). BiMathTutor – The program for further training of new tutors in University Mathematics. *INDRUM 2018 Proceedings*, pp. 458–459.

Jaworski, B. (1994). *Investigating Mathematics Teaching: A Constructivist Enquiry*. London: Falmer Press.

Klein, F. (2016). *Elementary Mathematics From a Higher Standpoint. Volume I: Arithmetic, Algebra, Analysis*. Heidelberg: Springer-Verlag (Original work published 1933)

Leontiev, A. N. (1978). *Activity, Consciousness and Personality*. Englewood Cliffs, NJ: Prentice Hall.

Leplat, J. (1997). *Regards Sur l'activité en situation de travail [About the Activity in a Job Situation]*. Paris: PUF.

Ma, L. (1999). *Knowing and Teaching Elementary Mathematics: Teachers' Understandings of Fundamental Mathematics in China and the United States*. Mahwah, NJ: Lawrence Erlbaum Associates.

Mangiante-Orsola, C., Perrin-Glorian, M.-J., & Strømskag, H. (2018). Theory of didactical situations as a tool to understand and develop mathematics teaching practices. *Annales de Didactique et de Sciences Cognitives, Volume Spécial English-French*, 145–174.

Martìn-Molina, V. (2016). The challenge of being a mathematics teacher. *INDRUM 2016 Proceedings*, pp. 431–432.

Mason, J. (2002). *Researching Your Own Practice: The Discipline of Noticing*. London: Routledge.

Mathieu-Soucy, S., Corriveau, C., & Hardy, N. (2018). Exploration of new post-secondary mathematics teachers' experiences: Preliminary results of a narrative inquiry. *INDRUM 2018 Proceedings*, pp. 403–411.

Murata, A. (2011). Introduction: Conceptual overview of Lesson Study. In L. Hart, A. Alston, & A. Murata (Eds.), *Lesson Study Research and Practice in Mathematics Education* (pp. 1–12). Dordrecht: Springer.

Nardi, E. (2008). *Amongst Mathematicians: Teaching and Learning Mathematics at University Level*. New York: Springer.

Nardi, E., Ryve, A., Stadler, E., & Viirman, O. (2014). Commognitive analyses of the learning and teaching of mathematics at university level: The case of discursive shifts in the study of Calculus. *Research in Mathematics Education*, 16, 182–198.

Neumann, R. (2001). Disciplinary Differences and University Teaching. *Studies in Higher Education*, 26(2), 135–146.

Petropoulou, G., Jaworski, B., Potari, D., & Zachariades, T. (2016). Addressing large cohorts of first year mathematics students in lectures. *INDRUM 2016 Proceedings*, pp. 390–399.

Poteaux, N. (2013). Pédagogie de l'enseignement supérieur en France: état de la question [Pedagogy of higher education in France: state of the art]. *Distances et Médiations des Savoirs. Distance and Mediation of Knowledge*, 1(4).

Rabardel, P. (2002). *Les Hommes et les Technologies: Approche Cognitive des Instruments Contemporains [People and Technology: A Cognitive Approach to Contemporary Instruments]*. Paris: Armand Colin. Retrieved from http://ergoserv.psy.univ-paris8.fr/Site/default.asp?Act_group=1 (Original work published 1995)

Robert, A., & Hache, C. (2013). Why and how to understand what is at stake in a mathematics class. In F. Vandebrouck (Ed.). *Mathematics Classroom: Students' Activities and Teacher's Practices* (pp. 23–73). Rotterdam: Sense.

Rogalski, J. (2013). Theory of Activity and developmental frameworks for an analysis of teachers' practices and students' learning. In F. Vandebrouck (Ed.). *Mathematics Classroom: Students' Activities and Teacher's Practices* (pp. 3–22). Rotterdam: Sense.

Romo, A., Barquero, B., & Bosch, M. (2016). Study and research paths in online teacher professional development. *INDRUM 2016 Proceedings*, pp. 400–410.

Rowland, T., Huckstep, P., & Thwaites, A. (2005). Elementary teachers' mathematics subject knowledge: The knowledge quartet and the case of Naomi. *Journal of Mathematics Teacher Education*, 8(3), 255–281.

Ruiz-Olarría, A. (2015). *La formación matemático-didáctica del profesorado de secundaria: De las matemáticas por enseñar a las matemáticas para la enseñanza [Mathematical-didactical training of secondary teachers: From mathematics to teach to mathematics for teaching]*. (PhD dissertation). Universidad Autónoma de Madrid.

Schmidt-Thieme, B. (2018). Getting language awareness: a curriculum for language and language teaching for pre-service studies for teachers of mathematics. *INDRUM 2018 Proceedings*, pp. 460–461.

Sfard, A. (2008). *Thinking as Communicating: Human Development, The Growth of Discourses, and Mathematizing*. New York: Cambridge University Press.

Shulman, L. (1987). Knowledge and teaching: Foundations of the new reform. *Harvard Educational Review*, 57(1), 1–23.

Sierra, T. A. (2006). *Lo matemático en el diseño y análisis de organizaciones didácticas. Los sistemas de numeración y la medida de magnitudes [Mathematics in the design and analysis of didactic organizations. Numbering systems and the measurement of magnitudes]*. (Doctoral dissertation, Universidad Complutense de Madrid, Spain).

Steen, L. A. (1999). Theories that gyre and gimble in the wabe. *Journal for Research in Mathematics Education*, 30(2), 235–241.

Strømskag Måsøval, H. (2013). Shortcomings in the milieu for algebraic generalisation arising from task design and vagueness in mathematical discourse. In C. Margolinas (Ed.), *Task design in mathematics education*. Proceedings of ICMI Study22 (pp. 231–239). Oxford: The International Commission on Mathematical Instruction.

Strømskag, H. (2017). A methodology for instructional design in mathematics – with the generic and epistemic student at the centre. *ZDM Mathematics Education*, 49, 909–921.

Strømskag, H. (2018). A model for instructional design in mathematics implemented in teacher education. *INDRUM 2018 Proceedings*, pp. 422–431.

Syrdalen, E. (2018). A university lecturer's role and interactions in a flipped classroom. *INDRUM 2018 Proceedings*, pp. 462–463.

Tabchi, T. (2018). University teachers-researchers' practices: The case of teaching discrete mathematics. *INDRUM 2018 Proceedings*, pp. 432–441.

Thoma, A., & Nardi, E. (2016). A commognitive analysis of closed-book examination task and lecturer's perspectives. *INDRUM 2016 Proceedings*, pp. 411–420.

Trouche, L. (2004). Managing the complexity of human/machine interactions in computerized learning environments: Guiding students' command process through instrumental orchestrations. *International Journal of Computers for Mathematical Learning*, 9(3), 281–307.

Vergnaud, G. (1998). Toward a cognitive theory of practice. In A. Sierpinska & J. Kilpatrick (Eds.), *Mathematics Education as a Research Domain: A Search for Identity* (pp. 227–240). Dordrecht: Kluwer.

Viirman, O. (2014). The functions of function discourse-university mathematics teaching from a commognitive standpoint. *International Journal of Mathematical Education in Science and Technology*, 45(4), 512–527.

Viirman, O. (2015). Explanation, motivation and question posing routines in university mathematics teachers' pedagogical discourse: a commognitive analysis. *International Journal of Mathematical Education in Science and Technology*, 46(8), 1165–1181.

Viirman, O. (2018). Characteristics of participation – A mathematician and a mathematics educator collaborating on a developmental research project. *INDRUM 2018 Proceedings*, pp. 442–451.

Viirman, O., Pettersson, I., Björklund, J., & Boustedt, J. (2018). Programming in mathematics teacher education – a collaborative teaching approach. *INDRUM 2018 Proceedings*, pp. 464–465.

Vygotsky, L. (1986). *Thought and Language*. Cambridge MA: MIT Press. (Original work published 1962)

Wenger, E. (1998). *Communities of Practice: Learning, Meaning, and Identity*. Cambridge, UK: Cambridge University Press.

Winsløw, C., Matheron, Y., & Mercier, A. (2013). Study and research courses as an epistemological model for didactics. *Educational Studies in Mathematics*, 83, 267–284.